SOURCES OF QUANTUM MECHANICS

CLASSICS OF SCIENCE SERIES

under the General Editorship of

Gerald Holton

Professor of Physics, Harvard University

ALREADY PUBLISHED

CLASSICS IN THE THEORY OF CHEMICAL COMBINATION
edited by O. Theodor Benfey

THE DISCOVERY OF RADIOACTIVITY AND TRANSMUTATION
edited by Alfred Romer

THE DEVELOPMENT OF HIGH-ENERGY ACCELERATORS
edited by M. Stanley Livingston

CLASSICS IN COORDINATION CHEMISTRY, PART I
edited by George B. Kauffman

SOURCES OF QUANTUM MECHANICS
edited by B. L. van der Waerden

RADIOCHEMISTRY AND THE DISCOVERY OF ISOTOPES
edited by Alfred Romer

Classics of Science, Volume V

SOURCES OF QUANTUM MECHANICS

Edited with a Historical Introduction by

B. L. VAN DER WAERDEN
University of Zurich

DOVER PUBLICATIONS, INC., NEW YORK

Published in Canada by General Publishing Company, Ltd., 30 Lesmill Road, Don Mills, Toronto, Ontario.

Published in the United Kingdom by Constable and Company, Ltd., 10 Orange Street, London WC 2.

This Dover edition, first published in 1968, is an unabridged republication of the work published in 1967 by North-Holland Publishing Co., Amsterdam. Reprinted by special arrangement with North-Holland Publishing Co.

International Standard Book Number: 0-486-61881-1

Library of Congress Catalog Card Number: 68-12916

Manufactured in the United States of America
Dover Publications, Inc.
180 Varick Street
New York, N.Y. 10014

PREFACE

The idea of collecting the most important early papers on Quantum Mechanics in a Source Publication is due to Max Born, and he intended to include 15 papers written by himself, Jordan, Heisenberg, Dirac and Pauli, and published during the years 1924–1926.

Dr. Paul Rosbaud, who until his death in 1963 acted as a scientific consultant to various publishing houses, took it upon him to discuss Born's project with several physicists. They all were in agreement that it would be best if all German papers were to be translated into English in order to make the sources available to all physicists and historians of Science. It was Pauli who recommended to Dr. Rosbaud the inclusion of earlier papers by Ladenburg, Kramers and Heisenberg which have prepared the way towards Quantum Mechanics.

When Dr. Rosbaud asked me to act as the editor of this volume, I discussed the list of papers with Born, Heisenberg, Heitler, Hund, Jordan, Kronig and others. They all agreed that several earlier papers written between 1917 and 1924 by Bohr, Einstein, Ehrenfest and Kuhn ought to be included, because these are necessary for a good understanding of the 'turning point' of the year 1925. At Jordan's advise, a paper by Van Vleck, which had strongly influenced Born's and Jordan's ideas on the interaction between matter and radiation, was also included in the project. However, a line had to be drawn somewhere. Therefore the following principles were adopted:

1. Papers on Quantum Theory were included only when they were judged indispensable for a proper understanding of the development of Quantum Mechanics. Thus, Bohr's great 1918 paper, in which the Principle of Correspondence was exposed, was included, but not Bohr's earlier papers. On the same principle, Planck's fundamental 1900 paper on the Law of Radiation was excluded, because a more fundamental

derivation of the same Law was given in Einstein's 1917 paper on the emission and absorption of radiation (no. 1 of the present collection). The papers of Debye and Sommerfeld, however important for the development of classical Quantum Theory, had to be omitted. Historians of Science, who want to learn more about pre-1925 Quantum Theory, will have to consult Sommerfeld's 'Atombau und Spektrallinien' and Pauli's article 'Quantentheorie' in Geiger and Scheel's Handbuch der Physik, Vol. 23.

2. Papers on the Zeeman Effect, Spin and Statistics were left aside, because these closely related subjects have been dealt with in the Pauli Memorial Volume (Interscience Publishers 1960), p. 199–244, by B. L. van der Waerden.

3. Papers on Wave Mechanics were not included. We hope to assemble them in a second volume.

4. The papers of John von Neumann, who gave Quantum Mechanics a rigorous mathematical foundation, were also reserved for the second volume.

5. Papers on Quantum Field Theory are outside the scope of this collection.

Even within these limits, only the most important papers could be included. Related papers are mentioned by title at the end of each paper.

Obvious misprints in the original papers have been tacitly corrected.

The papers collected in this volume naturally fall into two groups:

(i) *Towards Quantum Mechanics*. Papers 1–11, by Einstein, Ehrenfest, Bohr, Ladenburg, Kramers, Slater, Born, Van Vleck, Heisenberg, and Kuhn. The German papers of this group were translated by G. Field.

(ii) *Matrix Mechanics*. Papers 12–17, by Heisenberg, Born, Jordan, Dirac and Pauli. The German papers were translated by a team consisting of E. Sheldon, D. Robinson, G. Field and B. L. van der Waerden.

These two groups of papers are preceded by a historical introduction. In this introduction, use is made of letters of Heisenberg and others, which cast a new light on the history of Quantum Mechanics. Parts of these letters are reproduced in the original language.

I feel great gratitude to all who have helped me in selecting the papers and who gave me additional information. I am especially indebted to Born, Dirac, Heisenberg, Hund, Jordan, Kronig, Th. Kuhn

and Wigner, and to Mrs. Pauli, who had the great kindness of showi
me letters of Heisenberg to Pauli.

Thanks are also due to the Gesellschaft der Wissenschaften in Gö
tingen, which has borne the considerable cost of translating the Germa
papers into English.

B. L. van der Waerden

Zürich, September 1966

TABLE OF CONTENTS

INTRODUCTION

B. L. VAN DER WAERDEN

PART I. TOWARDS QUANTUM MECHANICS

Max Planck

Quantum theory was born on December 14, 1900, when Max Planck delivered his famous lecture before the Physikalische Gesellschaft, which was afterwards printed in Verhandlungen der Deutschen physikalischen Gesellschaft *2*, p. 237 under the title 'Zur Theorie des Gesetzes der Energieverteilung im Normalspektrum'.

In this paper, Planck assumed that the emission and absorption of radiation always takes place in discrete portions of energy, or 'energy quanta' $h\nu$, where ν is the frequency of the emitted or absorbed radiation. Starting with this assumption, Planck arrived at his famous formula for the density of black-body radiation at temperature T:

$$\rho = \frac{\alpha\nu^3}{\exp(h\nu/kT) - 1}.$$

An excellent commentary to Planck's paper was given by Martin Klein in Vol. 1 of the Archive for History of Exact Sciences, p. 459 (1962). We shall not reproduce Planck's paper here, because another derivation of Planck's law, which gives a better insight into the establishment of the equilibrium between the radiation and the emitting and absorbing molecules, was given by Einstein in paper **1**.

Rutherford

In order to explain the scattering of alpha particles by atoms, Rutherford assumed the atom to contain a charge $+Ne$ or $-Ne$ at its center surrounded by a sphere of electrification containing a charge $-Ne$ or $+Ne$ uniformly distributed throughout a sphere of radius R, e

being the fundamental unit of charge, and R being of the order of the radius of an atom, viz. 10^{-8} cm [1]. From these assumptions, he deduced the angular distribution of the scattered particles. The experimental results obtained by Geiger in 1910 were found to be in substantial agreement with Rutherford's theory, whereas they could not be explained by earlier theories.

The deductions from the theory are independent of the sign of the central charge, and Rutherford concludes: 'it has not been found possible to obtain definite evidence to determine whether it be positive or negative'. However, the drawings in the paper are made for the case of a positive central charge, and Rutherford himself seems to favour this assumption, for he writes:

> If the central charge be positive, it is easily seen that a positively charged mass if released from the centre of a heavy atom, would acquire a great velocity in moving through the electric field. It may be possible in this way to account for the high velocity of expulsion of α particles without supposing that they are initially in rapid motion within the atom.

Rutherford also considers the possibility that the negative charge, instead of being uniformly distributed throughout a sphere of radius R, is located in N rotating electrons. This hypothesis was considered by Nagaoka in 1904 (Phil. Mag. *7*, p. 445). Nagaoka had considered the properties of a 'Saturnian' atom, consisting of a central mass and Saturnian rings of rotating electrons. Rutherford notes that the angular distribution of scattered α-particles would be practically the same, whether the atom is considered to be a disk or a sphere, because large deviations are mainly due to the central charge.

Niels Bohr

The synthesis of Rutherford's atom model with Planck's quantum hypothesis was the great achievement of Niels Bohr. He supposed the atom to consist of a nucleus with positive charge Ze and Z electrons with charge $-e$ each, moving according to the laws of classical mechanics. In his papers of 1913, 1914 and 1915 (Phil. Mag. *26*, *27*, *29* and *30*) he introduced a set of assumptions concerning the stationary states of an atom and the frequency of the radiation emitted or absorbed when the atom passes from one such state to another,

[1] E. Rutherford: The Scattering of α and β Particles by Matter and the Structure of the Atom. Phil. Mag. *21* (sixth series) p. 669. Dated April 1911.

and he showed that it is possible in this way to obtain a simple interpretation of the main laws governing the line spectra of the elements, and especially to deduce the Balmer formula for the hydrogen spectrum.

We need not enter into details here, because the main principles of Bohr's theory are fully explained in Bohr's 1918 paper **3**.

Bohr's ideas were further developed and applied to more complicated spectra by Sommerfeld, Debye and others. For this development we must refer the reader to Pauli's excellent article 'Quantentheorie' in Geiger–Scheel's Handbuch der Physik, 1st edition, Vol. 23 (1926), reprinted in Pauli's Collected Scientific Papers, Vol. 1. The historian should also consult Sommerfeld's Atombau und Spektrallinien (preferably 4th ed., 1924), because it was mainly from this book that the young physicists who created Quantum Mechanics in 1925–26 learnt Quantum Theory.

Einstein

In Bohr's theory, the interaction between matter and radiation remained mysterious. Why does not the atom emit radiation, when it is in its ground state? What really happens when an atom passes from one stationary state to another? What laws determine the probabilities of these transitions?

The first one to bring more light into the darkness was Einstein (paper **1**). Einstein starts with Bohr's assumption that a molecule can only exist in a discrete set of states with energies ε_1, ε_2, ... If such molecules belong to a gas at temperature T, Einstein assumes, by analogy to the Boltzmann–Gibbs canonical distribution, the relative frequency W_n of a state Z_n to be [1]

$$W_n = p_n \exp(-\varepsilon_n/kT),$$

p_n being an integer called 'statistical weight' of the state.

In a radiation field, a molecule in state Z_n with energy ε_n may absorb radiation of frequency ν and pass to a state Z_m with higher energy ε_m. The probability for this process to happen during the time $\mathrm{d}t$ is assumed to be

$$\mathrm{d}W = B_n^m \rho \,\mathrm{d}t$$

[1] This assumption was already introduced in an earlier paper of Einstein: Verh. der D. Physik. Ges. *16*, p. 820 (1914). See also Einstein's paper on Specific Heat in Annalen der Physik (4) *22*, p. 180 (1907).

where ρ is the radiation density for frequency ν. Just so, a molecule in state Z_m may emit radiation and pass into a state Z_n with lower energy. The probability for this process is assumed to be

$$\mathrm{d}W = (A^n_m + B^n_m\rho)\mathrm{d}t.$$

If the radiation is in equilibrium with the molecular distribution of states at temperature T, the following condition must hold:

$$p_n \exp(-\varepsilon_n/kT)B^m_n\rho = p_m \exp(-\varepsilon_m/kT)(B^n_m\rho + A^n_m).$$

From this condition and from Wien's displacement law, Einstein derives Planck's radiation law and Bohr's frequency condition

$$\varepsilon_m - \varepsilon_n = h\nu.$$

Einstein next assumes that in an elementary process of emission or absorption only *directed* radiation bundles are emitted or absorbed. He says: 'Outgoing radiation in the form of spherical waves does not exist'. For the elementary processes, he postulates the conservation of momentum and energy. The momentum of a directed radiation bundle carrying an energy $h\nu$ is supposed to be $h\nu/c$, and the direction of the radiation bundle emitted from a molecule to be determined by chance. It is shown that the recoil momenta transferred from the radiation field to the molecules never disturb the thermodynamic equilibrium. This result is regarded by Einstein as a justification of his initial assumptions, for if one of these assumptions were to be changed, the result would not come out.

All subsequent research on absorption emission and dispersion of radiation was based upon Einstein's paper **1**.

The Adiabatic Hypothesis

Two important heuristic principles have guided quantum physicists during the period 1913–1925, viz. Ehrenfest's *Adiabatic Hypothesis* and Bohr's *Principle of Correspondence.*

The Adiabatic Hypothesis, first formulated by Ehrenfest [1] in 1913, says:

[1] P. Ehrenfest: Bemerkung betreffs der spezif. Wärme zweiatomiger Gase. Verh. D. physik. Ges. *15*, p. 451 (1913). A Theorem of Boltzmann and its connection with the theory of quanta. Proc. Kon. Akad. Amsterdam *16*, p. 591 (1913).

'If a system be affected in a reversible adiabatic way, allowed motions are transformed into allowed motions.'

The name Adiabatic Hypothesis is due to Einstein, as Ehrenfest states in paper **2**. Bohr calls the hypothesis 'Principle of mechanical transformability' (see **3**, § 1).

In paper **2**, Ehrenfest explains what he means by a 'reversible adiabatic affection', and he formulates the Adiabatic Hypothesis as sharply as possible, at the same time showing what is wanting in sharpness. He next demonstrates the importance of adiabatic invariants, and he indicates the difficulties that arise in the application of the hypothesis in singular cases. Finally, Ehrenfest shows that the adiabatic hypothesis is closely connected with the Second Law of Thermodynamics.

The Principle of Correspondence

Bohr's fundamental 1918 paper 'The quantum theory of line spectra' consists of three parts; the fourth part announced in the introduction never appeared. Part I 'On the general theory' will be reproduced here as paper **3**. Part II deals with the hydrogen spectrum. Part III, which was not published until 1922, contains a preliminary discussion of the spectra of other elements.

In Part I, Bohr once more enunciates the two fundamental assumptions of Quantum Theory:

I. That an atomic system can only exist permanently in a discontinuous series of 'stationary states',

II. That the radiation absorbed or emitted during a transition between two stationary states possesses a frequency ν given by

$$E' - E'' = h\nu. \tag{1}$$

Since these assumptions imply that no emission of radiation takes place in the stationary states, it follows that the ordinary laws of electrodynamics cannot be applied to these states without radical alterations. In many cases, however, the effect of that part of the electrodynamical forces which is connected with the emission of radiation will at any moment be very small as compared with the Coulomb forces. Therefore Bohr assumes that a close approximation of the motion in the stationary states can be obtained by retaining only the Coulomb forces and calculating the motions of the particles by ordinary mechanics.

Next, Bohr considers a transition between two stationary states. He remarks that in the limiting region of slow vibrations, it has been possible to account for the phenomenon of temperature radiation by ordinary electrodynamics. Hence '*we may expect that any theory capable of describing this phenomenon in accordance with observation will form some sort of natural generalisation of the ordinary theory of radiation.*'

If we analyse this argument, we see that it consists of two parts. First, Bohr expresses an experience from earlier investigations concerning the limiting region of slow vibrations, and next an expectation for future research. At this stage of Bohr's exposition, his expectation is formulated only in a qualitative form: the future theory of radiation must be a 'natural generalisation' of the classical theory.

Next, Bohr considers another limiting case, viz. the case of high quantum numbers. Once more, the starting point is a conclusion drawn from earlier research:

> 'We shall show ... that the conditions which will be used to determine the values of the energy in the stationary states are of such a type that the frequencies calculated by (1), in the limit where the motions in successive stationary states differ very little from each other, will tend to coincide with the frequencies to be expected on the ordinary theory of radiation from the motion of the system in the stationary states' (§ 1 of Bohr's paper **3**).

Immediately after this conclusion from research already carried out, Bohr formulates an expectation for future research. He first reminds us of his earlier conclusion from the limiting case of slow vibrations:

'In order to obtain the necessary relation to the ordinary theory of radiation in the limit of slow vibrations we are therefore led directly to certain conclusions about the probability of transition between two stationary states in this limit.'

These rather vague words in § 1 are a preliminary announcement of a much more definite expectation or claim, which Bohr formulates in § 2 for the case of one degree of freedom and again in § 3 for several degrees of freedom, and which he himself later called *Principle of Correspondence*. For the moment we note two things. At first, Bohr had only said 'we may expect ...'. Now, he uses the expression 'necessary relation', which is much stronger, and he repeats it in § 2.

Secondly, we may note that Bohr, in his preliminary announcement of 'certain conclusions' in § 1, speaks about the probability of transition 'in this limit', i.e. in the limit of *slow vibrations*. In his more definite statement of the Principle of Correspondence in § 2 and § 3, he considers another limiting case, viz. the case of *large quantum numbers*. This is not just the same thing. In many cases the two limits $n \to \infty$ and $\nu \to 0$ coincide, but in the case of the harmonic oscillator the frequency ν remains constant while n goes to infinity. It seems that Bohr, in his preliminary announcement in § 1, did not clearly distinguish between the two limiting cases.

In § 2, after having expanded the displacements of the particles in Fourier series (14) with coefficients C_τ, Bohr notes that, as far as the frequencies are concerned, there exists a close relation between the ordinary theory of radiation and the new theory for large quantum numbers n. Bohr now proceeds:

> In order to obtain the necessary connection, mentioned in the former section, to the ordinary theory of radiation in the limit of slow vibrations, we must further claim that a relation, as that just proved for the frequencies, will, in the limit of large n, hold also for the intensities of the different lines in the spectrum.

In this very remarkable sentence, Bohr formulates not only an expectation, but even a necessity, a claim which the future theory has to fulfill, and he concludes:

'Since now on ordinary electrodynamics the intensities of the radiations ... are directly determined from the coefficients C_τ in (14), we must therefore expect that for large values of n these coefficients will on the quantum theory determine the probability of spontaneous transition from a given stationary state for which $n=n'$ to a neighbouring state for $n=n''=n'-\tau$.'

This expectation or necessary connection between the classical and the future theory in the limit of large quantum numbers, is called the *Principle of Correspondence*. The name 'Korrespondenzprinzip' is found for the first time in a later paper of Bohr: Z. Phys. *2*, p. 423 (1920).

Three years later, in Z. Phys. *13* (1923), Bohr discussed anew the fundamental principles of Quantum Theory in connection with the Principle of Correspondence (paper **3h**, to be quoted at the end of paper **3**).

History of the Correspondence Principle

The first enunciation of an assumption akin to the Principle of Correspondence can be found already in Bohr's first paper in Phil. Mag. *26*,

p. 1 (1913). Bohr speaks of the constant in the expression

$$\nu = \frac{2\pi^2 me^4}{h^3}\left(\frac{1}{\tau_2^2} - \frac{1}{\tau_1^2}\right) \tag{4}$$

– the number (4) is Bohr's – and he proceeds:

From the above consideration it will follow that, taking the starting point in the form of the law of the hydrogen spectrum and assuming that the different lines correspond to a homogeneous radiation emitted during the passage between different stationary states, we shall arrive at exactly the same expression for the constant as that given in (4), if we only assume 1) that the radiation is sent out in quanta $h\nu$, and 2) that the frequency of the radiation emitted during the passing of the system between successive stationary states will coincide with the frequency of revolution of the electron in the region of slow vibrations.

For a fuller discussion of the development and meaning of the Principle of Correspondence in Bohr's papers we may refer to K. M. Meyer-Abich: Korrespondenz, Individualität und Komplementarität, Dissertation Hamburg 1964.

Applications of the Correspondence Principle

In § 2 of his fundamental paper **3**, Bohr applies the Principle of Correspondence to the special case in which certain coefficients C_τ are zero. In this case

we are led to expect that no transition will be possible for which $n'-n''$ is equal to one of these values τ.

Quite similar is Bohr's conclusion in the case of several degrees of freedom (see the end of Bohr's paper).

A successful application of this point of view to the intensity and polarization of the Stark components in hydrogen-like spectra was given by Kramers in his papers of 1919 and 1920, to be quoted at the end of Bohr's paper **3**.

Systematic guessing

The research work during the years 1919–1925 that finally led to Quantum Mechanics may be described as *systematic guessing, guided by the Principle of Correspondence.*

An important step in this direction was made in 1921 by Ladenburg in paper **4**. Kramers, in his papers **6** and **8**, improved on Ladenburg's results. The three papers **4**, **6** and **8** are not easy to understand for a modern reader, because the classical theory of dispersion and absorption of radiation is presupposed. Therefore, I shall first explain this theory[1], following Pauli's excellent exposition[2].

Classical dispersion theory

The polarization $\boldsymbol{P}$ caused by a vibrating electrical field is assumed to be proportional to the field:

$$\boldsymbol{P} = \alpha \boldsymbol{E}. \tag{2}$$

The factor α is connected with the diffraction index and hence measurable. It becomes large when the field frequency ν is near to an absorption frequency ν_i of the atom. Empirically it can well be represented by a formula of the type

$$\alpha = \sum_i \frac{e^2}{4\pi^2 m} \frac{f_i}{\nu_i^2 - \nu^2}. \tag{3}$$

This formula can be derived from classical theory on the assumption that the atom contains a certain number of oscillators, whose frequencies are equal to the absorption frequencies ν_i. If such an oscillator is treated as a particle with charge e and mass m, its electrical moment vector $\boldsymbol{P}$ satisfies the equation

$$\ddot{\boldsymbol{P}} + \gamma \dot{\boldsymbol{P}} + (2\pi\nu_0)^2 \boldsymbol{P} = \frac{e^2}{m} \boldsymbol{E}, \tag{4}$$

ν_0 being the frequency of the oscillator and γ the damping factor. For $\boldsymbol{E} = \boldsymbol{E}_0 \exp(2\pi i \nu t)$ a solution of the equation is $\boldsymbol{P} = \alpha \boldsymbol{E}$, where α is the complex number

$$\alpha = \frac{e^2/m}{4\pi^2(\nu_0^2 - \nu^2) + 2\pi\nu\gamma i}. \tag{5}$$

As long as ν is not very near to ν_0, the imaginary term in the denominator can be neglected. Taking the real part of $\boldsymbol{P}$ and summing over all oscillators, we obtain the formulae (2) and (3). The factor f_i was interpreted by Drude as the 'number of dispersion electrons'

[1] The classical theory of dispersion is mainly due to Drude. See P. Drude, Zur Geschichte der elektromagnetischen Dispersionsgleichungen. Annalen der Physik (4. Folge) *1*, p. 437, received Jan. 15, 1900.

[2] W. Pauli: Quantentheorie. Handbuch der Physik *23* (1926) p. 87. Reprinted in Pauli's Collected Papers (Interscience, New York 1964) p. 357.

pro atom; Pauli calls it the 'strength' of the oscillator. If (3) is assumed to be valid, the numbers f_i can be determined empirically by measuring anomalous dispersion in the neighbourhood of absorption lines.

The same numbers f_i also determine, according to classical theory, the strength of the absorption lines. In quantum theory, the strength of an absorption line is proportional to Einstein's absorption coefficient B_m^n and hence also to A_m^n because of Einstein's formula (see paper **1**)

$$A_m^n = \frac{8\pi h\nu^3}{c^3} B_m^n. \tag{6}$$

Following Pauli, we shall denote the probability coefficient A for a spontaneous passage of the atom from state P to state Q by A_{PQ}, and the corresponding coefficient f_i by f_{PQ}. Ladenburg's 'number of dispersion electrons' $\mathfrak{N}$ is a simple multiple of Pauli's f_{PQ}.

Ladenburg's paper 4

The central idea of this paper is, to equate the classical expression for the strength of an absorption line with the quantum-theoretical expression. Since the former is proportional to f_{PQ} and the latter to A_{PQ}, a formula of the form

$$A_{PQ} = \gamma_0 f_{PQ} \tag{7}$$

is obtained. An easy calculation shows that the coefficient γ_0 is equal to the classical damping factor

$$\gamma_0 = \frac{8\pi^2 e^2 \nu^2}{3mc^3}. \tag{8}$$

Our formulae (7) and (8) are Pauli's formulae (126) and (66). Our formula (7) is equivalent to Ladenburg's formula (7).

In formula (7), A_{PQ} can be measured by emission or absorption measurements. On the right-hand side, the coefficient f_{PQ} has a sense only in classical theory, but the same coefficients f_i also figure in (3), and the α in (3) can be determined from dispersion measurements. Hence, if the $f_i = f_{PQ}$ are solved from (7) and substituted into (3), a relation between measurable quantities is obtained.

By the principle of correspondence, this relation must hold in the limit for large quantum numbers. Hence, Ladenburg's guess is, that it be generally true. In fact, Ladenburg found a reasonable agreement between the f_i computed from absorption measurements, from dispersion measurements, and from classical theory.

'Virtual oscillators'

From the derivation given by Pauli and reproduced here, it follows that Drude's formula (3) is valid for a set of classical harmonic oscillators. Ladenburg does not write out this derivation, but it was generally known at his time, and we may safely assume that he had such a derivation in mind. *Hence we may say that Ladenburg replaced the atom, as far as its interaction with the radiation field is concerned, by a set of harmonic oscillators with frequencies equal to the absorption frequencies* ν_i *of the atom.*

This idea is not explicitly formulated in Ladenburg's paper **4**, but it is implicitly contained in it, and Ladenburg's contemporaries realized this, for in paper **5** the authors (Bohr, Kramers and Slater) write:

> The correspondence principle has led to comparing the reaction of an atom on a field of radiation with the reaction on such a field which, according to the classical theory of electro-dynamics, should be expected from a set of 'virtual' harmonic oscillators with frequencies equal to those determined by equation (1) for the various transitions between stationary states .... Such a picture has been used by Ladenburg in an attempt to connect the experimental results on dispersion qualitatively with considerations on the probability of transitions ....

In this passage, the oscillators of frequency ν_i are called *virtual harmonic oscillators*. It seems that this term was coined by Bohr in his paper **3h**.

Slater's idea

A closely related idea is that of a *virtual field of radiation* emitted by the virtual oscillators. This idea was developed by J. C. Slater in a most remarkable letter to the editor of Nature, dated January 28, 1924, and published in Nature *113*, p. 307 under the title 'Radiation and Atoms'.

In order to understand this letter we must bear in mind that Einstein, as early as 1905, in a highly interesting paper 'Ueber einen die Erzeugung und Verwandlung des Lichtes betreffenden heuristischen Gesichtspunkt' (Annalen d. Phys. *17*, p. 132), had ventured the hypothesis that the energy of light of frequency ν consists of discrete energy quanta $h\nu$, which are localized in single points moving in space, and which can be absorbed or generated only as a whole.

The essential part of Slater's letter to Nature runs thus (the italics are mine):

Any atom may, in fact, be supposed to communicate with other atoms all the time it is in a stationary state, by means of a *virtual field of radiation* originating from oscillators having the frequencies of possible quantum transitions and the function of which is to provide for *statistical conservation of energy and momentum* by determining the probabilities for quantum transitions. The part of the field originating from the given atom itself is supposed to induce a probability that that atom lose energy spontaneously, while radiation from external sources is regarded as inducing additional probabilities that it gain or lose energy, much as Einstein has suggested ....

The idea of the activity of the stationary states presented here suggested itself to me in the course of an attempt to combine the elements of the theories of electrodynamics and of light quanta by setting up a field to guide discrete quanta, which might move, for example, along the direction of Poynting's vector. But when the idea with that interpretation was described to Dr. Kramers, he pointed out that it scarcely suggested the definite coupling between emission and absorption processes which light quanta provide, but rather indicated a much greater independence between transition processes in distant atoms than I had perceived. The subject has been discussed at length with Prof. Bohr and Dr. Kramers, and a joint paper with them will shortly be published in the *Philosophical Magazine*, describing the picture more fully, and suggesting possible applications in the development of the quantum theory of radiation.

When Slater wrote this letter, he was in Copenhagen as a Sheldon Fellow of Harvard University. The discussions with Kramers and Bohr, from which the joint paper **5** resulted, took place in Copenhagen in Bohr's institute. Heisenberg, who visited Copenhagen at the end of March, 1924, told me that the ideas of the Bohr-Kramers-Slater paper played a very important role in the discussions among physicists concerning radiation, transition probabilities, and conservation of energy.

Bohr, Kramers and Slater

In the joint paper **5** of the three authors, we may distinguish three fundamental ideas, viz.:

1) Slater's idea of a 'Virtual radiation field',

2) statistical conservation of energy and momentum,

3) statistical independence of the processes of emission and absorption in distant atoms.

From our present point of view, i.e. from the point of view of Dirac's quantum-mechanical theory of emission and absorption, Slater's idea 1) is perfectly correct, but 2) and 3) are not. In fact, if we start with an atom in a stationary state and a field in a state φ_0 (e.g. in the vacuum state) at time $t=0$, the state of the field at any time t will be a linear combination

$$a_0\varphi_0 + a_1\varphi_1 + \ldots$$

and the coefficients a_1, a_2, ... will, in Dirac's first approximation, perform harmonic oscillations with frequencies ν_i that may be computed from Bohr's law

$$E' - E'' = h\nu.$$

On the other hand, Dirac's theory implies *strict* (not only statistical) conservation of energy and momentum, and strict dependence of emission and absorption of light quanta by distant atoms. Hence 2) and 3) are not acceptable to us, and in fact the assumption 3) was disproved by coincidence experiments of Geiger and Bothe in 1924 (Z. Phys. *32*, p. 639).

The ideas 3) and 2) are closely connected. From Slater's letter to Nature, it is perfectly clear that 3) is due to Kramers, but it is not quite clear to whom 2) is due. Therefore I asked Slater:

When you first had the idea of a 'virtual field' and explained it to Kramers, was the idea of *statistical conservation of energy and momentum* already in your mind, or was it the result of your discussion with Kramers?

From Slater's answer, dated Nov. 4, 1964, I quote:

As you suspected, the idea of statistical conservation of energy and momentum was put into the theory by Bohr and Kramers, quite against my better judgment. I had gone to Copenhagen with the idea that the field of the oscillators would be used to determine the behavior of photons, which I preferred to regard as real entities, satisfying conservation as we now know that they did, and I wished to introduce probability only in so far as the waves determined the probability of the photon's being at a given place at a given time. Bohr and Kramers opposed this view so vigorously that I saw that the only way to keep peace and get the main part of the suggestion published was to go along with them with the statistical idea.

Slater's statement is confirmed by a note published in Nature *116*, p. 278 (1925). In this note Slater discusses the possible existence of a field, obeying Maxwell's equations, the function of which whould be to guide corpuscular quanta, and he writes:

The theory in this form was developed in England, under the guidance of Mr. R. H. Fowler, to whom my sincerest thanks are due. The essential feature was the emission of the field before the ejection of the corpuscle: that is, during the stationary state before the transition .... When this view was presented to Prof. Bohr and Dr. Kramers, they pointed out that the advantages of this essential feature would be kept, although rejecting the corpuscular theory, by using the field to induce a probability of transition rather than by guiding corpuscular quanta. On reflection, it appeared that no phenomena at that time known demanded the existence of corpuscles. Under their suggestion, I became persuaded that the simplicity of the mechanism obtained by rejecting a corpuscular theory more than made up for the loss involved in discarding conservation of energy and rational causation, and the paper already quoted was written.

Kramers' dispersion theory

It was necessary to distinguish clearly between the correct idea 1) and the ideas 2) and 3) that proved incorrect afterwards, because Kramers' important papers **6** and **8** on the law of dispersion are based upon the idea 1), but not upon 2) and 3). Kramers' ideas can most easily be explained if we start with the formulae (2) and (3) derived from Drude's classical theory. Substituting (3) into (2), and writing P and E instead of $\boldsymbol{P}$ and $\boldsymbol{E}$, we obtain the fourth formula of Kramers' first paper:

$$P = E \sum_i f_i \frac{e^2}{m} \frac{1}{4\pi^2(\nu_i^2 - \nu^2)}. \tag{9}$$

Kramers notes that a formula of this type, in which the ν_i are equal to the absorption frequencies of the atom, represents the results of experiment with considerable accuracy. However, the formula does not satisfy the condition, required by the correspondence principle, that in the region of large quantum numbers the interaction between the atom and the radiation field tends to coincide with that expected on classical theory. To satisfy this condition, Kramers proposes another formula containing also negative terms corresponding to emission frequencies. If the absorption and emission frequencies are denoted by ν_i and ν_j, the formula proposed by Kramers can be written as

$$P = E \frac{e^2}{4\pi^2 m}\left(\sum \frac{f_i}{\nu_i^2 - \nu^2} - \sum \frac{f_j}{\nu_j^2 - \nu^2}\right),$$

the coefficients f_i and f_j being the same as our f_{PQ} defined by (7) and (8).

Thus, if the reaction of the atom against the incident radiation is compared with that of a set of virtual oscillators, we have to assume, according to Kramers, not only 'positive virtual oscillators' of strength $+f$ corresponding to absorption frequencies, but also 'negative virtual oscillators' of strength $-f$ corresponding to emission frequencies.

In his paper **6**, Kramers gives no derivation of his dispersion formula. In his second paper **8** (which was an answer to a note by G. Breit published in the same issue of Nature in August 1924), Kramers gives an outline of a derivation. A full derivation of the same formula is contained in the joint paper **9** of Kramers and Heisenberg. As Heisenberg told me, the redaction of this joint paper was entirely due to Kramers, and at the end of the derivation it is stated that a short outline of it was given in paper **8**. In fact, the main idea, viz. the replacement of a differential quotient by a difference quotient, is already formulated in paper **8**.

Max Born

The next important step towards Quantum Mechanics was made by Max Born in paper **7**. Born's idea is, to treat the interaction between two mechanical systems by the same methods by which Kramers treated the interaction between a radiation field and an atom.

In § 1 of his paper, Born gives an exposition of classical perturbation theory. The Hamiltonian is assumed to have the form

$$H = H_0 + \lambda H_1,$$

the perturbation H_1 being given by a Fourier series. In § 2, Born shows that classical dispersion theory can be treated as a special case of classical perturbation theory.

In § 3 Born states his problem: to pass from the classical formulae to those of 'Quantum Mechanics'. This word occurs for the first time in the introduction of Born's paper. Just like Kramers, Born starts with Slater's fundamental idea 1), and he expressly remarks that he will not make use of the 'statistical' ideas 2) and 3) of paper **5**, which are still debated.

Next, Born assumes that an atom in a stationary state n may be replaced, as far as the calculation of emission, absorption and dispersion is concerned, by a set of 'virtual oscillators' of frequencies

$$\nu(n, n') = \frac{1}{h} [W(n) - W(n')].$$

To every 'virtual resonator' corresponds, in the sense of Bohr's

Principle of Correspondence, a term in the Fourier series of the motion in state n, computed classically. Born now remarks that the frequency

$$(\nu\tau) = \nu_1\tau_1 + \dots + \nu_f\tau_f$$

of this term is to the quantum theoretic frequency $\nu(n, n')$ just as a differential coefficient is to a difference quotient. Therefore, in order to obtain quantum formulae from classical formulae, one has to replace all differential coefficients by the corresponding difference quotients or 'linear averages'. This principle is applied to the fundamental formula (16a) derived from classical perturbation theory, and thus Born obtains his 'quantum-mechanical' formula (33). In § 4 he notes that the same process, applied to the classical dispersion formula, yields the dispersion formula of Kramers.

At the time when Born's paper was written, in the summer of 1924, Heisenberg was Born's assistant in Göttingen. In a footnote, Born mentions the fact that some of the calculations were made by Heisenberg. Born also discussed the paper with Bohr, and by these discussions the fundamental notions were made more clear, as Born states in the introduction of his paper.

Kramers and Heisenberg

Heisenberg spent the following winter 1924/25 in Copenhagen, working with Bohr and Kramers. During this winter, the joint paper **10** of Kramers and Heisenberg on the refraction of radiation by atoms (received January 5, 1925) was written. The methods of this paper are closely related to those of Born's paper **7**. The same kind of multiple Fourier series is used, and the same method of replacing differential quotients by difference quotients is applied.

The final form of the main formula of paper **10** was suggested by Heisenberg, but the redaction of the paper is entirely due to Kramers, as Heisenberg told me in an oral communication.

We must return to the year 1924 and discuss an important American contribution.

Van Vleck

A clarification and extension of the Principle of Correspondence was given by J. H. Van Vleck, first in a short expository paper **9a** 'A Correspondence Principle for Absorption' (dated April 7, 1924), and more fully in paper **9** (printed Oct. 1924).

Van Vleck's idea is: If we want to estimate the absorption by means of the Principle of Correspondence, we have to compare the absorption, computed classically, with the difference between absorption and induced emission, computed from Einstein's formulae. In the limit of high quantum numbers, this difference must become equal to the classical absorption.

The importance of Van Vleck's investigation is stressed by Jordan in a letter to me dated December 1, 1961. I had asked Jordan to give me some information about the genesis of his joint papers **13** (with Born) and **15** (with Born and Heisenberg). From his answer I quote:

Die zwei bezeichneten Arbeiten sind hervorgegangen aus den damals sehr aktuellen Bestrebungen unter dem Thema 'Verschärfte Anwendung des Korrespondenzprinzips'. Es waren eine Reihe von Verfassern darauf gekommen, dass man aus dem Bohrschen Korrespondenzprinzip, welches ja zunächst nur eine etwas vage, diffuse Empfehlung zur Beurteilung quantentheoretischer Probleme aussprach, durch sinnvolles Erraten für mancherlei Spezialfragen exakte Antworten bekommen konnte. So wurden z.B. damals die der Quantenmechanik entsprechenden Intensitätsformeln für Multipletts und für Zeeman-Effekte in exakter Form auf der Grundlage korrespondenzmässiger Analogie von verschiedenen Verfassern nach und nach erarbeitet. Andererseits nahm Van Vleck folgendes Problem in Angriff: nach Einstein erforderte ja das Plancksche Gesetz bestimmte Proportionalitäten zwischen den Prozessen positiver Absorption, negativer Absorption und spontaner Emission. Bohr hatte diese Einsteinschen Ergebnisse lange Zeit recht skeptisch beurteilt, aber nun zeigte Van Vleck, dass man auch diese Einsteinschen Gesetze durch verschärfende Anwendung des Korrespondenzprinzips rechtfertigen konnte.

Heisenberg unternahm nun den Versuch, die Intensitätsgesetze der Balmer-Serie in ihrer exakten Form zu erraten auf Grund von Korrespondenz-Betrachtungen; das sollte also ähnlich gemacht werden, wei bei den Multipletts und Zeeman-Effekten. Das Problem erwies sich aber als zu schwierig, und er zog es dann vor, in mehr grundsätzlicher Weise über die Frage einer exakt formulierten Quantenmechanik nachzudenken. Hieraus entstand seine berühmte Arbeit [**12**].

Born und ich waren dafür sehr aufgeschlossen, weil wir zusammen uns gerade damit beschäftigt hatten, die erwähnte amerikanische Untersuchung weiterzuführen; die darin angewandten recht mühsamen Berechnungen konnten wir durch viel einfachere ersetzen. Born hatte in diesem Zusammenhang auch schon die 'Multiplikation' von Schemata von 'Übergangs-Amplituden' erwogen, doch wurde uns erst aus brieflichen Nachrichten Heisenbergs deutlich, welche Idee man mit dieser von Heisenberg ebenfalls entdeckten Möglichkeit der Multiplikation solcher Schemata durchführen konnte: die förmliche Formulierung von mechanischen Bewegungsgleichungen für solche Schemata, anstatt für klassische Koordinaten.

Heisenberg's symbolic multiplication will be discussed presently. We first have to mention an important formula for the sum of the strengths of the absorption lines corresponding to transitions from one state to all other states. This formula was discovered in May 1925 by W. Kuhn and rediscovered in June 1925 by W. Thomas.

The sum rule of Kuhn and Thomas

In Kuhn's paper **11**, the sum rule is obtained from the Principle of Correspondence. Kuhn starts, just as Ladenburg and Kramers did, with the fundamental formula of classical dispersion theory (9). The coefficients $f_i = f_{PQ}$ in this formula are called p_i in Kuhn's paper; we shall follow W. Thomas and call the coefficients f_a, because they correspond to absorption frequencies. When the frequencies ν_i are small as compared with ν, (9) becomes

$$P = -E \frac{e^2}{4\pi^2 m} \frac{1}{\nu^2} \Sigma f_a . \tag{10}$$

If the state of the atom is not the ground state, one has to modify this formula just as Kramers modified Ladenburg's formula by adding terms with opposite sign, thus obtaining

$$P = -E \frac{e^2}{4\pi^2 m} \frac{1}{\nu^2} (\Sigma f_a - \Sigma f_e) \tag{11}$$

(a = absorption, e = emission).

From this formula, Kuhn calculates the dispersion and compares it with a theoretical formula due to J. J. Thomson. The two expressions coincide, if $\Sigma f_a - \Sigma f_e = 1$. Hence Kuhn assumes that the formula

$$\Sigma f_a - \Sigma f_e = 1 \tag{12}$$

holds quite generally.

In paper **11a**, written at Breslau in June 1925 by W. Thomas, formula (12) is given in generalized form as

$$\Sigma f_a - \Sigma f_e = \text{Periodizitätsgrad.} \tag{13}$$

Thomas does not say how he obtained this formula.

PART II. THE BIRTH OF QUANTUM MECHANICS

Werner Heisenberg

We are now approaching the turning point, as Kronig calls it in his contribution to the Pauli Memorial Volume, that is, the point at which Heisenberg discovered a small path that led from the darkness towards the light of a new physics.

Heisenberg (born 1901) was a pupil of Sommerfeld in München. He first visited Göttingen at the occasion of Bohr's lectures in the summer of 1922, the so-called 'Bohr-Festspiele'. During the winter 1922–23 he studied in Göttingen and worked with Born; two papers **7b** and **7c**, in which the perturbation methods of the astronomers were applied to atomic systems, resulted from this collaboration. In the summer of 1923 Heisenberg returned to München to write his PhD-Thesis on a problem of hydrodynamics. In October 1923 he became Born's assistant at Göttingen.

With Born's permission I shall quote a passage from Born's Recollections concerning Heisenberg:

He looked like a simple peasant boy, with short, fair hair, clear bright eyes and a charming expression. He took his duties as an assistant more seriously than Pauli and was a great help to me. His incredible quickness and acuteness of apprehension has always enabled him to do a colossal amount of work without much effort: he finished his hydrodynamical thesis, worked on atomic problems partly alone, partly in collaboration with me, and helped me to direct my research students.

Born, Franck, Pauli and Heisenberg

To characterize the scientific atmosphere at Göttingen during the years 1923–25, some more quotations from Born's Recollections may be helpful:

Meanwhile James Franck pursued, with a great school of collaborators and pupils, his experimental investigations of quantum effects in the interaction of atoms and molecules which I followed with great interest. From our daily discussion there sprang a joint publication on the application of quantum theory to the kinetics of chemical reactions (Z.f. Phys. 1925).

My main interest during this period was directed however towards the quantum theory of the electronic structure of atoms. The question was, how far Bohr's theory did actually account for the facts, and to find its limitations.

Bohr assumed that an atom existed in stationary states which were described by special solutions of the equations of ordinary mechanics selected by certain quantum rules .... This principle worked exceedingly well in the case of one-electron system, like the hydrogen atom or helium ion. But could it be generalised for systems with several electrons like the neutral helium atom?

To decide this was the first point of our programme. It meant an adaptation of the classical perturbation methods of the astronomers to atomic systems. This problem was tackled in several papers in collaboration first with Pauli [**7a**, 1922] and later with Heisenberg [**7b** and **7c**, 1923]. Applied to helium these methods gave, as we rather expected, results which did not agree with spectroscopic measurements.

On the other hand, the qualitative results showed a fair agreement with the general properties of matter. That was shown in a particularly striking way by Heisenberg and myself in a paper [**7d**, 1924] where we derived the main properties of molecules with our perturbation methods, using the square root of a certain ratio (mass of the electron to mass of a nucleus) as an expansion parameter.

We became more and more convinced that a radical change of the foundations of physics was necessary, i.e. a new kind of mechanics for which we used the term quantum mechanics.

This word appears for the first time in physical literature in a paper of mine [**7**, 1924] in which an essential step towards the establishment of this new doctrine was made.

The fundamental problem of this paper **7** and of the earlier papers **4** (by Ladenburg) and **6** (by Kramers) is formulated by Born as follows:

We knew that frequencies of vibrations were proportioned to energy differences between two stationary states. It slowly became clear that this was the main feature of the new mechanics: each physical quantity depends on two stationary states, not on one orbit as in classical mechanics. To find the laws for these 'transition quantities' was the problem.

Born notes that the dispersion formula of Kramers contains only 'transition quantities', and he proceeds:

One can say that Kramers, guided by Bohr's principle of correspondence, guessed the correct expression for the interaction between the electrons in the atom and the electro-magnetic field of the light wave. That is at least the way I regarded his results. It was the first step from the bright realm of classical mechanics into the still dark and unexplored underworld of the new quantum mechanics. I made the next step with the question: Can one not find, by a similar systematic guessing, the interaction between two electronic systems in terms of 'transition quantities'? Indeed, by a proper re-interpretation of the classical perturbation theory the corresponding quantum formula could be constructed [paper **7**]. It was later fully confirmed by quantum mechanics.

Born and Jordan

Next, Born gives some details about his collaboration with Pascual Jordan, which led to their joint paper **7e** in Z. Phys. *33* (received June 11, 1925):

Jordan became my collaborator in the next problem I attacked; it was concerned with Planck's theory of radiation which led him to the existence of quanta. An inspection of his work showed that he used classical mechanics for the interaction of light and matter, an odd discrepancy. We translated Planck's calculations in the language of quantum theory, introducing 'transition quantities' instead of the corresponding classical quantities. Our paper on aperiodic quantum processes appeared in Z.f. Phys. 1925.

We were struck by the fact that the 'transition quantities' appearing in our formulae always corresponded to squares of amplitudes of vibrations in classical theory. So it seemed very likely that the notion 'transition amplitudes' could be formed. We discussed this idea in our daily meetings in which Heisenberg often took part, and I suggested that these amplitudes might be the central quantities and be handled by some kind of symbolic multiplication. Jordan has confirmed that I spoke to him about this possibility.

The solution of the riddle of 'symbolic multiplication' was found by Heisenberg in the spring of 1925. How did he arrive at this solution?

Bohr and Heisenberg

Of decisive importance for the evolution of Heisenberg's ideas was, according to his own testimony, his discussion with Bohr at the 'Bohr-Festspiele' in Göttingen in the summer of 1922. In the Bohr Memorial Volume (North-Holland Publishing Co., Amsterdam 1967) Heisenberg records:

For the first time I met Niels Bohr in Göttingen in the summer of 1922, when Bohr held a series of lectures at the invitation of the faculty of exact sciences, which we liked to call the 'Bohr Festival'. Sommerfeld, my teacher in Munich, had taken me along to Göttingen, although I was at that time only a 20 year old student in my fourth semester. Sommerfeld was warmly interested in his students, and he had noticed how strongly Bohr and his atomic theory interested me. The first impression of Bohr still remains quite clearly in my memory. Full of youthful excitement, but a little self-conscious and shy, his head a little to one side, the Danish physicist stood on the platform in the auditorium, the strong Göttingen summer light streaming in through the open windows. He spoke softly and with some hesitation, but behind every carefully chosen word one could discern a long chain of thought, which eventually faded somewhere in the background into a philosophical viewpoint which fascinated me.

At the end of the second and third lecture Bohr spoke of a calculation, which his collaborator, Kramers from Holland, had carried out on the so-called 'quadratic' Stark-effect in the hydrogen atom, and Bohr concluded with the remark that in spite of all the internal difficulties of atomic theory at that time, one should assume that Kramers' results were correct and would be verified by experiment. I knew Kramers' work rather well, as I had reviewed it in Sommerfeld's seminar in Munich, and therefore I dared to dissent during the discussion afterwards. I did not believe that Kramer's results were perfectly correct, for the quadratic Stark-effect could be thought of as a limiting case of the scattering of light with very large wave length. But since one knew in advance that a calculation of scattering on a hydrogen atom by the methods of classical physics must lead to a wrong result – the characteristic resonance effect would occur with the electron's orbital frequency – Kramer's calculation could hardly be expected to give a correct result. Bohr answered that one should here take into account the reaction of the radiation on the atom, but he was obviously worried by this objection. When the discussion was over, Bohr came to me and suggested that we should go for a walk together on the Hainberg outside Göttingen. Of course, I was very willing. That discussion, which took us back and forth over Hainberg's wooded heights, was the first thorough discussion I can remember on the fundamental physical and philosophical problems of modern atomic theory, and it has certainly had a decisive influence on my later career. For the first time I understood that Bohr's view of his theory was much more sceptical than that of many other physicists – e.g. Sommerfeld – at that time, and that his insight into the structure of the theory was not a result of a mathematical analysis of the basic assumptions, but rather of an intense occupation with the actual phenomena, such that it was possible for him to sense the relationship intuitively rather than derive them formally.

Thus I understood: knowledge of nature was primarily obtained in this way, and only as the next step can one succeed in fixing one's knowledge in mathematical form and subjecting it to complete rational analysis. Bohr was primarily a philosopher, not a physicist, but he understood that natural philosophy in our day and age carries weight only if its every detail can be subjected to the inexorable test of experiment.

Let us examine a little more closely the objection raised by Heisenberg in his first discussion with Bohr. According to classical theory, resonance between a radiation field and a hydrogen atom ought to take place when the frequency of the field is equal to that of the orbital motion. In reality, it takes place at the frequency of the light which can be emitted or absorbed by the atom. Hence the calculation of the dispersion from classical theory necessarily leads to wrong results, even for long waves. The same thing ought to be expected, according to Heisenberg, in the limiting case of infinitely long waves, i.e. in the case of the quadratic Stark effect.

For a harmonic oscillator the frequency of the emitted or absorbed light is just the same as that of the oscillator, so in this case the difficulty raised by Heisenberg does not yet occur. This explains why Heisenberg, in his letters to Pauli and Kronig and in his paper **12**, always takes the example of the anharmonic oscillator.

The anharmonic oscillator

The first letter of Heisenberg to Pauli, in which the anharmonic oscillator and its interaction with the radiation field is investigated, is dated September 29, 1922. I quote from this letter:

Ich bin überzeugt, dass die Formel $u_1/h - u_2/h = \Delta\nu$ im Grenzfall grosser Quantenzahlen stets die klassische Abklingung liefert.[1] Einen allgemeinen Beweis habe ich mir allerdings noch nicht überlegt, doch werde ich Ihnen schnell den Fall des unharmonischen Oszillators vorrechnen, aus dem man den allgemeinen Fall vermutlich verallgemeinern kann.

Die Gleichung des Oszillators sei (δ sehr klein)

$$\ddot{x} + \omega^2 x + \delta x^2 = 0.$$

Wir versuchen

$$x = a e^{i\omega t} + a_2 e^{2i\omega t} + a_3 e^{3i\omega t} + \ldots.$$

Es ergibt sich

$$a_2 = \left(\frac{\delta \cdot a}{\omega^2}\right) a \cdot \frac{1}{3} \qquad a_3 = \left(\frac{\delta a}{\omega^2}\right)^2 \cdot a \cdot \frac{1}{12}$$

$$a_4 = \left(\frac{\delta a}{\omega^2}\right)^3 \cdot a \cdot \frac{1}{54} \quad \text{u.s.w.} \quad \text{allgemein}$$

$$a_n = \left(\frac{\delta a}{\omega^2}\right)^{n-1} \cdot a \cdot z, \text{ wo } z \text{ eine reine Zahl ist.}$$

Also

$$x = a\, e^{i\omega t} + \frac{1}{3}\left(\frac{\delta}{\omega^2}\right) a^2\, e^{2i\omega t} + \frac{1}{12}\left(\frac{\delta}{\omega^2}\right)^2 \cdot a^3\, e^{3i\omega t} + \ldots.$$

Ist nun der Oszillator noch gedämpft, d.h. wird die Strahlung berücksichtigt, so tritt an Stelle von a augenscheinlich $a \cdot e^{-\gamma t}$; also wird x

$$x = a\, e^{-\gamma t + i\omega t} + \frac{1}{3}\left(\frac{\delta}{\omega^2}\right) a^2\, e^{-2\gamma t + 2i\omega t} + \ldots.$$

Tatsächlich wird also die Dämpfung der τ. Oberschwingung τ mal so gross, wie die der Grundschwingung. Dies folgt bei dieser Rechnung, wie man sieht, aus einer Art Dimensionsgleichung, und ich glaube, dass sich damit der allgemeine Beweis führen lässt, wenngleich ichs nicht sicher sagen kann.

The letter to Kronig

The same equation

$$\ddot{x} + \omega^2 x + \delta x^2 = 0$$

[1] The exact meaning of this sentence is not clear to me, but the continuation of the letter can be understood without it.

reoccurs in Heisenberg's letter to Kronig (dated June 5, 1925), which is reproduced on p. 23–25 of Kronig's article in the Pauli Memorial Volume. Just as in his earlier letter, Heisenberg solves the equation by a Fourier series, and derives recursive relations for the Fourier coefficients a, a_2, a_3, ... In his earlier letter to Pauli, Heisenberg had introduced a damping factor $e^{-\gamma t}$ without investigating the mechanism of the damping, but in the letter to Kronig an attempt is made to calculate the interaction between the oscillator and the electromagnetic field. The Coulomb force at a point of the field gives rise to a term $(1+x^2/a^2)^{-1}$ which is expanded as a Fourier series

$$b_0 + b_1 \cos \omega t + b_2 \cos 2\omega t + \ldots$$

in which the b's are explicit quadratic functions of the a's, e.g.

$$b_1 = -\frac{2(a_0 a_1 + \frac{1}{2} a_1 a_2 + \ldots)}{a^2}.$$

Now Heisenberg replaces this formula by its 'quantum-theoretical reinterpretation'

$$b_1(n, n-1) = -\frac{1}{a^2}\,[a_0(n)\, a_1(n, n-1) + a_1(n, n-1)\, a_0(n-1) + \ldots].$$

Kronig rightly remarks: 'From this letter it appears how the law for the multiplication of quantum-mechanical quantities, later expressed in matrix terminology, was beginning to take shape'. In fact, a few lines further down, Heisenberg takes a simple example of a multiplication of Fourier terms in classical theory:

$$b_2\, e^{2i\omega t} = (a_1\, e^{i\omega t})^2$$

and gives its quantum-mechanical reinterpretation as:

$$b_2(n, n-2) = a_1(n, n-1)\, a_1(n-1, n-2).$$

This is, in principle, the rule of matrix multiplication. We may conclude that Heisenberg discovered it early in June.

From Heisenberg's own account of the development of his ideas in the spring of 1925, published in the Pauli Memorial Volume (p. 42) it appears that he first tried to arrive at intensity formulae for the hydrogen spectrum by studying the Fourier expansion of the Kepler orbit, hoping that ultimately it might be possible just to guess the correct intensity formulae. However, the Kepler problem proved too difficult for this purpose. In fact, the Fourier expansion is much easier to obtain in the case of the anharmonic oscillator.

'Fabricating Quantum Mechanics'

In June 1925, a heavy attack of hay fever forced Heisenberg to leave Göttingen and to stay on the island of Helgoland, where no grass grows, for nine or ten days. Here his vague ideas on Quantum Mechanics acquired a more definite shape. Many years later he told me:

In Helgoland war ein Augenblick, in dem es mir wie eine Erleuchtung kam, als ich sah, dass die Energie zeitlich konstant war. Es war ziemlich spät in der Nacht. Ich rechnete es mühsam aus, und es stimmte. Da bin ich auf einen Felsen gestiegen und habe den Sonnenaufgang gesehen und war glücklich.

On his way back to Göttingen, Heisenberg met Pauli at Hamburg. On June 21 he writes to Pauli:

Noch einmal herzlichen Dank für die freundliche Aufnahme und Bewirtung in Hamburg .... In meinen Versuchen, eine Quantenmechanik zu fabrizieren, geht es nur langsam weiter, aber ich werde mich gar nicht drum kümmern, wie weit ich mich von der Theorie der bedingt-periodischen Lösungen entferne. Z.B. bleibt, was ich neulich sagte, richtig, dass schon die Energie des Oszillators $(n+\frac{1}{2})\nu h$, ebenso die des Rotators $(m+\frac{1}{2})^2(h^2/8\pi^2A)$ sein müsse ....

The next letter, of June 24, already contains the main ideas of Heisenberg's paper **12**:

Ueber meine eigenen Arbeiten hab ich fast keine Lust zu schreiben, weil mir selbst alles noch unklar ist und ich nur ungefähr ahne, wie es werden wird, aber vielleicht sind die Grundgedanken doch richtig. Grundsatz ist: Bei der Berechnung von irgendwelchen Grössen, als Energie, Frequenz usw. dürfen nur Beziehungen zwischen prinzipiell kontrollierbaren Grössen vorkommen. (Insofern scheint mir z.B. die Bohrsche Theorie beim Wasserstoff viel formaler als die Kramerssche Dispersionstheorie). Also beim Oszillator heisst die Bewegungsgleichung

$$\ddot{q} + \omega^2 q = 0;$$

setzt man symbolisch an

$$q = a(n, n-1)\, e^{i\omega(n,\, n-1)t},$$

so kriegt man natürlich

$$\omega(n, n-1) = \omega_0;$$

beim anharmonischen Oszillator bekommt man z.B.

$$\ddot{q} + \omega^2 q + \lambda q^2 = 0.$$

$$a_2(n, n-2)\left(-\omega^2(n, n-2) + \omega_0^2\right) + \lambda a_1(n, n-1)\, a_1(n-1, n-2) = 0 \text{ usw.}$$

Das Wichtigste ist aber die Festlegung der Konstante, d.h. die Quantenbedingung:

Klassisch heisst sie:

$$J = 2\pi m \, \Sigma \, a_\tau^2(\omega\tau)\cdot\tau, \quad \text{wo} \quad q = \Sigma \, a_\tau \, e^{i\omega\tau\cdot t}; \quad p = m\dot{q} = \Sigma \, mia_\tau\cdot(\tau\omega) \, e^{i\omega\tau\cdot t}$$

oder

$$1 = 2\pi m \, \Sigma \, \tau \frac{\delta}{\delta J}\cdot a_i^2(\omega\tau),$$

dies wird quantentheoretisch

$$h = 2\pi m \sum_\tau \{a^2(n+\tau, n) \, \omega(n+\tau, n) - a^2(n, n-\tau) \, \omega(n, n-\tau)\}.$$

Also beim Oszillator ($q = \frac{1}{2}[a(n, n-1) \, e^{i\omega t} + a(n, n-1) \, e^{-i\omega t}]$) [1]

$$h = \pi m[a^2(n, n+1) - a^2(n, n-1)]\omega_0.$$

Durch diese Gleichung, könnte man meinen, sind die $a(n, n-1)$ nur bis auf eine additive Konstante ermittelbar. Dies ist aber nicht der Fall, denn es muss einen tiefsten Zustand geben, von dem aus keine Sprünge mehr möglich sind. Die Definition des Normalzustandes ist, dass die a (*nach abwärts*) *verschwinden.* Hierdurch ist die Konstante festgelegt. Numeriert man die Zahl n noch so, dass für den Normalzustand $n=0$ ist (dies ist keine physikalische Aussage), so ergibt sich:

$$a^2(n, n-1) = \frac{nh}{m\pi\omega_0}.$$

Die Energie ist (die Quadrate von $\dot{q}$ und q sind wieder symbolisch gemeint):

$$W = \frac{m}{2}(\dot{q}^2 + \omega_0^2 q^2) =$$

$$= \frac{m}{2}\left[-\omega_0^2 \frac{(a(n, n-1) \, e^{i\omega t} - a \, e^{i\omega t})^2}{4} + \omega_0^2 \frac{(e^{i\omega t} + e^{-i\omega t})^2}{4}\right]^2 =$$

$$= \frac{m}{2}\omega_0^2 \frac{a^2(n, n+1) + a^2(n, n-1)}{2} = \frac{(n+\frac{1}{2})\omega_0 h}{2\pi}.$$

Für den anharmonischen Oszillator ergibt sich im wesentlichen Ihre Form 'B': $E = (n+\frac{1}{2})h\nu + \beta(n^2+n+\frac{1}{2})$.

Ich wäre Ihnen sehr dankbar, wenn Sie mir schreiben könnten, welche Argumente zu ungunsten dieser Formel sprechen. Abgesehen von der Formulierung der Quantenbedingung bin ich mit dem ganzen Schema noch nicht recht zufrieden. Der stärkste Einwand scheint mir der, dass die Energie, als Funktion der $\dot{q}$ und q geschrieben, im allgemeinen [3] keine Konstante zu werden braucht, auch wenn die Bewegungsgleichungen erfüllt sind. Es liegt dies letzten Endes daran, dass das Produkt zweier Fourierreihen doch nicht eindeutig definiert ist – ... Genauer durchgerechnet habe ich den Rotator, da kann man die Kronigschen und die Kembleschen Formeln bekommen ... Auch ich würde gerne verstehen, was eigentlich die Bewegungsgleichungen bedeuten, wenn man sie als Relation zwischen den Uebergangswahrscheinlichkeiten auffasst.

A few editor's notes may be helpful:

1 The second $a(n, n-1)$ should read $a(n, n+1)$.

2 This formula should read:

$$W = \frac{m}{2}(\dot{q}^2 + \omega_0^2 q^2) = \frac{m}{2}\left[-\omega_0^2 \frac{(a(n, n-1)\,\mathrm{e}^{\mathrm{i}\omega t} - a(n, n+1)\,\mathrm{e}^{-\mathrm{i}\omega t})^2}{4} + \omega_0^2 \frac{(a(n, n-1)\,\mathrm{e}^{\mathrm{i}\omega t} + a(n, n+1)\,\mathrm{e}^{-\mathrm{i}\omega t})^2}{4}\right].$$

The errors are probably copying errors only, the final result being correct. If $\dot{q}^2$ and $\omega_0^2 q^2$ are computed by regular matrix multiplication and added, the result is a constant diagonal matrix. It seems that this was the computation Heisenberg made late in the night in Helgoland: 'Ich rechnete es mühsam aus und es stimmte'.

3 The words 'im allgemeinen' ought to be stressed. In special cases such as the oscillator, Heisenberg had found the energy to be constant.

On June 29, Heisenberg writes:

... Inzwischen bin ich etwas, aber nicht viel weitergekommen und ich bin im Herzen wieder überzeugt, dass diese Quantenmechanik schon richtig ist, weshalb Kramers mich des Optimismus anklagt ...

The final text of paper 12

Ten days later, Heisenberg finished his paper 'Quantentheoretische Umdeutung ...' and sent the manuscript to Pauli, asking for his opinion. From his letter of July 9 I quote:

Wenn Sie glauben, dass ich Ihren Brief mit Hohngelächter gelesen hätte, so täuschen Sie sich sehr; im Gegenteil ist meine Meinung über die Mechanik seit Helgoland von Tag zu Tag radikaler ... Es ist wirklich meine Ueberzeugung, dass eine Interpretation der Rydberg-Formel im Sinne von Kreis und Ellipsenbahnen in klassischer Geometrie nicht den geringsten physikalischen Sinn hat und meine ganzen kümmerlichen Bestrebungen gehen dahin, den Begriff der Bahnen, die man doch nicht beobachten kann, restlos umzubringen und geeignet zu ersetzen. Deshalb getraue ich mich auch, Ihnen einfach das Manuskript meiner Arbeit kurzerhand zuzuschicken, weil ich glaube, dass sie, wenigstens im kritischen, d.h. negativen Teil wirkliche Physik enthält. Zwar habe ich ein sehr schlechtes Gewissen, weil ich Sie bitten muss, die Arbeit mir in 2–3 Tagen wiederzusenden, da ich sie noch in den letzten Tagen meines Hierseins entweder fertig machen oder verbrennen möchte. Meine eigene Meinung über das Geschreibsel, über das ich gar nicht sehr glücklich bin, ist die: dass ich von dem negativen heuristischen Teil fest überzeugt bin, dass ich aber den positiven für reichlich formal und dürftig halte; aber vielleicht können Leute, die mehr können, etwas vernünftiges daraus machen. Also lesen Sie bitte hauptsächlich die Einleitung ....

Summary of paper 12

The leading ideas of Heisenberg's paper may be summarized as follows:

1. In the atomic range, classical mechanics is no longer valid. In Heisenberg's own words: '... auch bei den einfachsten quantentheoretischen Problemen kann an eine Gültigkeit der klassischen Mechanik nicht gedacht werden.'

This was the common opinion of theoretical physicists of the schools of Copenhagen and Göttingen. What they were looking for, was a new mechanics, which Born had already called 'Quantenmechanik'.

2. One of the main conditions which this new theory had to fulfill was Bohr's 'Correspondence Principle': For large quantum numbers the results obtained from the new theory should converge to those obtained from classical mechanics.

This principle had guided physicists during the transition period from 1918 to 1925. A number of important results had been obtained by systematic guessing, based upon the Principle of Correspondence. In Heisenberg's paper, every formula of quantum mechanics is motivated by a corresponding classical formula.

3. An important device for satisfying the requirements of the Principle of Correspondence is: to replace differential quotients occurring in classical formulae by difference quotients. This device had already been employed in the papers of Born (**7**) and of Kramers and Heisenberg (**10**).

4. Heisenberg felt that the difficulties, to which the rules of quantization had given rise, were not mainly due to a deviation from classical mechanics, but rather to a breakdown of the *kinematics* underlying this mechanics. In the introduction to his paper, he motivates this as follows: 'The Einstein–Bohr frequency condition (which is valid in all cases) already represents such a complete departure from classical mechanics, or rather (from the viewpoint of wave theory) from the kinematics underlying this mechanics, that even for the simplest quantum-theoretical problems the validity of classical mechanics just cannot be maintained.'

In this sentence, the words 'from the viewpoint of wave theory' were obscure to me. I therefore asked Heisenberg what he had meant by this phrase and whether he had thought of light waves or of De Broglie waves. From his answer (dated Oct. 8, 1963) I quote:

Ich habe sicher die Wellentheorie des Lichtes gemeint und ausdrücken wollen, dass diese Lichtwellen doch von einer Strahlungsquelle erzeugt sein müssen, die die richtige Frequenz hat und nicht etwa die Frequenz der Elektronen in den Bohrschen Bahnen des Atoms. Dieser Punkt hatte mich ja schon lange aufgeregt (vergleiche mein Gespräch mit Bohr bei den Bohr-Festspielen 1922 in Göttingen) und ich habe der Wellentheorie des Lichtes wenigstens so weit vertraut, dass ich mir sagte: also muss im Atom irgend etwas mit dieser richtigen Frequenz schwingen und das muss doch bedeuten, dass man für das Elektron irgendeine seltsame Kinematik einführen muss, die diese Frequenz durch Kombination zweier stationärer Zustände liefert.

The feeling that 'something in the atom must vibrate with the right frequency' was shared by all those who replaced the atom by a set of 'virtual oscillators', i.e. by Ladenburg, Bohr, Kramers, Slater, Born and others. However, the idea to introduce a new kind of *kinematics* rather than a new kind of mechanics, is Heisenberg's own.

5. In his search for this new kinematics, Heisenberg had a completely new idea. He assumed that the equation of motion of an electron

(i) $$\ddot{x} + f(x) = 0$$

can be retained, and that only the kinematical interpretation of the quantity x as a location depending on time has to be rejected.

This fundamental idea occurs already in the letter to Kronig quoted before.

6. The question is now: What kind of quantities are to be substituted for x in the equation of motion?

In the classical case of a periodic motion the function $x(t)$ can be expanded in a Fourier series

(ii) $$x(t) = \sum_{-\infty}^{\infty} a_\alpha \, e^{i\alpha\omega t}.$$

In quantum theory, the coefficients a_α and the frequency depend on a quantum number n; therefore Heisenberg writes instead of (ii)

(iii) $$x(t) = \sum_{-\infty}^{\infty} a_\alpha(n) \, e^{i\alpha\omega_n t}.$$

Heisenberg's idea is now, to replace the terms of the Fourier series (iii) by a new kind of terms

(iv) $$a(n, n - \alpha) \, e^{i\omega(n, n-\alpha)t}$$

corresponding to the transition from n to $n-\alpha$, the time factor $\omega(n, n-\alpha)$ being 2π times the frequency of the light emitted in this transition.

The idea of introducing 'transition quantities' $a(n, n\pm\alpha)$ into the equation of motion occurs already in the letter to Kronig. Born too had considered the introduction of 'transition amplitudes', as we have seen.

Why did Born and Heisenberg want to introduce quantities like $a(n, m)$ depending on two states n and m into the theory?

Born's line of thought is clearly explained in his Recollections quoted before. He noted that the coefficients $f_i = f_{PQ}$ in the formulae of Ladenburg and Kramers, which are proportional to Einstein's emission probabilities A_{PQ}, always correspond to *squares of amplitudes* in the classical theory. He now conjectured that in Quantum Mechanics too 'transition amplitudes' could be formed, the implication being (of course) that the transition probabilities A_{PQ} be proportional to the squares of the transition amplitudes.

From Heisenberg's letters and from the introduction to his paper **12** we know that for him too the main problem was the calculation of the intensity of the radiation emitted in a transition P→Q. He knew that this intensity is proportional to Einstein's emission probability A_{PQ}, and he assumed, as we shall see presently, this probability to be proportional to $|a(n, n-\alpha)|^2$. He motivated the introduction of the $a(n, n-\alpha)$ by saying that the intensities and hence the $|a(n, n-\alpha)|$ are observable, in contrast to the classical functions $x(t)$.

7. Now the question arises: What exactly is, in Heisenberg's line of thought, the connection between $a(n, n-\alpha)$ and the intensity of the emitted radiation?

In § 1 of Heisenberg's paper, this connection is not clearly indicated. Instead of the $a(n, n-\alpha)$ certain quantities $\mathfrak{A}(n, n-\alpha)$ are introduced, which are said to be complex vectors determining polarization and phase of the emitted light. The part of the radiation corresponding to the transition $n \to n-\alpha$ is said to be represented by the expression

$$\operatorname{Re}\{\mathfrak{A}(n, n-\alpha)\, e^{i\omega(n,n-\alpha)t}\} \tag{1}$$

analogous to the classical expression

$$\operatorname{Re}\{\mathfrak{A}_\alpha(n)\, e^{i\omega(n)\cdot\alpha t}\}. \tag{2}$$

As Th. Kuhn pointed out in a letter to me dated December 22, 1965, quantities $\mathfrak{A}_a$ and $\mathfrak{A}_e$, called characteristic amplitudes for absorption and emission transitions, also occur in paper **10** by Kramers and Heisenberg, to which paper **12** ties. The $\mathfrak{A}_a$ and $\mathfrak{A}_e$ are complex amplitudes of electrical moments of virtual oscillators, from which absorption and emission may be calculated by classical electrodynamics. The $\mathfrak{A}_e$ have all the properties which Heisenberg claims for his $\mathfrak{A}(n, n-\alpha)$, so we may safely assume, as Kuhn did in his letter, that Heisenberg's $\mathfrak{A}(n, n-\alpha)$ are just the $\mathfrak{A}_e$ of paper **10**, i.e. the amplitudes of the electrical moments of the virtual oscillators corresponding to transitions $n \to n-\alpha$.

Later on, in § 2, the $\mathfrak{A}_\alpha(n)$ in (2) are replaced by $a_\alpha(n)$, which are no longer vectors, but amplitudes in the Fourier expansion of the coordinate x of the electron. The relations between the $\mathfrak{A}_\alpha$ and the a_α, or between the $\mathfrak{A}(n, n-\alpha)$ and the $a(n, n-\alpha)$ are not specified, which is rather confusing.

Still, by an analysis of Heisenberg's arguments, we can establish a definite numerical connection between the squares $|a(n, n\pm\alpha)|^2$ and the radiation emitted or absorbed in the transitions $n \to n + \alpha$.

At the end of § 2, Heisenberg writes the main formula of the dispersion theory of Kramers as follows. Let $E \cos 2\pi\nu t$ be the primary wave and M the moment induced by this wave. Then

$$M = e^2E \cos 2\pi\nu t \cdot \frac{2}{h} \sum_\alpha \left\{ \frac{|a(n, n+\alpha)|^2 \, \nu(n, n+\alpha)}{\nu^2(n, n+\alpha) - \nu^2} - \frac{|a(n, n-\alpha)|^2 \, \nu(n, n-\alpha)}{\nu^2(n, n-\alpha) - \nu^2} \right\}.$$

In Kramers' original formula, the same denominators $\nu_i^2 - \nu^2$ occur, but the numerators contain the factors called f_i or f_{PQ} in our previous account. Heisenberg's formula coincides with that of Kramers if we put

$$\text{(v)} \qquad |a(n, n+\alpha)|^2 \, \nu(n, n+\alpha) = \frac{h}{8\pi^2 m} f_{n+\alpha, n}$$

and

$$\text{(vi)} \qquad |a(n, n-\alpha)|^2 \, \nu(n, n-\alpha) = \frac{h}{8\pi^2 m} f_{n, n-\alpha} \, .$$

From this we may conclude that Heisenberg assumed (v) and (vi) to hold, even if he did not expressly say so. This conclusion is confirmed by a footnote to formula (16) of Heisenberg's paper, which says that

(16) is equivalent to the sum rule of Kuhn and Thomas. In fact, this equivalence holds exactly when (v) and (vi) are assumed.

Einstein's probability coefficients A_{PQ} and B_{PQ} for emission and absorption are proportional to the f_{PQ}, as we have seen. Therefore, by (v) and (vi), they can be calculated from the absolute values $|a(n, n+\alpha)|$. This explains Heisenberg's statement: 'Equations (11) and (16), taken together, contain, if they can be solved, a complete determination not only of the frequencies and energies, but also of the quantum theoretical transition probabilities'.

It is quite clear that Heisenberg, when he wrote his paper, considered the formulae (v) and (vi) as evident. In fact, they can easily be derived from classical formulae by Heisenberg's 're-interpretation', as follows.

The fundamental idea underlying the Kramers–Heisenberg paper **10** (and also the earlier papers **4–9**) was, to calculate the interaction of an atom with a radiation field by replacing the atom by a set of 'virtual oscillators' having frequencies

$$\nu = \frac{1}{h}(\varepsilon_n - \varepsilon_m)$$

and by applying classical electrodynamics to these oscillators. Now the coordinate x of a classical harmonic oscillator of frequency ν is

$$x(t) = a_1\, \mathrm{e}^{2\pi i\nu t} + \bar{a}_1\, \mathrm{e}^{-2\pi i\nu t}.$$

The energy emitted per unit of time is, according to classical theory,

$$\text{(vii)} \qquad -\frac{\mathrm{d}E}{\mathrm{d}t} = \frac{4e^2}{3c^3}(2\pi\nu)^4|a_1|^2.$$

In quantum theory, the left-hand side of this formula has to be replaced by the emitted energy $h\nu$, multiplied by the emission probability $A_n^{n-\alpha}$:

$$A_n^{n-\alpha}h\nu = \frac{8\pi^2e^2\nu^2}{3mc^3} f_{n,n-\alpha}h\nu.$$

To the right-hand side of (vii) we have to apply Heisenberg's method of re-interpretation, replacing a_1 by $a(n, n-\alpha)$. Thus, formula (vi) is obtained. An analogous calculation for absorption would yield (v). Another possibility to obtain (v) would be, to make use of the Hermitean symmetry of the matrix $a(n, m)$.

8. An idea which Heisenberg stresses very much is the postulate 'to establish a quantum-theoretical mechanics based entirely upon relations between observable quantities.'

Heisenberg regards the location of an electron as non-observable. This was an error, because according to the fully developed quantum mechanics the three coordinates x, y, z of an electron are observable. However, this error was extremely fruitful, because it stimulated Heisenberg to look out for other, directly observable quantities.

The first quantities introduced by Heisenberg are the products

$$\mathfrak{A}(n, n-\alpha)\, e^{i\omega(n,n-\alpha)t} \tag{viii}$$

analogous to the terms

$$\mathfrak{A}_\alpha(n)\, e^{i\omega(n).\alpha t}$$

of a classical Fourier series. The classical quantity represented by this Fourier series seems to be the electrical moment $-e\mathfrak{x}$ of the atom, and the $\mathfrak{A}(n, n-\alpha)$ may be interpreted accordingly, as we have said already, as the amplitudes of electrical moments of virtual oscillators. These amplitudes are, as Heisenberg says, complex vectors with 6 real coordinates each, and they determine intensity, polarization and phase of the radiation. Since the intensity and polarization of the radiation are directly observable, the quantities $\mathfrak{A}(n, n-\alpha)$ are in fact observable apart from a common phase factor $e^{i\varphi}$.

The fundamental quantities of Heisenberg's mechanics are not the $\mathfrak{A}(n, n-\alpha)$ but the $a(n, n-\alpha)$ and $a(n, n+\alpha)$ which determine the matrix elements of the coordinate x. Now what is the relation between the $\mathfrak{A}$ and the a?

Let us start with the definition of the electrical moment of an electron

$$\mathfrak{M} = -e\mathfrak{x}.$$

If this is interpreted as a matrix equation, we obtain for the single matrix elements

$$\mathfrak{A}(n, m)\, e^{i\omega(n,m)} = -e\mathfrak{x}(n, m).$$

Hence the first coordinate of the vector $\mathfrak{A}(n, n-\alpha)$ should be, if our interpretation of the $\mathfrak{A}$ is correct, $-ea(n, n-\alpha)$.

In the letters to Kronig and Pauli, only the $a(n, n\pm\alpha)$ occur, not the $\mathfrak{A}$. In § 2 of Heisenberg's paper, the notation is just the same as in the letters, and the main results of § 2 are already present in the letters. Formulae (3)–(8) of § 1 also occur in the letters, but with a instead of $\mathfrak{A}$ and b instead of $\mathfrak{B}$. Logically this is more correct, for the quantities occurring in (3)–(8) are not vectors. If one rewrites these formulae in the original notation and forgets all about the $\mathfrak{A}$, it is much easier to understand Heisenberg's line of thought. It seems that Heisenberg introduced the $\mathfrak{A}$ just because of his observability principle, because they are more directly connected with the radiation field than the a.

9. Formulae (7) and (8) contain Heisenberg's multiplication law, which was afterwards found to be identical with the law of matrix multiplication. In Heisenberg's paper as well as in his letters to Kronig and Pauli, this rule was formulated first for the case of a square x^2, and afterwards for an arbitrary product xy. The classical rule for finding the square of a Fourier series (ii) is given by formula (3) of Heisenberg's paper. The quantum-theoretical analogue of this formula is Heisenberg's formula (7). As Heisenberg rightly remarks, this type of combination results 'almost necessarily' from the frequency-combination relations. In fact, these relations require

$$\omega(n, n-\alpha) + \omega(n-\alpha, n-\beta) = \omega\,(n, n-\beta),$$

therefore the term $a(n, n{-}\alpha)\mathrm{e}^{\mathrm{i}\omega(n,n-\alpha)t}$ has to be multiplied by a term like

$$a(n-\alpha, n-\beta)\ \mathrm{e}^{\mathrm{i}\omega(n-\alpha,n-\beta)t}$$

in order to obtain $\mathrm{i}\omega(n, n{-}\beta)t$ in the exponent. This idea is clearly expressed in Heisenberg's letter to Kronig.

As we have seen, Born too had thought of a symbolic multiplication of 'transition amplitudes', and discussed this idea with Jordan and Heisenberg. It is possible that the development of Heisenberg's ideas was influenced by this suggestion. However, his main line of thought was different from Born's. As we know from his letter to Kronig, Heisenberg started with the problem of calculating the force exerted by an anharmonic oscillator upon the surrounding field, and by a suitable modification of the classical expression of this force by a Fourier series he found his symbolic multiplication.

The next step was now obvious. Inserting for x in (i) the set of quantities $a(n, n{-}\alpha)$, and using symbolic multiplication to calculate $f(x)$, Heisenberg at once obtained his equation of motion.

10. The old theory was based upon the equation of motion and the

quantum condition. Hence, the next step to be taken was, to find an equivalent of the quantum condition within the framework of the new theory.

In the case of one degree of freedom (the only case considered in Heisenberg's paper) the quantum condition of Bohr's theory can be written as

$$\int m\dot{x}\,\mathrm{d}x = J = nh$$

integrated over a full period of the motion. Substituting for x the Fourier series (iii), one would obtain

$$\text{(ix)} \qquad nh = 2\pi m \sum_{\alpha=-\infty}^{\infty} |a_\alpha(n)|^2 \alpha^2 \omega_n.$$

Heisenberg replaces this formula by another one, obtained by differentiation with respect to n, which he regards as more natural from the point of view of the Principle of Correspondence, viz. his formula (15):

$$\text{(x)} \qquad h = 2\pi m \sum_{\alpha=-\infty}^{\infty} \alpha \frac{\mathrm{d}}{\mathrm{d}n} (\alpha\omega_n |a_\alpha|^2).$$

Since the expression $\alpha\omega_n|a_\alpha|^2$ is defined for integer n only, one might object to this differentiation. Yet it is quite harmless, because Heisenberg uses (x) only as an intermediate step towards a better formula, in which the derivative is replaced by a difference:

$$\text{(xi)} \quad h = 4\pi m \sum_{\alpha=0}^{\infty} \{|a(n, n+\alpha)|^2 \omega(n, n+\alpha) - |a(n, n-\alpha)|^2 \omega(n, n-\alpha)\}.$$

This is Heisenberg's quantum condition. It is equivalent to the sum rule of Kuhn and Thomas, as Heisenberg notes in a footnote.

11. In formula (xi), only differences occur on the right-hand side. One more condition is necessary to determine $a(n, n-\alpha)$ for the lowest possible value of n, i.e. for the ground state. It is very easy to find this condition. As we have seen, Heisenberg assumed the squares $|a(n, n-\alpha)|^2$ to be proportional to the probabilities of the transitions $n \to n-\alpha$. For the ground state, no transition to a lower level is possible, hence one has to put

$$a(n, n-\alpha) = 0$$

if n is the quantum number of the ground state.

Born's reaction to Heisenberg's paper

On July 11 or 12 Heisenberg gave Born the final version of his paper, asking him to decide whether it was worth publishing. Born writes in his Recollections:

Meanwhile Heisenberg pursued some work of his own, keeping its idea and purpose somewhat dark and mysterious. Towards the end of the summer semester, in the first days of July 1925, he came to me with a manuscript and asked me to read it and to decide whether it was worth publishing ... He added that though he had tried hard, he could not make any progress beyond the simple considerations contained in his paper, and he asked me to try myself, which I promised.

I remember that I did not read this manuscript at once because I was tired after the term ... But when, after a few days, I read it I was fascinated. Heisenberg had taken up the idea of transition amplitudes and developed a calculus for them, by following up the correspondence with the coefficients of the classical expansion of a vibrating quantity into its harmonic components (Fourier series). If two such expansions are multiplied, one obtains a new expansion for the product ... Now Heisenberg suggested forgetting everything about the series and considering the set of coefficients which represented the physical quantity in question; then one has a multiplication rule for those sets of coefficients. Yet each set is still attached to a single classical orbit or stationary state while in quantum theory every observable effect depends on the transition between states. Using our previous experience about the correspondence between the concepts of classical and quantum theory, Heisenberg defined transition amplitudes and their multiplication in analogy to the sets of classical Fourier coefficients and their multiplication. His most audacious step consists in the suggestion of introducing the transition amplitudes of the coordinates q and momenta p in the formulae of mechanics ...

I was deeply impressed by Heisenberg's considerations, which were a great step forward in the programme which we had pursued.

Born must have studied Heisenberg's manuscript before July 15, for on July 15 he writes in a letter to Einstein:

... Heisenberg's neue Arbeit, die bald erscheint, sieht sehr mystisch aus, ist aber sicher richtig und tief ...

Born's conjecture on $pq - qp$

On July 19, Born took the train to Hannover to attend the meeting of the Deutsche Physikalische Gesellschaft. His own account, confirmed by Jordan's testimony, runs thus:

After having sent Heisenberg's paper to the Zeitschrift für Physik for publication, I began to ponder about his symbolic multiplication, and was soon so involved in it that I thought the whole day and could hardly sleep at night. For I felt there was something fundamental behind it ... And one morning ... I suddenly saw light: Heisenberg's symbolic multiplication was nothing but the matrix calculus, well known to me since my student days from the lectures of Rosanes in Breslau.

I found this by just simplifying the notation a little: instead of $q(n, n+\tau)$... I wrote $q(n, m)$, and re-writing Heisenberg's form of Bohr's quantum conditions I recognised at once its formal significance. It meant that the two matrix products $\boldsymbol{pq}$ and $\boldsymbol{qp}$ are not identical. I was familiar with the fact that matrix multiplication is not commutative; therefore I was not too much puzzled by this result. Closer inspection showed that Heisenberg's formula gave only the value of the diagonal elements $(m=n)$ of the matrix $\boldsymbol{pq}-\boldsymbol{qp}$: it said that they were all equal and had the value $h/2\pi i$. But what were the other elements $m \neq n$?

Here my own constructive work began. Repeating Heisenberg's calculation in matrix notation, I soon convinced myself that the only reasonable value of the non-diagonal elements should be zero, and I wrote down the strange equation

$$\boldsymbol{pq} - \boldsymbol{qp} = \frac{h}{2\pi i}\,\mathbf{1}$$

where $\mathbf{1}$ is the unit matrix. But this was only a guess, and my attempts to prove it failed.

Pauli's reaction

On July 19, in the train from Göttingen to Hannover, Born met Pauli. Born relates:

I joined him in his compartment, and absorbed by my new discovery, I at once told him about the matrices and my difficulties in finding the value of those non-diagonal elements. I asked him whether he would like to collaborate with me in this problem. But instead of the expected interest, I got a cold and sarcastic refusal. 'Yes, I know you are fond of tedious and complicated formalism. You are only going to spoil Heisenberg's physical ideas by your futile mathematics', and so on ...

The same emotional distinction between Heisenberg's 'physical ideas' and the 'formal mathematics' of the school of Göttingen is made in a letter from Pauli to Kronig dated October 9, 1925, and reproduced in the Pauli Memorial Volume, p. 25–26. In this letter, Pauli writes:

Die Heisenbergsche Mechanik hat mir wieder Lebensfreude und Hoffnung gegeben. Die Lösung des Rätsels bringt sie zwar nicht, aber ich glaube, dass es jetzt wieder möglich ist, vorwärts zu kommen. Man muss zunächst versuchen, die Heisenbergsche Mechanik noch etwas mehr vom Göttinger formalen Gelehrsamkeitsschwall zu befreien und ihren physikalischen Kern noch besser blosszulegen.

At another point of his narrative, Born returns to Pauli's refusal:

Concerning Pauli's refusal to collaborate with me, he explained later to other people that he had not seriously pondered about Heisenberg's ideas and did not wish to interfere with his plans.

Jordan's proof and the joint paper 13

The next day, i.e. July 20, Born asked his pupil Jordan to help him in his work.

Jordan accepted and after only a few days brought me the solution of the problem: He showed that the canonical equations of motion, applied to the matrices $\boldsymbol{p}$ and $\boldsymbol{q}$, led to the result that the time derivative of $\boldsymbol{pq}-\boldsymbol{qp}$ must vanish, hence the matrix itself must be diagonal. Then we began to write a joint paper ...

This joint paper of Born and Jordan is paper **13** of the present collection; it was received on September 27, 1925. Born characterizes its contents as follows:

This paper by Jordan and myself contains the formulation of matrix mechanics, the first printed statement of the commutation law quoted above, some simple applications to the harmonic and anharmonic oscillator, and another fundamental idea: the quantization of the electro-magnetic field (by regarding its components as matrices).

Analysis of the paper 13

The principal new ideas in the Born–Jordan paper are:

1) Interpretation of Heisenberg's symbolic multiplication as matrix multiplication,

2) The formula for $\boldsymbol{pq}-\boldsymbol{qp}$,

3) Jordan's proof of this formula, based upon the calculation of the time derivative of $\boldsymbol{pq}-\boldsymbol{qp}$,

4) Proof of the conservation of energy by the same method,

5) Proof of Bohr's Frequency Condition,

6) Quantization of the electromagnetic field by regarding its components as matrices,

7) Justification, based upon this quantization, of Heisenberg's assumption that the squares of the absolute values of the elements in the matrix representing the electrical moment of an atom determine the transition probabilities.

Of these fundamental ideas, the first two are due to Born, and the last five to Jordan. This is completely clear from Born's statements in his Recollections and from Jordan's statements in his letters to me, written in 1961, 1962, and 1964.

Jordan's proof of the formula for $\boldsymbol{pq}-\boldsymbol{qp}$ is not an exact mathematical proof. If the derivative $\dot{\boldsymbol{g}}$ of a matrix

$$\boldsymbol{g} = (g_{nm}\, \mathrm{e}^{2\pi i\nu(n,m)t})$$

is zero, it does not follow that $\boldsymbol{g}$ is diagonal, unless we assume, as Jordan does, that $\nu(n, m) \neq 0$ for $n \neq m$. This is just an assumption, not a consequence of the equations of motion.

Yet, Jordan's proof shows clearly that it is *reasonable* to assume that $\boldsymbol{pq}-\boldsymbol{qp}$ is diagonal: it is compatible with the equations of motion. This explains why Born was at once convinced by Jordan's argument.

Conservation of energy was proved in the same way by showing that the derivative $\dot{\boldsymbol{H}}$ is zero. Mathematically, both proofs were based upon a clever theory of 'symbolic differentiation' of matrix functions (§ 2 of paper **13**).

For the proof of Bohr's Frequency Condition

$$W_n - W_m = h\nu(n, m)$$

see § 4 of the paper. From Heisenberg's letter to Jordan, dated September 10, to be quoted presently, it is clear that this proof is due to Jordan.

The ideas 6) and 7) are highly important and original, but we shall not discuss them here, because Quantum Field Theory is outside the scope of this book. For the same reason, Chapter IV of the paper, which contains a first elaboration of the ideas 6) and 7), will not be included in our source publication. However, the last section of the 'three-men-paper' **14**, which deals with the same subject, will be included because of its fundamental importance for the history of Quantum Mechanics.

The remainder of the Born–Jordan paper is devoted to a mathematical elaboration of the fundamental ideas and their application to special cases.

This elaboration and the redaction of the paper are mainly due to Jordan. Born was in Silvaplana from the middle of August until the middle of September and he was very tired, as he repeatedly states in his Recollections. In a letter to me, dated October 8, 1964, Jordan gives the following account of the redaction of the paper:

Während des Aufenthaltes Borns in Silvaplana war ich in Hannover, in meinem Elternhause, und ich habe dort noch einen Teil des Stoffes überlegt, der dann in der Arbeit von Born und mir dargestellt wurde. Ich führte darüber einen Briefwechsel mit Born, dem ich natürlich meine Fortschritte berichtete. Ich erinnere mich noch, dass er nach einiger Zeit eine Pause des Briefwechsels für zweckmässig hielt, da ihm die zweifache Beanspruchung durch eine etwas strapaziöse Sanatoriums-Kur und durch unsere briefliche Unterhaltung über dieses spannende Thema nicht gut bekam. Es dürfte in der Tat so gewesen sein, wie Sie vermuten – nämlich, dass ich, als wir uns dann in Göttingen wiedersahen, die Arbeit im ersten Entwurf schon weitgehend aufgezeichnet hatte. (Genauere Erinnerung an diesbezügliche Einzelheiten habe ich freilich nicht mehr). Ich halte es für wahrscheinlich, dass das Eingangsdatum 27.9.25 durchaus exakt das Eintreffen unseres Manuskriptes bei der Zeitschrift angibt.

Jordan's statements are confirmed by three letters from Heisenberg to Jordan, from which I shall reproduce the essential passages:

München, 20.8.
... Von Born hörte ich, dass Sie in der Quantenmechanik grosse Fortschritte gemacht hätten und es würde mich natürlich riesig interessieren, über Ihre Rechnungen einiges zu erfahren. Da, wie Born mir schrieb, Sie gerne Korrekturen meiner Arbeit haben möchten, schicke ich Ihnen das Manuskript – Korrekturen habe ich leider keine mehr. Ich wäre Ihnen aber sehr dankbar, wenn Sie mir kurz über Ihre Rechnungen berichten könnten...

A new request was dispatched from Munich on September 10:

Lieber Jordan! Es hat mich sehr gefreut, zu hören, dass Sie den Beweis für die Frequenzbedingung gefunden zu haben glauben und ich bitte Sie sehr, ihn mir so bald, wie irgend möglich, nach Kopenhagen zu schreiben; denn von Uebermorgen an will ich wieder selbst Physik treiben ...

This time Jordan sent a summary of the first two chapters of the Born–Jordan paper **13**. Heisenberg answered at once:

Kopenhagen 13.9.25
Lieber Jordan! Vielen Dank für Ihren Brief und den Beweis. Ihre und Borns Arbeit scheint mir ein sehr grosser Fortschritt; der Kernpunkt ist offenbar die Bornsche Frequenzbedingung. Ich hoffe, Sie werden Ihre schönen Resultate jetzt bald mit Born publizieren, sie sind sicher reif dazu ...

Dirac's paper 14

The main ideas of the Born–Jordan paper **13** were re-discovered independently by P.A.M. Dirac, a young physicist at Cambridge, England. In an interview on June 26, 1961, Dirac told me how he hit upon the fundamental ideas of his paper **14**:

In July 1925 Heisenberg came to Cambridge and lectured at the Kapitza club, but I was not present at the lecture and did not know anything about it. The first I heard of Heisenberg's new ideas was in early September, when R. H. Fowler gave me the proof sheets of Heisenberg's paper [**12**]. At first I could not make much of it, but after about two weeks I saw that it provided the key to the problem of quantum mechanics. I proceeded to work it out by myself. I had previously learnt the Transformation Theory of Hamiltonian Mechanics from lectures by R. H. Fowler and from Sommerfeld's book Atombau und Spektrallinien.

Heisenberg's lecture at the Kapitza club was delivered on July 28. The title was 'Termzoologie und Zeemanbotanik'. In private talks with other physicists, Heisenberg also explained his ideas on Quantum Mechanics. Fowler must have participated in one of these talks and asked Heisenberg to send him proof sheets.

In paper **14**, Dirac first gives a summary of Heisenberg's ideas. Simplifying Heisenberg's notation just as Born simplified it, he writes the multiplication rule in the form

$$xy(nm) = \Sigma_k\, x(nk)\, y(km).$$

Dirac also introduces a sum of two quantum variables x and y defined by

$$\{x + y\}(nm) = x(nm) + y(nm).$$

In the matrix algebra defined by these two operations, Dirac considers operations $\mathrm{d}/\mathrm{d}v$ satisfying the conditions

$$\frac{\mathrm{d}}{\mathrm{d}v}(x + y) = \frac{\mathrm{d}}{\mathrm{d}v}x + \frac{\mathrm{d}}{\mathrm{d}v}y$$

$$\frac{\mathrm{d}}{\mathrm{d}v}(xy) = \frac{\mathrm{d}}{\mathrm{d}v}x \cdot y + x \cdot \frac{\mathrm{d}}{\mathrm{d}v}y.$$

He finds that the most general operation having this property is defined by

$$\frac{\mathrm{d}x}{\mathrm{d}v} = xa - ax.$$

Next Dirac asks, what corresponds to $xy-yx$ in classical theory. By an ingenious application of the Principle of Correspondence, he finds that the most reasonable assumption is that $xy-yx$ corresponds to $ih/2\pi$ times the Poisson bracket

$$[x, y] = \Sigma_r \left\{ \frac{\partial x}{\partial q_r} \frac{\partial y}{\partial p_r} - \frac{\partial y}{\partial q_r} \frac{\partial x}{\partial p_r} \right\},$$

summed over the u degrees of freedom of the mechanical system.

Hence Dirac makes the fundamental assumption

$$xy - yx = (ih/2\pi) \cdot [x, y]. \tag{11}$$

Special cases are

$$\begin{aligned} q_r q_s - q_s q_r &= 0, \\ p_r p_s - p_s p_r &= 0, \\ q_r p_s - q_s p_r &= \delta_{rs} ih/2\pi. \end{aligned} \tag{12}$$

In a letter dated July 28, 1961, I have asked Dirac:

How did you find formula (11)? Did you follow the road indicated in your paper? Did you first find the general formula (11) and next the special case (12), or conversely?

The answer, dated August 21, 1961, was:

I did get the general Poisson bracket formula before the special cases. I got the general formula on the lines indicated in the paper, by working with the case of large quantum numbers. At that time I was expecting some kind of connection between the new mechanics and Hamiltonian dynamics (as Hamiltonian dynamics was used so much with Sommerfeld's development of the Bohr theory) and it seemed to me that this connection should show up best with large quantum numbers.

In § 5 of his paper, again starting with classical mechanics, Dirac writes the equations of motion as

$$\dot{x} = [x, H]$$

and the conditions that variables P_r, Q_r shall be canonical as

$$[Q_r, Q_s] = 0, \qquad [P_r, P_s] = 0, \qquad [Q_r, P_s] = \delta_{rs}.$$

In § 6, he derives Bohr's relation connecting the frequencies with the energy differences:

$$H(nn) - H(mm) = (h/2\pi) \cdot \omega(nm).$$

The 'three men's paper' 15

Paper **15**, by Born, Heisenberg and Jordan, gives a logically consistent exposition of Matrix Mechanics. In Heisenberg's letters, this paper is usually called 'Dreimännerarbeit'. We shall try to trace the history of this highly important paper.

We shall restrict ourselves to the main part of the paper, which deals with mechanical systems having a finite number of degrees of freedom, and leave aside the last part (Chapter 4, § 3), which is entirely due to Jordan and which deals with the quantization of the electromagnetical field and the calculation of fluctuations in this field.

The history of the paper begins on September 12, 1925, when Heisenberg, upon his arrival in Copenhagen, received a letter from Jordan containing a summary of the theory of Born and Jordan. This letter is lost, but an account of its contents was given in a letter from Heisenberg to Pauli (dated September 18), to be reproduced presently. Apparently Jordan's letter contained just a summary of the Born–Jordan paper **13**, the only difference being that Jordan, in his letter, seems to have used Gothic letters for quantum variables. In a letter to Pauli, Heisenberg makes an ironical remark on the use of Gothic letters 'damit es vornehmer aussieht'. In the printed paper **13**, the Gothic letters were replaced by heavy type.

Perturbation theory and canonical transformation

Heisenberg answered Jordan on September 13. The first three sentences of his answer have been reproduced already (see the end of the section 'Analysis of the paper **13**'). The letter continues thus:

In den zwei Tagen, in denen ich hier bin, hab ich mir noch nicht viel überlegen können. Doch ist mir ein sehr einfacher Beweis der Kramersschen Dispersiontheorie und der Formeln von Kramers und mir auf Grund des neuen Formalismus eingefallen:

Sei die Energie $\mathfrak{H}=\mathfrak{H}_0+\lambda\mathfrak{H}_1$ (der Einfachheit halber seien sowohl $\mathfrak{H}_0$ wie $\mathfrak{H}_1$ in $\mathfrak{p}$ u. $\mathfrak{q}$ 'separiert':

Ferner:
$$\left.\begin{array}{ll} \mathfrak{H}_0 = \sum a_s\mathfrak{p}^s + b_s\mathfrak{q}^s & \mathfrak{H}_1 = \sum \alpha_s\mathfrak{p}^s + \beta_s\mathfrak{q}^s. \\ \mathfrak{q} = \mathfrak{q}_0 + \lambda\mathfrak{q}_1 & \mathfrak{p} = \mathfrak{p}_0 + \lambda\mathfrak{p}_1. \end{array}\right\} \quad (1)$$

Dann heisst das mit λ proportionale Glied der Bewegungsgleichungen:

$$(2)\quad \left\{\begin{array}{l} \dot{\mathfrak{q}}_1 = \sum s a_s(\mathfrak{p}_0^{s-2}\mathfrak{p}_1 + \mathfrak{p}_0^{s-3}\mathfrak{p}_1\mathfrak{p}_0 + \mathfrak{p}_0^{s-4}\mathfrak{p}_1\mathfrak{p}_0^2 + \\ \qquad \dots + \mathfrak{p}_1\mathfrak{p}_0^{s-2}) + \sum s\alpha_s\mathfrak{p}_0^{s-1} \\ \dot{\mathfrak{p}}_1 = -\sum s b_s(\mathfrak{q}_0^{s-2}\mathfrak{q}_1 + \mathfrak{q}_0^{s-3}\mathfrak{q}_1\mathfrak{q}_0 + \dots) - \sum s\beta_s\mathfrak{q}_0^{s-1}. \end{array}\right.$$

Ich behaupte nun, dass diese Gleichungen die Lösung besitzen:

$$(3)\quad \begin{cases} \mathfrak{q}_1 = \dfrac{2\pi i}{h}(\mathfrak{F}\mathfrak{q}_0 - \mathfrak{q}_0\mathfrak{F}) \\[2ex] \mathfrak{p}_1 = \dfrac{2\pi i}{h}(\mathfrak{F}\mathfrak{p}_0 - \mathfrak{p}_0\mathfrak{F}), \end{cases}$$

wobei $\mathfrak{F}$ definiert ist durch

(3a) $\dot{\mathfrak{F}} = \mathfrak{H}_1$ ($\mathfrak{H}_1$ habe zunächst kein konstantes Glied).

Dann wird nämlich aus (2):

$$(4)\quad \begin{cases} \dot{\mathfrak{q}}_1 = \dfrac{2\pi i}{h}\sum s a_s(\mathfrak{F}\mathfrak{p}_0^{s-1} - \mathfrak{p}_0^{s-1}\mathfrak{F}) + \sum s\alpha_s\mathfrak{p}_0^{s-1} \\[2ex] \dot{\mathfrak{p}}_1 = -\dfrac{2\pi i}{h}\sum s b_s(\mathfrak{F}\mathfrak{q}_0^{s-1} - \mathfrak{q}_0^{s-1}\mathfrak{F}) - \sum s\beta_s\mathfrak{q}_0^{s-1}. \end{cases}$$

Setzt man nun die linke Seite aus (3) (3a) ein und berücksichtigt rechts die Bewegungsgleichung (Nullter Ordnung) so folgt:

$$(5)\quad \begin{cases} \dot{\mathfrak{q}}_1 = \dfrac{2\pi i}{h}(\mathfrak{F}\dot{\mathfrak{q}}_0 - \dot{\mathfrak{q}}_0\mathfrak{F} + \mathfrak{H}_1\mathfrak{q}_0 - \mathfrak{q}_0\mathfrak{H}_1) = \dfrac{2\pi i}{h}(\mathfrak{F}\dot{\mathfrak{q}}_0 - \dot{\mathfrak{q}}_0\mathfrak{F}) + \sum s\alpha_s\mathfrak{p}_0^{s-1} \\[2ex] \dot{\mathfrak{p}}_1 = \dfrac{2\pi i}{h}(\mathfrak{F}\dot{\mathfrak{p}}_0 - \dot{\mathfrak{p}}_0\mathfrak{F} + \mathfrak{H}_1\mathfrak{p}_0 - \mathfrak{p}_0\mathfrak{H}_1) = \dfrac{2\pi i}{h}(\mathfrak{F}\dot{\mathfrak{p}}_0 - \dot{\mathfrak{p}}_0\mathfrak{F}) - \sum s\beta_s\mathfrak{q}_0^{s-1} \end{cases}$$

Diese Gleichungen (5) aber lassen sich sofort verifizieren, wenn man links für $\mathfrak{H}_1$ die Entwicklung (1) ansetzt.

Schwierigkeiten machen mir noch die Fälle, wo $\mathfrak{H}_1$ ein konstantes Glied enthält. Aber abgesehen davon gibt Gleichung (3) die Kramersschen sowie die von Kramers und mir aufgestellten Formeln exakt wieder.

What Heisenberg develops here, is a first-order perturbation theory. The theory is still restricted to the case in which the perturbation H_1 contains no constant term, but a few days later Heisenberg managed to remove this restriction and to develop a complete perturbation theory for systems having one degree of freedom. On September 16, he writes Jordan:

Ueber das, was ich im letzten Brief schrieb, hinausgehend, habe ich jetzt eine vollständige Störungstheorie der Probleme mit einem Freiheitsgrad, bis in beliebiger Näherung und auch für den Fall, dass $\overline{H}_1$, $\overline{H}_2$ ... nicht Null sind. Für $\overline{H}_2$ bzw. W_2 kommt natürlich Born's Formel heraus. Viele Grüsse!

Heisenberg's perturbation theory was explained in a letter to Born, which is lost, and also in a letter to Pauli, dated September 18. The letter begins with an exposition of the Born–Jordan theory:

Die mir z.Z. sehr am Herzen liegende Quantenmechanik hat inzwischen, hauptsächlich durch Born und Jordan, entschiedene Fortschritte gemacht, die ich

Ihnen, schon um mir selbst klar zu werden was ich glauben soll, im Folgenden erzählen werde. Man kann tatsächlich jetzt ziemlich allgemein Energiesatz, Frequenzbedingung und Dispersionsformel auf Grund der damaligen Annahmen beweisen.

1.) Annahmen. Unter einer Grösse $\mathfrak{p}$ (deutsch geschrieben damit es vornehmer aussieht) verstehe ich die Gesamtheit von Grössen der Form $p_{ik}e^{2\pi i \nu_{ik} \cdot t}$; für die ν_{ik} werden die Kombinationsrelationen als erfüllt angesehen

$$\nu_{ik} + \nu_{kl} = \nu_{il} \quad \text{oder} \quad \nu_{ik} = \frac{W_i - W_k}{h} \tag{1}$$

wobei aber natürlich *nicht* vorausgesetzt wird, dass W die Energie sei ...

Als Produkt wird definiert

$$\mathfrak{a} \cdot \mathfrak{b} = \mathfrak{c} \quad \text{bedeutet:} \quad c_{ik} = \sum_l a_{il} b_{lk}$$

Diese Produktbildung ist distributiv und genügt der Relation $\mathfrak{a}(\mathfrak{b}\mathfrak{c}) = (\mathfrak{a}\mathfrak{b})\mathfrak{c}$; sie ist aber nicht kommutativ ...

Die Energie habe zunächst die Form $\mathfrak{H}(\mathfrak{p}, \mathfrak{q}) = \mathfrak{H}_a(\mathfrak{p}) + \mathfrak{H}_b(\mathfrak{q})$ ausserdem sei sie nach Potenzen von $\mathfrak{p}$ und $\mathfrak{q}$ entwickelbar (negative Exponenten sind *auch* zugelassen):

$$\mathfrak{H}_a = \sum_s a_s \mathfrak{p}^s; \qquad \mathfrak{H}_s = \sum_s b_s \mathfrak{q}^s. \tag{2}$$

2.) Die Bewegungsgleichungen lauten:

$$\left\{\begin{aligned} \dot{\mathfrak{p}} &= -\frac{\partial \mathfrak{H}}{\partial \mathfrak{q}} = -\sum s \cdot b_s \mathfrak{q}^{s-1}. \\ \dot{\mathfrak{q}} &= \frac{\partial \mathfrak{H}}{\partial \mathfrak{p}} = +\sum s \cdot a_s \mathfrak{p}^{s-1}. \end{aligned}\right. \tag{3}$$

(hierbei ist unter $\dot{\mathfrak{a}}$ die Gesamtheit $2\pi i \nu_{ik} a_{ik}$ verstanden).

3.) Nun kommt gleich die in dieser Mechanik entscheidende Quantenbedingung und es war eine sehr gescheite Idee von Born, dass er sie in der Form schrieb:

$$\mathfrak{p}\mathfrak{q} - \mathfrak{q}\mathfrak{p} = \frac{h}{2\pi i}\mathfrak{e} \tag{4}$$

($\mathfrak{e}$ sei die 'Einheitsdeterminante', d.h. $\mathfrak{e}_{ii} = 1$, $\mathfrak{e}_{ik} = 0$ für $k \neq i$). Was zunächst die *periodischen* Glieder von (4) betrifft, so folgt aus den Bewegungsgleichungen leider, dass sie Null sind:

$$\frac{d}{dt}(\mathfrak{p}\mathfrak{q} - \mathfrak{q}\mathfrak{p}) = \overbrace{\dot{\mathfrak{p}}\mathfrak{q} - \mathfrak{q}\dot{\mathfrak{p}}}^{=0} + \overbrace{\mathfrak{p}\dot{\mathfrak{q}} - \dot{\mathfrak{q}}\mathfrak{p}}^{=0} = 0$$

Für das konstante Glied von (4) findet man durch Einsetzen der Fourierkoeffizienten leicht, dass es der Kuhnschen Formel entspricht.

Wir nehmen also (4) als richtig an, dann folgen (durch geeignete Multiplikation mit $\mathfrak{p}$, $\mathfrak{q}$ und Addition) die weiteren wichtigen Relationen:

$$\text{(5)}\qquad \left.\begin{array}{l} \mathfrak{p}^s\mathfrak{q} - \mathfrak{q}\mathfrak{p}^s = s\,\dfrac{h}{2\pi i}\,\mathfrak{p}^{s-1} \\[2ex] \mathfrak{p}\mathfrak{q}^s - \mathfrak{q}^s\mathfrak{p} = s\,\dfrac{h}{2\pi i}\,\mathfrak{q}^{s-1} \end{array}\right\}\ \text{für alle Exponenten } s.$$

Denkt man sich nun eine Grösse $\mathfrak{W}$, die definiert ist durch

$$\text{(6)}\qquad W_{ik} = \begin{cases} W_i & i = k, \\ 0 & i \neq k \end{cases}$$

so sieht man, dass man $\dot{\mathfrak{q}}$ bzw. $\dot{\mathfrak{p}}$ (6) ersetzen kann durch $2\pi i/h(\mathfrak{W}\mathfrak{q}-\mathfrak{q}\mathfrak{W})$ bezw. $2\pi i/h(\mathfrak{W}\mathfrak{p}-\mathfrak{p}\mathfrak{W})$. (Ueberhaupt treten in dieser Mechanik stets solche schiefsymmetrische Bildungen an Stelle von Differentialquotienten.)
Nun folgt erstens aus (3) und (6):

$$\text{(7)}\qquad \begin{cases} \mathfrak{W}\mathfrak{q} - \mathfrak{q}\mathfrak{W} = \dfrac{h}{2\pi i}\sum\limits_s s a_s \mathfrak{p}^{s-1}. \\[2ex] \mathfrak{W}\mathfrak{p} - \mathfrak{p}\mathfrak{W} = -\dfrac{h}{2\pi i}\sum\limits_s s b_s \mathfrak{q}^{s-1}. \end{cases}$$

Ferner folgt aus (5):

$$\text{(8)}\qquad \begin{cases} \mathfrak{H}\mathfrak{q} - \mathfrak{q}\mathfrak{H} = \dfrac{h}{2\pi i}\sum\limits_s s a_s \mathfrak{p}^{s-1} \\[2ex] \mathfrak{H}\mathfrak{p} - \mathfrak{p}\mathfrak{H} = -\dfrac{h}{2\pi i}\sum\limits_s s b_s \mathfrak{q}^{s-1} \end{cases}$$

Bildet man jetzt $\mathfrak{W}-\mathfrak{H}=\mathfrak{x}$, so wird aus (7) und (8)

$$\text{(9)}\qquad \begin{cases} \mathfrak{q}\mathfrak{x} = \mathfrak{x}\mathfrak{q}; \quad \mathfrak{p}\mathfrak{x} = \mathfrak{x}\mathfrak{p}; \quad \text{daraus} \\ \mathfrak{q}^s\mathfrak{x} = \mathfrak{x}\mathfrak{q}^s; \quad \mathfrak{p}^s\mathfrak{x} = \mathfrak{x}\mathfrak{p}^s; \quad \text{also auch} \\ \mathfrak{H}\mathfrak{x} = \mathfrak{x}\mathfrak{H} \quad \text{oder} \quad \mathfrak{H}\mathfrak{W} = \mathfrak{W}\mathfrak{H}; \quad \text{dies letztere bedeutet} \end{cases}$$

$$\text{(10)}\qquad \dot{\mathfrak{H}} = 0.$$

$H_{ik}=0$ für $i\neq k$; setzt man $H_{ii}=H_i$, so folgt aus (7) und (8)

$$\mathfrak{W}\mathfrak{q} - \mathfrak{q}\mathfrak{W} = \mathfrak{H}\mathfrak{q} - \mathfrak{q}\mathfrak{H} \quad \text{oder} \quad (W_i - W_k)q_{ik} = (H_i - H_k)q_{ik}$$

oder

$$\text{(11)}\qquad \frac{H_i - H_k}{h} = \nu_{ik}$$

Bis hierher stammt der ganze Beweis von Born und Jordan.

5.) Störungstheorie. Wir denken uns die Energiefunktion entwickelt nach λ:

$$\text{(12)}\qquad \mathfrak{H} = \mathfrak{H}_0 + \lambda\mathfrak{H}_1 + \lambda^2\mathfrak{H}_2 + \cdots$$

wobei wieder

$$\mathfrak{H}_i = \mathfrak{H}_{ia}(\mathfrak{p}) + \mathfrak{H}_{ib}(\mathfrak{q}).$$

Gesucht

$$(13)\qquad \begin{aligned} \mathfrak{q} &= \mathfrak{q}_0 + \lambda\mathfrak{q}_1 + \lambda^2\mathfrak{q}_2 + \dots \\ \mathfrak{p} &= \mathfrak{p}_0 + \lambda\mathfrak{p}_1 + \lambda^2\mathfrak{p}_2 + \dots \\ \mathfrak{W} &= \mathfrak{W}_0 + \lambda\mathfrak{W}_1 + \lambda^2\mathfrak{W}_2 + \dots \end{aligned}$$

Aus (11) folgt (umgekehrt, wie vorhin) auch dass die *Bewegungsgleichungen* immer dann gelten, wenn die Quantenbedingungen (4) erfüllt sind. Ich setze also jetzt $\mathfrak{q}$, $\mathfrak{p}$ in einer solchen Form an, dass die Bedingung (4) von selbst erfüllt ist und suche nachträglich die in diese Form eingehende unbekannte Funktion $\mathfrak{S}$ *so* zu bestimmen, dass die Energie $\mathfrak{H}=\mathfrak{W}$ wird. $\mathfrak{S}$ entspricht (bis auf einen Faktor $h/2\pi i$) dem klassischen S der kanonischen Transformation; überhaupt ist hier jede Transformation eine 'kanonische', die $\mathfrak{p}\mathfrak{q}-\mathfrak{q}\mathfrak{p}$ in sich überführt:

$$\mathfrak{p}\mathfrak{q} - \mathfrak{q}\mathfrak{p} = \mathfrak{p}'\mathfrak{q}' - \mathfrak{q}'\mathfrak{p}'.$$

Der Ansatz, der eine solche Transformation leistet heisst:

$$(14)\qquad \mathfrak{S} = \lambda\mathfrak{S}_1 + \lambda^2\mathfrak{S}_2 + \dots$$

$$\mathfrak{q} = \mathfrak{q}_0 + \lambda(\mathfrak{S}_1\mathfrak{q}_0 - \mathfrak{q}_0\mathfrak{S}_1) + \lambda^2(\tfrac{1}{2}(\mathfrak{S}_1^2\mathfrak{q}_0 - 2\mathfrak{S}_1\mathfrak{q}_0\mathfrak{S}_1^2 + \mathfrak{q}_0\mathfrak{S}_1^2) + \mathfrak{S}_2\mathfrak{q}_0 - \mathfrak{q}_0\mathfrak{S}_2) + \lambda^3\dots$$

$$\mathfrak{p} = \mathfrak{p}_0 + \lambda(\mathfrak{S}_1\mathfrak{p}_0 - \mathfrak{p}_0\mathfrak{S}_1) + \lambda^2(\tfrac{1}{2}(\mathfrak{S}_1^2\mathfrak{p}_1 - 2\mathfrak{S}_1\mathfrak{p}_0\mathfrak{S}_1 + \mathfrak{p}_0\mathfrak{S}_1^2) + \mathfrak{S}_2\mathfrak{p}_0 - \mathfrak{p}_0\mathfrak{S}_2) + \lambda^3\dots$$

Das Bildungsgesetz der $\mathfrak{q}_i$ bzw. $\mathfrak{p}_i$ werden Sie wohl sofort erkennen, es handelt sich stets um iterierte Bildungen der Form $\mathfrak{S}\mathfrak{q}-\mathfrak{q}\mathfrak{S}$. Die Gleichungen (14) haben, wie man leicht ausrechnet, zur Folge, dass auch für jede beliebige Funktion $\mathfrak{F}(\mathfrak{q})$ gilt

$$(15)\qquad \mathfrak{F}(\mathfrak{q}) = \mathfrak{F}(\mathfrak{q}_0) + \lambda(\mathfrak{S}_1\mathfrak{F} - \mathfrak{F}\mathfrak{S}_1) + \lambda^2(\tfrac{1}{2}(\mathfrak{S}_1^2\mathfrak{F} - 2\mathfrak{S}_1\mathfrak{F}\mathfrak{S}_1 + \mathfrak{F}\mathfrak{S}_1^2) + \mathfrak{S}_2\mathfrak{F} - \mathfrak{F}\mathfrak{S}_2) + \lambda^3\dots$$

und entsprechend für $\mathfrak{p}$.

Aus (15) folgt demnach für $\mathfrak{H}$:

$$\mathfrak{H} = \mathfrak{H}(\mathfrak{q}_0\mathfrak{p}_0) + \lambda(\mathfrak{S}_1\mathfrak{H} - \mathfrak{H}\mathfrak{S}_1) + \lambda^2(\tfrac{1}{2}(\mathfrak{S}_1^2\mathfrak{H} - 2\mathfrak{S}_1\mathfrak{H}\mathfrak{S}_1 + \mathfrak{H}\mathfrak{S}_1^2) + \mathfrak{S}_2\mathfrak{H} - \mathfrak{H}\mathfrak{S}_2) + \lambda^3\dots$$

Setzt man die Entwicklung (12) ein, und setzt $H=W$, so folgt nach dem berühmten Born–Paulischen Muster:

$$(16)\qquad \begin{cases} \mathfrak{H}_0 = \mathfrak{W}_0. \\ \mathfrak{S}_1\mathfrak{W}_0 - \mathfrak{W}_0\mathfrak{S}_1 + \mathfrak{H}_1 = \mathfrak{W}_1. \\ \mathfrak{S}_2\mathfrak{W}_0 - \mathfrak{W}_0\mathfrak{S}_2 + \tfrac{1}{2}(\mathfrak{S}_1^2\mathfrak{H}_0 - 2\mathfrak{S}_1\mathfrak{H}_0\mathfrak{S}_1 + \mathfrak{H}_0\mathfrak{S}_1^2) + \mathfrak{S}_1\mathfrak{H}_1 - \mathfrak{H}_1\mathfrak{S}_1 + \mathfrak{H}_2 = \mathfrak{W}_2. \end{cases}$$

Die Integration geht so vor sich, dass man zuerst die Mittelwerte bildet, z.B. $\mathfrak{H}_1=\mathfrak{W}_1$, dann den Rest durch Fourierreihen integriert. In der Tat bedeutet ja $\mathfrak{S}_i\mathfrak{W}_0-\mathfrak{W}_0\mathfrak{S}_i$ stets nur $-h\mathfrak{S}_{ik}^{(0)}S_{ik}^{(i)}$. Also kann man $\mathfrak{S}_1$, $\mathfrak{S}_2$ u.s.w. successiv bestimmen. Hängt $\mathfrak{H}$ explizit von der Zeit ab, so kommt, wie klassisch, auf der linken Seite von (16) noch ein $-\frac{h}{2\pi i}\frac{\partial \mathfrak{S}_1}{\partial t}, -\frac{h}{2\pi i}\frac{\partial \mathfrak{S}_2}{\partial t}$ u.s.w. hinzu, was die Integration dann allgemein möglich macht.

Die Gleichungen (14) und (16) enthalten sowohl die Kramerssche Dispersionstheorie, als auch die Bornsche Formel für die Energie W_2 und die von Kramers und mir aufgestellten Formeln.

Ich glaube, das ist alles, was man zunächst verlangen kann. Allerdings fehlt noch das wichtigste, nämlich die wirkliche Integration im Fall des Wasserstoffs, d.h. also der Fall mehrerer Freiheitsgrade und ähnliches. Allgemein aber hab ich nach dem Gelingen der obigen Beweise doch sehr grosses Zutrauen zur ganzen Theorie gefasst.

The first part of this letter, up to the sentence 'Bis hierher stammt der ganze Beweis von Born und Jordan', is just a summary of Chapter 2 of paper **13**. Next, Heisenberg develops his own perturbation theory. It is based upon the following definition of a 'canonical transformation': 'Quite generally every transformation which transforms $\mathfrak{p}\mathfrak{q}-\mathfrak{q}\mathfrak{p}$ into itself, is canonical'.

Just as in classical mechanics a canonical transformation is generated by a function $S(p, q)$, Heisenberg generates a canonical transformation by a matrix

$$\text{(A)} \qquad \mathfrak{S} = \lambda\mathfrak{S}_1 + \lambda^2\mathfrak{S}_2 + \dots.$$

He assumes the transformed $\mathfrak{p}$ and $\mathfrak{q}$ to be linear functions of the original $\mathfrak{p}_0$ and $\mathfrak{q}_0$. His formula for $\mathfrak{q}$ is

$$\mathfrak{q} = \mathfrak{q}_0 + \lambda(\mathfrak{S}_1\mathfrak{q}_0 - \mathfrak{q}_0\mathfrak{S}_1) + \lambda^2\{\tfrac{1}{2}(\mathfrak{S}_1^2\mathfrak{q}_0 - 2\mathfrak{S}_1\mathfrak{q}_0\mathfrak{S}_1 + \mathfrak{q}_0\mathfrak{S}_1^2) + \mathfrak{S}_2\mathfrak{q}_0 - \mathfrak{q}_0\mathfrak{S}_2\} + \lambda^3 \dots.$$

The terms of the series are, as Heisenberg himself says, the iterated commutators $\mathfrak{S}\mathfrak{q}-\mathfrak{q}\mathfrak{S}$. In fact, the series can be written as

$$\text{(B)} \quad \mathfrak{q} = \mathfrak{q}_0 + (\mathfrak{S}\mathfrak{q}_0 - \mathfrak{q}_0\mathfrak{S}) + \tfrac{1}{2}\{\mathfrak{S}(\mathfrak{S}\mathfrak{q}_0 - \mathfrak{q}_0\mathfrak{S}) - (\mathfrak{S}\mathfrak{q}_0 - \mathfrak{q}_0\mathfrak{S})\mathfrak{S}\} + \frac{1}{3!}\{\dots\} + \dots.$$

The same formulae hold for $\mathfrak{p}$ and also for $\mathfrak{H}=\mathfrak{H}(\mathfrak{p}, \mathfrak{q})$:

$$\text{(C)} \qquad \mathfrak{H} = \mathfrak{H}(\mathfrak{q}_0, \mathfrak{p}_0) + (\mathfrak{S}\mathfrak{H} - \mathfrak{H}\mathfrak{S}) + \tfrac{1}{2}\{\dots\} + \dots.$$

The relation

$$\mathfrak{p}_0\mathfrak{q}_0 - \mathfrak{q}_0\mathfrak{p}_0 = \frac{h}{2\pi i}\mathbf{1}$$

is in fact preserved by this transformation.

The problem is now, to find a matrix $\mathfrak{S}$ such that the transformed matrix $\mathfrak{H}$ becomes a diagonal matrix $\mathfrak{W}$. This problem can, at least formally, be solved by a successive calculation of the coefficient matrices $\mathfrak{S}_1$, $\mathfrak{S}_2$, ... in (A), as Heisenberg shows.

This perturbation theory formed the starting point of the three men's paper **15**. About September 15, Born and Jordan had returned to Göttingen, and 'an intense correspondence between Göttingen and Copenhagen' started, as Born writes in his Recollections. Of this correspondence, only Heisenberg's letters to Jordan are preserved. We have quoted one of these already. The next letter, dated September 21, begins thus:

Lieber Jordan! Vielen Dank für Ihren Brief; an Born schrieb ich schon eine Skizze der Störungstheorie, deshalb schreib ich sie Ihnen nicht nocheinmal, Sie können ja alles bei Born nachlesen ...

The letter to Born must have contained the same perturbation theory as the letter to Pauli quoted before. It seems that Born did not study this letter carefully. In Born's opinion (stated in his Recollections), Heisenberg's perturbation theory was wrong. For Heisenberg's canonical transformation, Born substituted a simpler and more elegant transformation viz.

(D) $$\boldsymbol{P} = \boldsymbol{S}\boldsymbol{p}\boldsymbol{S}^{-1}, \qquad \boldsymbol{Q} = \boldsymbol{S}\boldsymbol{q}\boldsymbol{S}^{-1}$$

This implies

$$\boldsymbol{PQ} - \boldsymbol{QP} = \boldsymbol{S}(\boldsymbol{pq} - \boldsymbol{qp})\boldsymbol{S}^{-1}$$

hence, if $\boldsymbol{pq}-\boldsymbol{qp}$ is a multiple of the unit matrix, $\boldsymbol{PQ}-\boldsymbol{QP}$ will be the same multiple of the unit matrix. This means: Born's transformation (D) is canonical in the sense of Heisenberg.

Heisenberg's transformation (B) is a special case of Born's. If one substitutes $S = e^{\mathfrak{S}}$, Born's transformation (D) becomes

(E) $$\boldsymbol{P} = e^{\mathfrak{S}}\boldsymbol{p}\, e^{-\mathfrak{S}}, \qquad \boldsymbol{Q} = e^{\mathfrak{S}}\boldsymbol{q}\, e^{-\mathfrak{S}},$$

and by expanding the latter formula in a power series, one obtains Heisenberg's formula (B).

Heisenberg soon recognized the equivalence of his formula (B) with Born's form (D). In a letter to Jordan, dated September 29, he writes

3.) Born's Form für die kanonische Transformation ist fast der Glanzpunkt der bisherigen Theorie, hoffentlich gelingt es bald, daraus Integrationsverfahren zu gewinnen. Auch würde mich sehr interessieren, ob die Bornsche Form die *allgemeinste* kanonische Transformation darstellt.

4.) Wegen des 'Rechenfehlers' in meiner Störungstheorie hab ich mich anfangs sehr geschämt und dies an Born geschrieben–besonders da es kein Rechenfehler, sondern nur eine falsch erratene Formel wäre; aber inzwischen habe ich mir doch überlegt, dass mein Leichtsinn mich diesmal doch gar nicht in die Irre geführt hat, sondern meine Formeln waren ebenso richtig, wie die Bornschen (nur ist bei Born die Begründung viel schöner und durchsichtiger, während man bei mir etwas durch vollständige Induktion 'leicht' beweisen müsste). Man kann nämlich, in der Bornschen Ausdrucksweise, statt

$$S = 1 + \lambda S_1 + \lambda^2 S_2 + \dots$$

auch den Ansatz umschreiben:

$$S = 1 + \lambda S_1 + \lambda^2(\tfrac{1}{2}S_1^2 + S_2) + \lambda^3(\tfrac{1}{6}S_1^3 + \tfrac{1}{2}(S_1S_2 + S_2S_1) + S_3) + \dots$$

das ist ja Geschmacksache. Bildet man dann $q = Sq_0S^{-1}$ u.s.w., so bleibt alles so, wie ich schrieb. Also: diesmal war alles noch gutgegangen.

A little later, Heisenberg also recognized that Born's form and his own are connected by the substitution $S = e^{\mathfrak{S}}$, for in a letter to Pauli, dated October 12, 1925, he writes:

... In der Quantenmechanik gibt es einige Neuigkeiten: Die allgemeine kanonische Transformation lautet

$$P_k = Sp_kS^{-1}$$

$$Q_k = Sq_kS^{-1}.$$

Setzt man $S = e^{1 + \lambda\varphi_1 + \lambda^2\varphi_2 \dots}$, so ergibt sich die Störungstheorie, von der ich damals schrieb. Ferner, die allgemeine Integrationsmethode lautet offenbar: man soll eine Transformation S finden die

$$\mathscr{H}(p_k, q_k) = S^{-1}\mathscr{H}(P_k, Q_k)S = W$$

diagonal macht ...

Transformation to Principal Axes

Born's form (D) of the Heisenberg–Born canonical transformation led directly to the heart of the new theory, viz. to the Principal Axes Transformation.

Heisenberg had already noted that his formula (B) holds just in the same form for any functions $\mathfrak{F}(\mathfrak{q})$ or $\mathfrak{F}(\mathfrak{p})$ and hence also for the energy $\mathfrak{H}$ (see his formula C). Just so, Born noted that for any function $\boldsymbol{f(P, Q)}$ the same transformation formula holds as for $\boldsymbol{P}$ and $\boldsymbol{Q}$:

$$\boldsymbol{f(P, Q) = Sf(p, q)S^{-1}}. \tag{18}$$

This is formula (18) of paper **15**. The redaction of this part of the paper is due to Born. Born's 'functions' $\mathfrak{F}(\boldsymbol{P}, \boldsymbol{Q})$ are, according to § 1 of the paper, defined by 'additions and multiplications of matrices', i.e. they are polynomials or power series in $\boldsymbol{P}$ and $\boldsymbol{Q}$.

From this formula (18), Born immediately proceeds to the Principal Axes Transformation:

The importance of the canonical transformation rests on the following statement: If any pair $\boldsymbol{p}_0$, $\boldsymbol{q}_0$ is given that satisfies equation (15) [i.e. the equation for $\boldsymbol{pq}-\boldsymbol{qp}$], the problem of the integration of the canonical equations can be reduced to the following problem: To determine a function $\boldsymbol{S}$ such that with

$$\boldsymbol{p} = \boldsymbol{S}\boldsymbol{p}_0\boldsymbol{S}^{-1}, \qquad \boldsymbol{q} = \boldsymbol{S}\boldsymbol{q}_0\boldsymbol{S}^{-1}$$

the function

$$\boldsymbol{H}(\boldsymbol{pq}) = \boldsymbol{S}\boldsymbol{H}(\boldsymbol{p}_0\boldsymbol{q}_0)\boldsymbol{S}^{-1} = W$$

becomes a diagonal matrix.

Born's great advantage over Heisenberg and Jordan was that he was familiar not only with Matrix Calculus, which he had learnt from Rosanes at Breslau, but also with Hilbert's theory of integral equations and quadratic forms in an infinite number of variables. In a footnote to Chapter 3 of paper **15** Born quotes Hilbert's 'Grundzüge einer allgemeinen Theorie der linearen Integralgleichungen' (Teubner, Leipzig 1912) and Hellinger's paper in Crelle's Journal reine u. ang. Math. *136* (1910), in which the case of the continuous spectrum is treated in detail. In § 3 of the same Chapter 3, Born gives an account of Hellinger's theory. He states that Hilbert and Hellinger had treated only the case of bounded quadratic forms (or of bounded operators in Hilbert space in the terminology of John von Neumann), but he is confident that essentially the same theorems will hold for non-bounded forms as well (second footnote to § 1, Chapter 3).

Three years later, von Neumann succeeded in proving that Born's confidence was fully justified, but the exact conditions and demonstrations were very hard to find. The spectral resolution of hypermaximal (or, as we now call them, self-adjoint) operators was one of von Neumann's most brilliant mathematical achievements [1]. Von Neumann told me that he had struggled with the problem for a long time, and that an idea of Carleman finally brought the solution.

Canonical transformation and diagonalization of the matrix $\boldsymbol{H}(\boldsymbol{p}, \boldsymbol{q})$

[1] J. von Neumann: Allgemeine Eigenwerttheorie Hermitescher Funktionaloperatoren. Math. Annalen *102*, p. 49 (received December 15, 1928).

are the central ideas of the three men's paper. Perturbation theory, as developed in §§ 4–5 of Chapter 1, in § 2 of Chapter 2 and in a more general form in § 2 of Chapter 3 is just a straightforward application of the central ideas.

Eigenvalues and Eigenvectors

The redaction of Chapter 3 of paper **15** is due to Born. Born knew from Hilbert and Hellinger that the problem of diagonalization of a Hermitean form is equivalent to an 'Eigenvalue Problem'. In § 1, he already calls the diagonal elements W_n *Eigenvalues*, and their totality *Spectrum* of the form $A(xx^*)$. In § 2 he mentions the well-known fact that the Principal Axes can be obtained by solving the set of linear equations

$$Wx_k - \sum H(kl)x_l = 0,$$

in which $H(kl)$ are the elements of the matrix $\boldsymbol{H}$, or the coefficients of the Hermitean form.

Born knew that the eigenvectors $(x_1, x_2, \ldots)$ have finite sums of squares

$$\sum x_k x_k^*,$$

i.e. that they are elements of a Hilbert space. However, he did not realize that the eigenvectors determine stationary states of the atom; he used them only as a mathematical aid in performing the transformation to principal axes. The physical significance of the eigenvectors was not made clear before Schroedinger.

General commutation relations

After having found his Perturbation Theory, Heisenberg next generalized the relation for $\mathfrak{p}\mathfrak{q}-\mathfrak{q}\mathfrak{p}$ to several variables. In a letter to Jordan, dated September 21, 1925, he writes:

... Was die Systeme mit mehreren Freiheitsgraden betrifft, so bin ich überzeugt, dass die Quantenbedingungen lauten

$$(1)\qquad \begin{cases} \mathfrak{p}_r\mathfrak{q}_s - \mathfrak{q}_s\mathfrak{p}_r = \begin{cases} \dfrac{h}{2\pi i} \text{ für } s = r \\ 0 \quad \text{für } s \neq r \end{cases} \\ \mathfrak{p}_r\mathfrak{p}_s - \mathfrak{p}_s\mathfrak{p}_r = 0 \\ \mathfrak{q}_r\mathfrak{q}_s - \mathfrak{q}_s\mathfrak{q}_r = 0. \end{cases}$$

Dass diese Bedingungen die Frequenzbedingung liefern, lässt sich leicht nach Ihrer Methode zeigen; soviel ich mich erinnere, schrieben Sie mir ähnliche Bedingungen in einem Brief, machten aber nachträglich Einwendungen? Im Fall der Separation reduziert sich (1) auf Ihre und Borns Formeln, im Fall der Nichtsepariertheit ist (1) nicht ohne weiteres aus der klassischen Theorie einzusehen, ich glaub aber doch wegen der Frequenzbedingung, dass (1), auch bei unperiodischen Systemen stets richtig ist.

Independently, the same relations (1) were proposed by *Pauli*, by *H. Weyl* and by *Dirac*. Pauli must have hit upon the relations (1) between September 18, when Heisenberg sent him a summary of the Born–Jordan paper, and September 24, when Heisenberg wrote him a postcard from Copenhagen:

Ihre Vermutung über die Quantenbedingung bei mehreren Freiheitsgraden ist natürlich richtig; ich hatte mir schon vor ein paar Tagen einen einfachen Beweis für die Frequenzbedingung überlegt, der so kurz ist, dass ich ihn hier auf der Karte schreiben will ...

Heisenberg now proves that the commutation relations (1) imply

$$\frac{2\pi i}{h}(\mathfrak{F}\mathfrak{p}_\varrho - \mathfrak{p}_\varrho\mathfrak{F}) = -\frac{\partial\mathfrak{F}}{\partial\mathfrak{q}_\varrho}$$

for every function $\mathfrak{F}$ that can be developed into a power series in $\mathfrak{q}$ and $\mathfrak{p}$. Applying this to the enrgy $\mathfrak{H}$, he obtains the frequency condition

$$\frac{H_n - H_m}{h} = \nu(nm).$$

The proof is the same as in paper **15**, Chapter 1, formula (14), and Chapter 2, formula (4). This derivation of the frequency condition was, for Heisenberg, a strong argument in favour of the commutation relations (1).

Born and Jordan had objections against the commutation relations. From Heisenberg's answer (Sept. 29) I quote:

Die Bedenken wegen der Quantenbedingung scheinen mir nirgends zwingend, wenn ich auch einräumen muss, dass ich keine entscheidenden Gründe für die Form (1) ... habe, als nur den der formalen Schönheit. Vgl. z.B. die Klammersymbole (die Jakobischen) der klassischen Mechanik ...

The analogy with the Poisson brackets also led Dirac to the commutation relations, as we have seen.

The third to discover the commutation relations – a few days after

Heisenberg and Pauli – was Hermann Weyl at Zürich. His letter to Born, dated September 27, 1925 begins thus:

Lieber Herr Born!
Ihr Ansatz zur Quantentheorie hat auf mich gewaltigen Eindruck gemacht ...

Next, Weyl considers the matrices $\boldsymbol{p}$ and $\boldsymbol{q}$ as infinitesimal generators of Lie groups $\{e^{\xi \boldsymbol{p}}\}$ and $\{e^{\eta \boldsymbol{q}}\}$, and writes Born's relation for $\boldsymbol{pq}-\boldsymbol{qp}$ as

$$e^{\xi \boldsymbol{p}}\, e^{\eta \boldsymbol{q}} = e^{h\xi\eta}\, e^{\eta \boldsymbol{q}}\, e^{\xi \boldsymbol{p}}.$$

By h, Weyl means $h/2\pi$. Using this formula, he gives a proof of the Conservation of Energy and proceeds:

Unbestimmt bleiben bei Ihrem Ansatz ja immer die Phasen, in dem Sinne, dass alle Gleichungen bestehen bleiben, wenn man $\boldsymbol{p}_{\mu\nu}$, $\boldsymbol{q}_{\mu\nu}$ ersetzt durch $\boldsymbol{p}_{\mu\nu}\, e^{i(\varphi_\mu-\varphi_\nu)}$ $\boldsymbol{q}_{\mu\nu}\, e^{i(\varphi_\mu-\varphi_\nu)}$; φ_μ beliebige Zahlen. Ihnen wird man aber wohl die physikalische Bedeutung absprechen. Im Hinblick darauf ist es, glaube ich, nicht zu gewagt, Ihren Ansatz für mehrere Freiheitsgrade so zu verallgemeinern:

$$\boldsymbol{p}_i\boldsymbol{q}_k - \boldsymbol{q}_k\boldsymbol{p}_i = h\delta_{ik} \qquad \left[\delta_{ik} = \begin{cases} 0 \ (i \neq k) \\ 1 \ (i = k) \end{cases}, \quad i, k = 1, 2, \dots f.\right]$$

Dann kann man mit den $\boldsymbol{p}$ und $\boldsymbol{q}$ vernünftig rechnen. Und bei mehreren unabhängigen Oszillatoren nebeneinander binden die hinzukommenden Kommutationsbedingungen wie $\boldsymbol{p}_1\boldsymbol{q}_2-\boldsymbol{q}_2\boldsymbol{p}_1=0$ nur die Phasen $\varphi^{(1)}$ und $\varphi^{(2)}$ aneinander, was physikalisch nichts zu sagen hat.

Physical Applications of the Theory

This is the title of Chapter 4 of the three men's paper. It consists of three sections.

The first section (§ 1) deals with the angular momentum $\boldsymbol{M}$. The commutation relation

$$\boldsymbol{M}_x\boldsymbol{M}_y - \boldsymbol{M}_y\boldsymbol{M}_x = (h/2\pi i)\boldsymbol{M}_z$$

is derived, and it is proved that the eigenvalues of $\boldsymbol{M}_z$ are $m\cdot(2\pi/h)$, where m is integer or half-integer, the transition rules for m and $j = \max m$ being

$$m \to m+1 \quad \text{or} \quad m \quad \text{or} \quad m-1,$$

$$j \to j+1 \quad \text{or} \quad j \quad \text{or} \quad j-1,$$

and $\boldsymbol{M}^2$ being a diagonal matrix with elements

$$j(j+1)(h/2\pi)^2.$$

From these results, the formulae for the intensities and polarization of the single Zeeman components of a spectral line are derived.

All three authors agree that this section is due to Jordan and Heisenberg. In a letter to me, dated July 23, 1962, Jordan gives the following more detailed information:

Bis ungefähr Seite 600 unten [369 middle] stammte der Inhalt von mir. Die wichtigen Formeln (18) und die daraus abgeleiteten Folgerungen, die zur Auswahlregel für j führten, stammen von Heisenberg; jedoch war Formel (24) sowie auch (25) wohl auch schon im älteren Grundstock des Manuskriptes von Born und mir vorhanden. Seite 604 [373–374] stammte wiederum von Heisenberg ...

The discussions between Jordan and Heisenberg about the angular momentum probably took place in Göttingen after October 23. In the many letters of Heisenberg to Jordan and Pauli, dated September 13 to October 23, the angular momentum is not yet mentioned.

The very short second section (§ 2 of Chapter 4) contains a preliminary discussion of the Zeeman effect, probably due to Heisenberg.

The last section (§ 3) deals with fluctuations in the radiation field. This historically important section is entirely due to Jordan.

The final redaction of paper 15

The redaction of the three men's paper was not an easy task, first because this kind of physics was completely new to each of the three authors, secondly because each was a highly original individual, whose way of looking at problems was quite different from that of the others in many cases.

A first draft was made by Born and Jordan while Heisenberg was at Copenhagen. In the first redaction of Chapter 1, Jordan defined differentiation with respect to a matrix by formula (4) of Chapter 1, as in the Born–Jordan paper **13**. Heisenberg was not satisfied by this definition and replaced it by his definition (3). Formula (4) was only mentioned, not used in the sequence. This was 'sicherlich eine zweckmässige Vereinfachung', as Jordan puts it in his letter to me quoted before. The arguments in favour of Heisenberg's definition of $\partial \boldsymbol{f}/\partial \boldsymbol{q}$ and $\partial \boldsymbol{f}/\partial \boldsymbol{p}$ were exposed in a letter to Born dated October 5, and in a letter to Jordan dated October 7.

According to an earlier letter, Heisenberg intended to return to Göttingen between October 18 and October 20. In fact, Heisenberg was still at Copenhagen on October 12, but on October 23 he was back at Göttingen. On this latter date he wrote Pauli:

Im Augenblick bin ich ausschliesslich mit der Quantenmechanik beschäftigt und ich zweifle noch stark, ob das Problem, diese 3-Männerbeit zu schreiben, überhaupt eine Lösung hat in endlicher Zeit. Aber ich möchte so gern Ihre Meinung hören über den Entwurf, der allerdings schrecklicherweise nur etwa $\frac{1}{3}$ des Endumfangs erreicht. Der wichtigste Teil der Arbeit scheint mir trotz des mathematischen Charakters die Hauptachsentransformation; sie ist natürlich nur als Integrationsmethode gedacht und hilft auch dafür bis jetzt nur wenig, aber sie ist doch formal sehr schön.

We may safely assume that Heisenberg's draft covered, roughly speaking, Chapters 1 and 2 of paper **15** (slightly more than one third of the whole paper).

Chapter 3 was written by Born. Starting with the theory of Hermitean forms in Hilbert space, Born developed a perturbation theory which is more generally applicable than that of Chapters 1 and 2, and mathematically equivalent to Schroedinger's perturbation theory. In § 3, the continuous spectrum is treated by Hellinger's method.

Chapter 3 was finished before Born's departure to America on October 28, 1925, but the whole paper was not yet in definite shape. From Jordan's letter to me, dated October 8, 1964, I quote:

Nach meiner sehr bestimmten Erinnerung ... kann ich mit Sicherheit sagen, dass die Arbeit zwar hinsichtlich des Materials, welches sie bringen sollte, vor Borns Abreise in der Hauptsache fertig war, dann aber nach seiner Abreise erst endgültig formuliert und zusammengestellt wurde. Für die endgültige Zusammenstellung war Heisenberg damals federführend; wir haben uns natürlich über die Einzelheiten noch laufend besprochen. ... Das amtliche Eingangsdatum 16.11 ist sicherlich so zu verstehen, dass unser Manuskript ein oder zwei Tage vorher von Göttingen aus abgeschickt wurde.

The Introduction to paper **15** is certainly due to Heisenberg. About this introduction and about his redactional activity in general he writes Pauli on November 16, 1925:

Ich habe mir alle Mühe gegeben, die Arbeit physikalischer zu machen, als sie war und ich bin so halb zufrieden damit. Aber ich bin immer noch ziemlich unglücklich über die ganze Theorie und war so froh, dass Sie mit der Ansicht über Mathematik und Physik so ganz auf meiner Seite stehen. Hier bin ich in einer Umgebung, die genau entgegengesetzt denkt und fühlt und ich weiss nicht, ob ich nur zu dumm bin, um Mathematik zu verstehen. Göttingen zerfällt in zwei Lager, die einen, die, wie Hilbert (oder auch Weyl in einem Brief an Jordan) von dem grossen Erfolg reden, der durch die Einführung der Matrizenrechnung in die Physik errungen sei, die anderen, die, wie Franck, sagen, dass man die Matrizen doch nie verstehen könne.

It appears from this letter that Heisenberg, although he himself had introduced matrices into physics and rediscovered the law of matrix multiplication, considered the application of matrix calculus to physics difficult to understand from the physical point of view. Heisenberg and Pauli both wanted to make the theory 'more physical', and Heisenberg tried to explain the physical principles underlying the theory as clearly as possible in the introduction to paper **15**.

Pauli's paper 16

In this paper Pauli shows that the hydrogen spectrum can be derived from the new theory.

Pauli's starting point is a method, due to Lenz, for integrating the classical equations of motion of a particle in a Coulomb field (see paper **16a**). If

$$\mathfrak{p} = m\mathfrak{v}$$

is the linear momentum of the particle and

$$\mathfrak{P} = m[\mathfrak{r}\mathfrak{v}]$$

the angular momentum, the vector

$$\mathfrak{A} = \frac{1}{Ze^2m}[\mathfrak{P}\mathfrak{p}] + \frac{\mathfrak{r}}{r}$$

can be shown to be constant in time. Scalar multiplication with $\mathfrak{r}$ yields

$$(\mathfrak{A}\mathfrak{r}) = -\frac{1}{Ze^2m}\mathfrak{P}^2 + r$$

and this is the equation for a conic.

Pauli now shows that the same calculations can, step by step, be performed in Quantum Mechanics. He first shows that $\mathfrak{P}$ is constant in time, and next that

$$\mathfrak{A} = \frac{1}{Ze^2m}\tfrac{1}{2}\{[\mathfrak{P}\mathfrak{p}] - [\mathfrak{p}\mathfrak{P}]\} + \frac{\mathfrak{r}}{r}$$

is constant in time. He shows that $\mathfrak{P}^2$ and $\boldsymbol{P}_z$ can be supposed to be diagonal matrices, the eigenvalues of $\boldsymbol{P}_z$ being $m(h/2\pi)$, as in paper **15**. The quantum number m assumes all integer values from $-k$ to k. Next,

$$\boldsymbol{A}_x\boldsymbol{A}_y - \boldsymbol{A}_y\boldsymbol{A}_x \quad \text{and} \quad \mathfrak{A}^2$$

are calculated, and the energy E is found to be

$$E = \frac{RhZ^2}{n^2}.$$

Finally, the perturbations caused by an electrical field and by crossed electrical and magnetic fields are investigated.

Pauli's paper was received on January 17, 1926, but the main result must have been obtained before November 3, 1925, for on that date Heisenberg writes Pauli:

... Ich brauche Ihnen wohl nicht zu schreiben, wie sehr ich mich über die neue Theorie des Wasserstoffs freue ...

Pauli's paper convinced most physicists that Quantum Mechanics is correct.

Dirac's paper 17

In Heisenberg's fundamental paper **12** and in the papers **13–16** following it, dynamical variables were assumed to be represented by matrices (x_{mn}). In paper **17**, Dirac starts with a more general assumption. He supposes the quantum variables x, y, ... to be elements of an algebra, which means that sums $x+y$ and products xy are defined which satisfy the ordinary laws of algebra, excluding the commutative law of multiplication. Among the elements of this algebra are the numbers of classical mathematics which are called *c-numbers*. All other elements are called *q-numbers*.

The Poisson bracket expression $[x, y]$ is defined by

$$xy - yx = ih[x, y],$$

h being $(2\pi)^{-1}$ times Planck's constant. The equations of motion are written as

$$\dot{x} = [x, H].$$

Starting with a given set of canonical variables q_r, p_r satisfying the Heisenberg–Dirac commutation relations, Dirac calls a set of variables Q_r, P_r canonical if one can deduce the equations

$$[Q_r, Q_s] = 0 \qquad [P_r, P_s] = 0$$
$$[Q_r, P_s] = 0 \quad (r \neq s) \qquad \text{or} \quad 1 \quad (r = s)$$

from the relations connecting the Q_r, P_r with the p_r, q_r.

He claims that these latter relations may be put in the form

$$Q_r = bq_rb^{-1}, \qquad P_r = bp_rb^{-1}.$$

This means that Dirac rediscovered Born's formula for the canonical transformation. For Born, this formula was of fundamental importance: Perturbation Theory and Transformation to Principal Axes were based upon it. Dirac does not lay much stress upon this form of the canonical transformation. He writes '...these formulae do not appear to be of great practical value'.

Dirac's attitude stands in a marked contrast to that of Born, Weyl and Hilbert, who all regarded the reduction of atomic dynamics to an eigenvalue problem as highly important.

Heisenberg hesitated between the two attitudes. We have already quoted his letter of October 23: 'Der wichtigste Teil der Arbeit scheint mir trotz des mathematischen Charakters die Hauptachsentransformation; sie ist natürlich nur als Integrationsmethode gedacht und hilft auch dafür bis jetzt nur wenig, aber sie ist doch formal sehr schön.' In a later letter (November 16) he writes '... Wie man wirklich integriert, haben Sie ja in Ihrer Wasserstoff-Arbeit gezeigt, und alles andere ist doch nur formaler Kram'.

In § 4 of his paper, Dirac considers multiply periodic systems. He shows that for such a system the q-numbers can be represented by matrices in such a way that the product xy is represented by the product matrix.

In our present terminology, multiply periodic systems are systems having a purely discrete energy spectrum. Strictly speaking, the matrix representation can be applied to this case only. Dirac's more general approach has the advantage that it has a meaning also when a part of the spectrum or the whole spectrum is continuous. Dirac clearly realized this (see the end of § 1 of his paper).

In § 5–7, Dirac gives a preliminary investigation of the hydrogen atom. Since the hydrogen problem was completely solved by Pauli five days before, we shall not reproduce this part of Dirac's paper.

PART I

TOWARDS QUANTUM MECHANICS

Received March 3, 1917

1

ON THE QUANTUM THEORY OF RADIATION

A. EINSTEIN

The formal similarity between the chromatic distribution curve for thermal radiation and the Maxwell velocity-distribution law is too striking to have remained hidden for long. In fact, it was this similarity which led W. Wien, some time ago, to an extension of the radiation formula in his important theoretical paper, in which he derived his displacement law

$$\varrho = \nu^3 f(\nu/T). \tag{1}$$

As is well known, he discovered the formula

$$\varrho = \alpha\nu^3 \exp(-h\nu/kT), \tag{2}$$

which is still accepted as correct in the limit of large values of ν/T (Wien's radiation formula). Today we know that no approach which is founded on classical mechanics and electrodynamics can yield a useful radiation formula. Rather, classical theory must of necessity lead to Rayleigh's formula

$$\varrho = \frac{k\alpha}{h}\nu^2 T. \tag{3}$$

Next, Planck in his fundamental investigation based his radiation formula

$$\varrho = \alpha\nu^3 \frac{1}{\exp(h\nu/kT) - 1} \tag{4}$$

on the assumption of discrete portions of energy, from which quantum theory developed rapidly. It was then only natural that Wien's argument, which had led to eq. (2), should have become forgotten.

Editor's note. This paper was published as Phys. Zs. **18** (1917) 121. It was first printed in Mitteilungen der Physikalischen Gesellschaft Zürich, No. 18, 1916.

Not long ago I discovered a derivation of Planck's formula which was closely related to Wien's original argument * and which was based on the fundamental assumption of quantum theory. This derivation displays the relationship between Maxwell's curve and the chromatic distribution curve and deserves attention not only because of its simplicity, but especially because it seems to throw some light on the mechanism of emission and absorption of radiation by matter, a process which is still obscure to us. By postulating some hypotheses on the emission and absorption of radiation by molecules, which suggested themselves from quantum theory, I was able to show that molecules with a quantum-theoretical distribution of states in thermal equilibrium, were in dynamical equilibrium with the Planck radiation; in this way, Planck's formula (4) could be derived in an astonishingly simple and general way. It was obtained from the condition that the internal energy distribution of the molecules demanded by quantum theory, should follow purely from an emission and absorption of radiation.

But if these hypotheses on the interaction between radiation and matter turn out to be justified, they must produce rather more than just the correct statistical distribution of the internal energy of the molecules: for there is also a momentum transfer associated with the emission and absorption of radiation; this produces, purely through the interaction between the radiation and the molecules, a certain velocity distribution for the latter. This must evidently be identical with the velocity distribution of the molecules which is entirely due to their collisions among themselves, i.e. it must agree with the Maxwell distribution. It has to be required that the mean kinetic energy of a molecule (per degree of freedom) should be equal to $\frac{1}{2}kT$ in a Planck radiation field of temperature T. This requirement should hold independently of the nature of the molecules under consideration and independently of the frequencies emitted or absorbed by them. We want to demonstrate in the present paper that this far-reaching requirement is in fact satisfied quite generally, thus lending new support to our simple hypotheses concerning the elementary processes of emission and absorption.

To obtain such a result however requires a certain extension of the hypotheses, which had been up to now solely concerned with an exchange

* Verh. d. Deutschen physikal. Gesellschaft **18** Nr. 13/14 (1916) 318. The arguments used in that paper are reproduced in the present discussion.

of *energy*. The question arises: does the molecule suffer an impulse when it emits or absorbs energy ε? As an example, let us consider the emission of radiation from the point of view of classical electrodynamics. When a body emits an energy ε, it has a recoil (momentum) ε/c, provided the whole of the radiation is emitted in the same direction. If, however, the emission process has spatial symmetry, such as in the case of spherical waves, no recoil is produced at all. This second possibility is also of importance in the quantum theory of radiation. If a molecule absorbs or emits energy ε in the form of radiation during its transition from one quantum-theoretically possible state to another, such an elementary process can be thought of as being partially or completely directional, or else symmetrical (non-directional). It will become apparent that we shall only then arrive at a theory which is free from contradictions, if we consider such elementary processes to be perfectly directional; this embodies the main result of the subsequent discussion.

1. Fundamental hypothesis of quantum theory. Canonical distribution of states

In quantum theory a molecule of a given kind can only exist in a discrete set of states $Z_1, Z_2, \ldots Z_n, \ldots$, with (internal) energies $\varepsilon_1, \varepsilon_2, \ldots \varepsilon_n, \ldots$, apart from its orientation and translatory motion. If such molecules belong to a gas at temperature T, the relative frequency W_n of such states Z_n is given by the formula

$$W_n = p_n \exp(-\varepsilon_n/kT), \tag{5}$$

which corresponds to the canonical distribution of states in statistical mechanics. In this formula, $k = R/N$ is the well-known Boltzmann constant, and p_n is a number, independent of T and characteristic for the molecule and its nth quantum state, which can be called the statistical 'weight' of this state. Formula (5) can be derived from Boltzmann's principle, or from purely thermodynamical considerations. It expresses the most extreme generalisation of Maxwell's velocity-distribution law.

The latest fundamental developments in quantum theory are concerned with a theoretical derivation of the quantum-theoretically possible states Z_n and their weights p_n. For the present basic investigation, a detailed determination of the quantum states is not required.

2. Hypotheses on the radiative exchange of energy

Let Z_n and Z_m be two quantum-theoretically possible states of the gas molecule, whose energies are ε_n and ε_m, respectively, and satisfy the inequality $\varepsilon_m > \varepsilon_n$. Let us assume that the molecule is capable of a transition from state Z_n into state Z_m with an absorption of radiation energy $\varepsilon_m - \varepsilon_n$; that, similarly, the transition from state Z_m to state Z_n is possible, with emission of the same radiative energy. Let the radiation absorbed or emitted by the molecule have frequency ν which is characteristic for the index combination (m, n) that we are considering.

For the laws governing this transition, we introduce a few hypotheses which are obtained by carrying over the known situation for a Planck resonator in classical theory to the as yet unknown one in quantum theory.

(a) *Emission of radiation.* According to Hertz, an oscillating Planck resonator radiates energy in the well-known way, regardless of whether or not it is excited by an external field. Correspondingly, let us assume that a molecule may go from state Z_m to a state Z_n and emit radiation energy $\varepsilon_m - \varepsilon_n$ with frequency μ, without excitation from external causes. Let the probability $\mathrm{d}W$ for this to happen during the time interval $\mathrm{d}t$, be

$$\mathrm{d}W = A_m^n \, \mathrm{d}t, \tag{A}$$

where A_m^n is a constant characterising the index combination under consideration.

The statistical law which we assumed, corresponds to that of a radioactive reaction, and the above elementary process corresponds to a reaction in which only γ-rays are emitted. It need not be assumed here that the time taken for this process is zero, but only that this time should be negligible compared with the times which the molecule spends in states Z_1, etc.

(b) *Absorption of radiation.* If a Planck resonator is located in a radiation field, the energy of the resonator is changed through the work done on the resonator by the electromagnetic field of the radiation; this work can be positive or negative, depending on the phases of the resonator and the oscillating field. We correspondingly introduce the following quantum-theoretical hypothesis. Under the influence of a radiation density ϱ of frequency ν, a molecule can make a transition from state Z_n to state Z_m by absorbing radiation energy

$\varepsilon_m - \varepsilon_n$, according to the probability law

$$\mathrm{d}W = B_n^m \varrho \, \mathrm{d}t. \tag{B}$$

We similarly assume that a transition $Z_m \to Z_n$, associated with a liberation of radiation energy $\varepsilon_m - \varepsilon_n$, is possible under the influence of the radiation field, and that it satisfies the probability law

$$\mathrm{d}W = B_m^n \varrho \, \mathrm{d}t. \tag{B'}$$

B_n^m and B_m^n are constants. We shall give both processes the name 'changes of state due to irradiation'.

We now have to ask ourselves what is the momentum transfer to the molecule for such changes of state. Let us first discuss the case of absorption of radiation. If a radiation bundle in a given direction does work on a Planck resonator, the corresponding energy is removed from the radiation bundle. To this transfer of energy there also corresponds a momentum transfer from radiation bundle to resonator, by momentum conservation. The resonator is thus acted upon by a force in the beam direction of the radiation bundle. If the energy transfer is negative, then the force acts on the resonator in the opposite direction. If the quantum hypothesis holds, we can obviously interpret the process in the following way. If the incident radiation bundle produces the transition $Z_n \to Z_m$ by absorption of radiation, a momentum $(\varepsilon_m - \varepsilon_n)/c$ is transferred to the molecule in the direction of propagation of the beam. For the absorption process $Z_m \to Z_n$, the momentum transfer has the same magnitude, but is in the opposite direction. For the case where the molecule is acted upon simultaneously by several radiation bundles, we assume that total energy $\varepsilon_m - \varepsilon_n$ associated with an elementary process is removed from, or added to, a *single* such radiation bundle. Thus here, too, the momentum transferred to the molecule is $(\varepsilon_m - \varepsilon_n)/c$.

For an energy transfer by emission of radiation in the case of a Planck resonator, no momentum transfer to the resonator takes place, since emission occurs in the form of a spherical wave, according to classical theory. As was remarked previously, a quantum theory free from contradictions can only be obtained if the emission process, just as absorption, is assumed to be directional. In that case, for each elementary emission process $Z_m \to Z_n$ a momentum of magnitude $(\varepsilon_m - \varepsilon_n)/c$ is transferred to the molecule. If the latter is isotropic, we shall have to assume that all directions of emission are equally probable.

If the molecule is not isotropic, we arrive at the same statement if the orientation changes with time in accordance with the laws of chance. Moreover, such an assumption will also have to be made about the statistical laws for absorption, (B) and (B′). Otherwise the constants B_n^m and B_m^n would have to depend on the direction, and this can be avoided by making the assumption of isotropy or pseudo-isotropy (using time-averages).

3. Derivation of the Planck radiation law

We now look for that particular radiation density ϱ, for which the exchange of energy between radiation and molecules in keeping with the probability laws (A), (B), and (B′) does not disturb the molecular distribution of states given by eq. (5). For this it is necessary and sufficient that the number of elementary processes of type (B) taking place per unit time should, on average, be equal to those of type (A) and (B′) taken together. From this condition one obtains from (5), (A), (B), (B′) the equation

$$p_n \exp(-\varepsilon_n/kT) B_n^m \varrho = p_m \exp(-\varepsilon_m/kT)(B_m^n \varrho + A_m^n)$$

for the elementary processes associated with the index combination (m, n).

If, in addition, ϱ tends to infinity with T, as will be assumed, the relation

$$p_n B_n^m = p_m B_m^n \tag{6}$$

has to hold between the constants B_n^m and B_m^n. We then obtain from our equation,

$$\varrho = \frac{A_m^n / B_m^n}{\exp[(\varepsilon_m - \varepsilon_n)/kT] - 1} \tag{7}$$

as the condition for dynamical equilibrium.

This expresses the temperature dependence of the radiation density according to Planck's law. From Wien's displacement law (1) it follows immediately that

$$\frac{A_m^n}{B_m^n} = \alpha \nu^3 \tag{8}$$

and

$$\varepsilon_m - \varepsilon_n = h\nu, \tag{9}$$

where α and h are universal constants. To compute the numerical value of the constant α, one would have to have an exact theory of electrodynamic and mechanical processes; for the present, one has to confine oneself to a treatment of the limiting case of Rayleigh's law for high temperatures, for which the classical theory is valid in the limit.

Eq. (9) is of course the second principal rule in Bohr's theory of spectra. Since its extension by Sommerfeld and Epstein, this may well be claimed to have become a safely established part of our science. It also contains implicitly the photochemical principle of equivalence, as has been shown by me.

4. A method for calculating the motion of molecules in the radiation field

We now turn to a discussion of the motion of our molecules under the influence of radiation. For this we shall make use of a method which is well known from the theory of Brownian motion, and which I employed on several occasions for numerical computations of motions in a radiation field. To simplify the calculation we shall only consider the case where the motions take place in just one direction, the X-direction of the coordinate system. Furthermore, we shall confine ourselves to a calculation of the average value of the kinetic energy of the progressive motion, and we shall thus not attempt to prove that such velocities v obey the Maxwell distribution law. The mass M of the molecule is assumed sufficiently large, so that higher powers of v/c can be neglected in comparison with lower ones; we can then apply the laws of ordinary mechanics to the molecule. Finally, no real loss of generality is introduced if we perform the calculations as if the states with index m and n were the only possible states for the molecule.

The momentum Mv of a molecule undergoes two different types of change during the short time interval τ. Although the radiation is equally constituted in all directions, the molecule will nevertheless be subjected to a force originating from the radiation, which opposes the motion. Let this be equal to Rv, where R is a constant to be determined later. This force would bring the molecule to rest, if it were not for the irregularity of the radiative interactions which transmit a momentum Δ of changing sign and magnitude to the molecule during time τ; such an unsystematic effect, as opposed to that previously mentioned, will sustain some movement of the molecule. At the end of the short time interval τ, the momentum of the molecule will have the value

$$Mv - Rv\tau + \Delta.$$

Since the velocity distribution is supposed to remain constant with time, the average of the absolute value of the above quantity must be equal to Mv; the mean values of the squares of these quantities, taken over a long time interval or over a large number of molecules, must therefore be equal:

$$\overline{(Mv - Rv\tau + \Delta)^2} = \overline{(Mv)^2}.$$

Since we were specifically concerned with the systematic effect of v on the momentum of the molecule, we shall have to neglect the average value $\overline{v\Delta}$. Expanding the left-hand side of the equation, one therefore obtains

$$\overline{\Delta^2} = 2RM\overline{v^2}\tau. \tag{10}$$

The mean square value $\overline{v^2}$, which the radiation of temperature T produces in our molecules by interacting with them, must be of the same size as the mean square value $\overline{v^2}$ obtained from the gas laws for a gas molecule at temperature T in the kinetic theory of gases. For the presence of our molecules would otherwise disturb the thermal equilibrium between the thermal radiation and an arbitrary gas held at the same temperature. We must therefore have

$$\tfrac{1}{2}\overline{Mv^2} = \tfrac{1}{2}kT. \tag{11}$$

Eq. (10) thus becomes

$$\overline{\Delta^2}/\tau = 2RkT. \tag{12}$$

The investigation is now continued as follows. For a given radiation $(\varrho(\nu))$, $\overline{\Delta^2}$ and R can be calculated, using our hypotheses on the interaction between radiation and molecules. If the results are inserted in eq. (12), this equation must become an identity, if ϱ is expressed as a function of ν and T, using Planck's equation (4).

5. Calculation of *R*

Consider a molecule of the kind discussed above, moving uniformly with velocity v along the X-axis of the coordinate system K. We wish to find the average momentum which is transferred from the radiation field to the molecule per unit time. In order to calculate it, we have to describe the radiation in a coordinate system K′ which is at rest relative to the molecule in question. For we had only formulated our hypotheses on emission and absorption for the case of stationary molecules. The transformation to the [coordinate] system K′ has been carried out in a number of places in the literature, particularly accurately in Mosengeil's Berlin dissertation. For completeness, however, I shall reproduce these simple arguments at this point.

Referred to K, the radiation is isotropic, i.e. the radiation of frequency range $d\nu$ per unit volume, associated with a given infinitesimal solid angle $d\varkappa$ relative to its direction of propagation, is given by

$$\varrho \, d\nu \, \frac{d\varkappa}{4\pi}, \tag{13}$$

where ϱ depends only on the frequency ν, but not on the direction. To this particular radiation there corresponds a particular radiation in K′ which is similarly characterised by a frequency range $d\nu'$ and a certain solid angle $d\varkappa'$. The volume density of this particular radiation is given by

$$\varrho'(\nu', \varphi') \, d\nu' \, \frac{d\varkappa'}{4\pi}. \tag{13'}$$

This defines ϱ'. It depends on the direction, which we shall define in the usual way by means of the angle φ' with the X'-axis and the angle ψ' which the projection in the $Y'Z'$-plane makes with the Y'-axis. To these angles correspond the angles φ and ψ, which determine the direction of $d\varkappa$ in K in an analogous manner.

First of all it is clear that the transformation law between (13) and (13′) must be the same as that for the squares of the amplitudes, A^2 and A'^2, of a plane wave with corresponding direction. We therefore find, to the desired approximation, that

$$\frac{\varrho'(\nu', \varphi') \, d\nu' \, d\varkappa'}{\varrho(\nu) \, d\nu \, d\varkappa} = 1 - 2 \frac{v}{c} \cos \varphi, \tag{14}$$

or

$$\varrho'(\nu', \varphi') = \varrho(\nu) \frac{d\nu}{d\nu'} \frac{d\varkappa}{d\varkappa'} \left(1 - 2 \frac{v}{c} \cos \varphi\right). \tag{14'}$$

The theory of relativity further gives the following formulae, valid to the desired approximation,

$$\nu' = \nu\left(1 - \frac{v}{c}\cos\varphi\right), \tag{15}$$

$$\cos\varphi' = \cos\varphi - \frac{v}{c} + \frac{v}{c}\cos^2\varphi, \tag{16}$$

$$\psi' = \psi. \tag{17}$$

With the same approximation, we have from (15)

$$\nu = \nu'\left(1 + \frac{v}{c}\cos\varphi'\right).$$

Therefore, again to the same approximation,

$$\varrho(\nu) = \varrho\left(\nu' + \frac{v}{c}\nu'\cos\varphi'\right)$$

or

$$\varrho(\nu) = \varrho(\nu') + \frac{\partial\varrho}{\partial\nu}(\nu')\cdot\frac{v}{c}\nu'\cos\varphi'. \tag{18}$$

Moreover, from (15), (16) and (17),

$$\frac{d\nu}{d\nu'} = 1 + \frac{v}{c}\cos\varphi',$$

$$\frac{d\varkappa}{d\varkappa'} = \frac{\sin\varphi\, d\varphi\, d\psi}{\sin\varphi'\, d\varphi'\, d\psi'} = \frac{d(\cos\varphi)}{d(\cos\varphi')} = 1 - 2\frac{v}{c}\cos\varphi'.$$

By means of these two relations and (18), we can write (14′) in the form

$$\varrho'(\nu', \varphi') = \left[(\varrho)_\nu + \frac{v}{c}\nu'\cos\varphi'\left(\frac{\partial\varrho}{\partial\nu}\right)_\nu\right]\left(1 - 3\frac{v}{c}\cos\varphi'\right). \tag{19}$$

Using (19) and our hypothesis on the emission and absorption of radiation by the molecule, we can easily calculate the average momentum transferred to the molecule per unit time. Before doing so, however, we shall have to say a few words in justification of this approach. It could be objected that eqs. (14), (15), (16) are based on Maxwell's theory of the electromagnetic field which cannot be reconciled with quantum theory. But this objection relates more to the form than to the real essence of the matter. For whatever the shape of a future theory of the electromagnetic processes, the Doppler principle and the aberration law will at all events remain preserved, and hence also eqs. (15) and (16). Furthermore, the validity of the energy relation (14) certainly extends beyond wave theory; according to the theory of relativity, this transformation law also holds, e.g., for the energy density of a mass moving with (almost) the velocity of light and having infinitesimally small rest density. Eq. (19) can therefore lay claim to being valid for any theory of radiation.

According to (B), the radiation associated with the solid angle $\mathrm{d}\varkappa'$ would give rise to $B_n^m \varrho'(\nu', \varphi')\,\mathrm{d}\varkappa'/4\pi$ elementary absorption processes of the type $Z_n \rightarrow Z_m$ per second, if the molecule were to be restored to the state Z_n immediately after each such elementary process. But in reality, the time for remaining in state Z_n per second is equal to $S^{-1} p_n \exp(-\varepsilon_n/kT)$ from (5), where the abbreviation

$$S = p_n \exp(-\varepsilon_n/kT) + p_m \exp(-\varepsilon_m/kT) \tag{20}$$

has been used. The number of such processes per second is thus really

$$\frac{1}{S} p_n \exp(-\varepsilon_n/kT)\, B_n^m \varrho'(\nu', \varphi') \frac{\mathrm{d}\varkappa'}{4\pi}.$$

For each such elementary process a momentum $[(\varepsilon_m - \varepsilon_n)/c] \cos \varphi'$ gets transmitted to the atom in the direction of the positive X'-axis. Analogously we find, starting from (B'), that the corresponding number, per second, of elementary processes for an absorption of the type $Z_m \rightarrow Z_n$ is

$$\frac{1}{S} p_m \exp(-\varepsilon_m/kT)\, B_m^n \varrho'(\nu', \varphi') \frac{\mathrm{d}\varkappa'}{4\pi},$$

and in such a process a momentum $-[(\varepsilon_m - \varepsilon_n)/c] \cos \varphi'$ is transferred to the molecule. The total momentum transfer to the molecule produced by the absorption of radiation is therefore, per unit time,

$$\frac{h\nu'}{cS} p_n B_n^m [\exp(-\varepsilon_n/kT) - \exp(-\varepsilon_m/kT)] \int \varrho'(\nu', \varphi') \cos \varphi' \frac{\mathrm{d}\varkappa'}{4\pi}.$$

This follows from (6) and (9), and the integration extends over all elementary solid angles. On integrating, one obtains from (19)

$$-\frac{h\nu}{c^2 S}\left(\varrho - \tfrac{1}{3}\nu \frac{\partial \varrho}{\partial \nu}\right) p_n B_n^m [\exp(-\varepsilon_n/kT) - \exp(-\varepsilon_m/kT)] v.$$

Here the effective frequency is again denoted by ν (instead of ν').

But this expression represents the whole of the average momentum transferred per unit time to a molecule moving with velocity v. For it it clear that the elementary radiative emission processes, which take place in K′ without interaction with the radiation field, have no preferred direction, so that they cannot transmit any momentum to the molecule, on average. The final result of our discussion is therefore

$$R = \frac{h\nu}{c^2 S}\left(\varrho - \tfrac{1}{3}\nu \frac{\partial \varrho}{\partial \nu}\right) p_n B_n^m [\exp(-\varepsilon_n/kT)][1 - \exp(-h\nu/kT)]. \qquad (21)$$

6. Calculation of $\overline{\Delta^2}$

It is much simpler to calculate the effect of the irregularity of the elementary processes on the mechanical behaviour of the molecule, because the calculation can be based on a molecule at rest, to the degree of approximation to which we had restricted ourselves from the beginning.

Consider an arbitrary event, causing a momentum transfer λ to a molecule in the X-direction. This momentum can be assumed of different sign and magnitude in different cases. Nevertheless λ is supposed to satisfy a certain statistical law, such that its average value vanishes. Now let $\lambda_1, \lambda_2, \ldots$ be the momenta transmitted to the molecule due to a number of mutually independent causes, so that the total momentum transfer Δ is given by

$$\Delta = \sum \lambda_v.$$

Then, if the averages $\overline{\lambda_v}$ of the individual λ_v vanish,

$$\overline{\Delta^2} = \overline{\sum \lambda_v^2}. \qquad (22)$$

If the mean square values $\overline{\lambda_v^2}$ of the individual momenta are all equal ($\overline{\lambda_v^2} = \overline{\lambda^2}$), and if l is the total number of events producing these momenta, the relation

$$\overline{\Delta^2} = l\overline{\lambda^2} \qquad (22a)$$

holds.

According to our hypotheses, a momentum $\lambda = (h\nu/c)\cos\varphi$ is transferred to the molecule for each absorption and emission process. Here, φ denotes the angle between the X-axis and a randomly chosen direction. One therefore obtains

$$\overline{\lambda^2} = \tfrac{1}{3}(h\nu/c)^2. \tag{23}$$

Since we assume that all the elementary processes which occur can be regarded as mutually independent events, we are allowed to use (22a). Then l is the number of elementary events which occur in time τ. This is twice the number of absorption processes $Z_n \to Z_m$ taking place in time τ. We therefore have

$$l = \frac{2}{S} p_n B_n^m \exp(-\varepsilon_n/kT)\varrho\tau \tag{24}$$

and from (23), (24) and (22),

$$\frac{\overline{\Delta^2}}{\tau} = \frac{2}{3S}\left(\frac{h\nu}{c}\right)^2 p_n B_n^m \exp(-\varepsilon_n/kT)\varrho. \tag{25}$$

7. Conclusion

We now have to show that the momenta transferred from the radiation field to the molecule according to our basic hypotheses, never disturb the thermodynamic equilibrium. For this, we need only insert the values for Δ^2/τ and R determined by (25) and (21), after replacing in (21) the expression

$$\left(\varrho - \tfrac{1}{3}\nu\frac{\partial\varrho}{\partial\nu}\right)[1 - \exp(-h\nu/kT)]$$

by $\varrho h\nu/3kT$, from (4). It is then seen immediately that our basic equation (12) is identically satisfied.

We have now completed the arguments which provide a strong support for the hypotheses stated in § 2, concerning the interaction between matter and radiation by means of absorption and emission processes, or in- or outgoing radiation. I was led to these hypotheses by my endeavour to postulate for the molecules, in the simplest possible manner, a quantum-theoretical behaviour that would be the analogue of the behaviour of a Planck resonator in the classical theory. From the general quantum assumption for matter, Bohr's second postulate (eq. 9) as well as Planck's radiation formula followed in a natural way.

Most important, however, seems to me to be the result concerning the momentum transfer to the molecule due to the absorption and emission of radiation. If one of our assumptions about the momenta were to be changed, a violation of eq. (12) would be produced; it seems hardly possible to maintain agreement with this relation, imposed by the theory of heat, other than on the basis of our assumptions. The following statements can therefore be regarded as fairly certainly proved.

If a radiation bundle has the effect that a molecule struck by it absorbs or emits a quantity of energy $h\nu$ in the form of radiation (ingoing radiation), then a momentum $h\nu/c$ is always transferred to the molecule. For an absorption of energy, this takes place in the direction of propagation of the radiation bundle, for an emission in the opposite direction. If the molecule is acted upon by several directional radiation bundles, then it is always only a single one of these which participates in an elementary process of irradiation; this bundle alone then determines the direction of the momentum transferred to the molecule.

If the molecule undergoes a loss in energy of magnitude $h\nu$ without external excitation, by emitting this energy in the form of radiation (outgoing radiation), then this process, too, is directional. Outgoing radiation in the form of spherical waves does not exist. During the elementary process of radiative loss, the molecule suffers a recoil of magnitude $h\nu/c$ in a direction which is only determined by 'chance', according to the present state of the theory.

These properties of the elementary processes, imposed by eq. (12), make the formulation of a proper quantum theory of radiation appear almost unavoidable. The weakness of the theory lies on the one hand in the fact that it does not get us any closer to making the connection with wave theory; on the other, that it leaves the duration and direction of the elementary processes to 'chance'. Nevertheless I am fully confident that the approach chosen here is a reliable one.

There is room for one further general remark. Almost all theories of thermal radiation are based on the study of the interaction between radiation and molecules. But in general one restricts oneself to a discussion of the *energy* exchange, without taking the *momentum* exchange into account. One feels easily justified in this, because the smallness of the impulses transmitted by the radiation field implies that these can almost always be neglected in practice, when compared with other effects causing the motion. For a *theoretical* discussion, however, such small effects should be considered on a completely equal footing with more conspicuous effects of a radiative *energy* transfer, since energy and momentum are linked in the closest possible way. For this reason a theory can only be regarded as justified when it is able to show that the impulses transmitted by the radiation field to matter lead to motions that are in accordance with the theory of heat.

Related papers

1a A. Einstein, *Über einen die Erzeugung und Verwandlung des Lichtes betreffenden heuristischen Gesichtspunkt.* Ann. d. Phys. **17** (1905) 132.

1b A. Einstein, *Zur Theorie der Lichterzeugung und Lichtabsorption.* Ann. d. Phys. **20** (1906) 199.

1c A. Einstein, *Beiträge zur Quantentheorie.* Verh. der D. Physikal. Ges. **16** (1914) 820.

1d A. Einstein, *Strahlungs-Emission und Absorption nach der Quantentheorie.* Verh. der D. Physikal. Ges. **18** (1916) 318.

Communicated in the meeting of June 24, 1916

2

ADIABATIC INVARIANTS AND THE THEORY OF QUANTA

P. EHRENFEST

Introduction

In the treatment of a continually increasing number of physical problems, use is at the same time made of the principles of classical mechanics and electrodynamics, and of the hypothesis of the quanta, which is in conflict with them. Through the study of these problems it is hoped to arrive at some general point of view which may trace the boundary between the 'classical region' and the 'region of the quanta'.

One fundamental law stands amidst the theory of quanta, which is wholly derived from classical foundations: the *Displacement Law* of W. Wien on the change of the distribution of energy over the spectrum involved by a reversible adiabatic compression of radiation. This fact deserves our attention. It might be possible that also in more general cases, when we do not restrict ourselves to harmonic motions, the reversible adiabatic transformations should be treated in a classical way, whereas in the calculation of other processes (e.g. an isothermal addition of heat) the quanta come into play.

From this point of view I started in some papers in which on the one hand I studied Planck's hypothesis of energy elements*, and on the other tried to extend this hypothesis to more general motions.†

Editor's note. This paper was published as Phil. Mag. **33** (1917) 500–513. It is an abridged translation of a paper published in Verslagen Kon. Akad. Amsterdam **25** (1916) 412–433 under the title 'Over adiabatische veranderingen van een stelsel in verband met de theorie der quanta'. An English translation of the full text was published in Proc. Acad. Amsterdam **19** (1917) 576–597 under the title 'On Adiabatic Changes of a System in Connection with the Quantum Theory'. A German translation was published in Ann. Physik **51** (1916) 327–352.

* P. Ehrenfest, Ann. d. Phys. **36** (1911) 91–118 (quoted as paper A).

† P. Ehrenfest, Verh. d. D. phys. Ges. **15** (1913) 451 (quoted as B). P. Ehrenfest, 'A Theorem of Boltzmann and its connexion with the theory of quanta', Proc. Acad. Amsterdam (quoted as C).

In these researches I especially made use of the following hypothesis, to which Einstein* gave the name 'Adiabatenhypothese'.

If a system be affected in a reversible adiabatic way, allowed motions are transformed into allowed motions.†

Suppose that for some class of motions we, for the first time, introduce the quanta. In some cases the hypothesis fixes completely which special motions are to be considered as allowed: this occurs if the new class of motions can be derived by means of an adiabatic transformation from some class for which the allowed motions are already known (especially if the new motions can be derived from harmonic motions of one degree of freedom).**

In other cases the hypothesis gives restrictions to the arbitrariness which exists otherwise in the introduction of the quanta.

In these applications of the adiabatic hypothesis the socalled '*adiabatic invariants*' are of great importance, i.e. those quantities which may have the same values before and after the adiabatic affection. Especially I have shown before†† that arbitrary periodic motions (of one or more degrees of freedom) possess the adiabatic invariant

$$\overline{2T}/\nu \tag{1}$$

(ν, frequency; $\overline{T}$, mean with respect to time of the kinetic energy), which in the case of harmonic motions of one degree of freedom reduces to‡

$$\varepsilon/\nu \tag{2}$$

The object of the considerations of this paper is:

(1) To formulate as sharply as possible the adiabatic hypothesis, at the same time showing what is wanting in sharpness, especially for non-periodic motions.

(2) To demonstrate the importance of the '*adiabatic invariants*' for the theory of quanta. In this respect the discussion of the invariant $\overline{2T}/\nu$ mentioned above gives the connexion between the adiabatic hypothesis

* A. Einstein, 'Beiträge z. Quantentheorie', Verh. d. D. phys. Ges. **16** (1914) 826.

† For the definitions of the expressions used here comp. § 1, 2.

**Examples: C, § 3; this paper, § 7, 8.

†† Paper B, § 1.

‡ Comp. A, § 2; C, § 2. The existence of this adiabatic invariant may be considered as the root of Wien's displacement law.

and the formulae by which Planck, Debye, Bohr, Sommerfeld and others have introduced the quanta.
(3) To indicate the difficulties which arise in the application of the hypothesis, if the reversible adiabatic transformation leads through singular motions.
(4) To indicate the connexion between the adiabatic problems and the statistical-mechanical roots of the Second Law of Thermodynamics. Boltzmann's deduction of this law is based upon a statistical principle which has been destroyed by the introduction of the quanta. At the present time we possess a statistical deduction of this law for some special systems (e.g. for systems with simple harmonic motions) but not for general systems.*

I take the liberty of publishing my considerations, in the hope that others may overcome the difficulties which I could not solve. Perhaps on closer examination it will appear that the adiabatic hypothesis is not generally valid; in any case, the correctness of Wien's displacement law seems to indicate that the reversible adiabatic processes take a prominent place in the theory of quanta – it seems that they may be treated in a 'classical' way.

1. Definition of a reversible adiabatic affection of a system. Motions $\beta(a)$ and $\beta(a')$ which are adiabatically related to each other

Let the coordinates of the system be denoted by $q_1 \ldots q_n$. The potential energy Φ may contain besides the coordinates q certain 'parameters' $a_1, a_2 \ldots$, the values of which can be altered infinitely slowly. The kinetic energy T may be a homogeneous quadratic function of the velocities $\dot{q}_1 \ldots \dot{q}_n$, the coefficients of which are functions of the q and may be of the $a_1, a_2 \ldots$. By changing the parameters from the values $a_1, a_2 \ldots$ to the values $a_1', a_2' \ldots$ in an infinitely slow way, a given motion $\beta(a)$ may be transformed into another motion $\beta(a')$. This special type of influencing upon the system may be called '*a reversible adiabatic affection*', the motions $\beta(a)$ and $\beta(a')$ '*adiabatically related to each other*'.

Remarks. A. If some of the motions considered are distinctly

* P. Ehrenfest, Phys. Z. **15** (1914) 657 (paper D). Also § 8 of this paper.

non-periodic (e.g. the hyperbolic motion in the case of a Newtonian attraction), the addition '*reversible*' loses its original meaning.

B. The definition given above must be generalized in a suitable manner, if the system is affected by an (infinitely slowly increasing) magnetic field (Zeeman effect), or if the mechanical system is replaced by an electrodynamical one (reversible adiabatic compression of radiation).

2. Formulation of the adiabatic hypothesis for systems with periodical or quasi-periodical motions

Consider the system first when the parameters have some given values $a_{10}, a_{20}\ldots$. The theory of quanta will not allow every motion $\beta(a_0)$ which is possible with these values of the parameters according to the equations of the classical mechanics, but only some distinct special motions.* Consequently we speak of the 'allowed' motions $B\{a_0\}$ belonging to the values $a_{10}, a_{20}\ldots$ of the parameters. To any other values $a_1, a_2\ldots$ belong other 'allowed' motions $B\{a\}$. Now our hypothesis asserts:

For *general* values $a_1, a_2\ldots$ of the parameters, those and only those motions are allowed which are adiabatically related to the motions which were allowed for the special values $a_{10}, a_{20}\ldots$ (i.e. which can be transformed into them, or may be derived from them in an adiabatic reversible way).

Remarks. A. Whether it be possible to extend the hypothesis to non-periodic motions, and how this should be done, I am not able to tell on account of some difficulties, which are mentioned in § 9.

B. Some forms of adiabatic affections may be realized physically – for instance, the strengthening of an electric or a magnetic field surrounding an atom (Stark and Zeeman effect). Others have more the character of a mathematical fiction (e.g. the change of a central field of force).

* In the newer form of his radiation theory, Planck speaks only of 'critical' motions, besides which other motions are 'allowed' too. In order not to become not diffuse, we will leave this form of the theory of quanta out of consideration. The suitable adaptation of the considerations given in this paper is easily to be found.

3. The adiabatic invariants and their application

Each application of the adiabatic hypothesis forces us to look for '*adiabatic invariants*' – that is, for quantities which retain their value during the transformation of a motion $\beta(a)$ into a motion $\beta(a')$ related adiabatically to the former. Indeed, from the hypothesis follows immediately the property:

If an adiabatic invariant Ω for the 'allowed' motions $B\{a_0\}$, belonging to the *special* values $a_{10}, a_{20}\ldots$, possesses the distinct numerical values $\Omega', \Omega''\ldots$, *it possesses exactly the same values* for the 'allowed' motions belonging to *arbitrary* values of the parameters $a_1, a_2\ldots$.

4. The adiabatic invariant $\overline{2T}/\nu$ for periodic motions and ε/ν especially for harmonic motions*

Suppose that the system under consideration possesses the following properties: For arbitrarily fixed values of the parameters $a_1, a_2\ldots$, all the motions that have to be considered are *periodic*, independently of the phases $(q_{10}\ldots q_{n0}, \dot{q}_{10}\ldots\dot{q}_{n0})$ the motion starts with. The period P may depend in some way or other on the $a_1, a_2\ldots$ and on the beginning phase.

Then the integral with respect to time of twice the kinetic energy, extended over one period, is an adiabatic invariant:

$$\delta' \int_0^P 2T \, \mathrm{d}t = 0. \tag{3}$$

In this formula δ' denotes the difference in value for two infinitely near, adiabatically, related motions of the system. (For the demonstration of form. 3 the reader is referred to the original paper†, Proc. Acad. Amsterdam **19** (1917) 576). Putting the reciprocal of the period P equal to the frequency ν, and denoting the mean of T with respect to the time by $\overline{T}$, form. 3 expresses:

$$\overline{2T}/\nu = \textit{adiabatic invariant}. \tag{4}$$

* Comp. paper C, § 1, 2. *Other instances of adiabatic invariants:* If the system possesses cyclic coordinates, the cyclic momenta are invariants. If the rotation of a ring of electrons is affected by an increasing magnetic field, the *sum of the moment of momentum and of the electrokinetic moment* is an invariant (Zeeman effect, magnetization). If an increasing electric field acts on a hydrogen atom of Bohr, then the component of the moment of momentum parallel to the lines of force is an invariant. For changes of a central field of force it is the moment of momentum.

† Comp. the Editor's note at the bottom of the first page of the present paper.

In the case of a simple harmonic motion of one degree of freedom we know that the mean of the kinetic energy is equal to the mean of the potential energy; hence both of them are equal to half the total energy. So here we have

$$\varepsilon/\nu = \textit{adiabatic invariant}. \tag{5}$$

5. Geometrical interpretation of the adiabatic invariant $\overline{2T}/\nu$ in the phase-space (p-q space)

To get a connexion with the formulae used by Planck, Debye, Bohr, Sommerfeld and others to introduce the quanta, we will avail ourselves of a transformation of the *integral of Action,* to which Sommerfeld has drawn attention:*

$$\int_0^P 2T\,\mathrm{d}t = \int_0^P \sum_h p_h \dot{q}_h\,\mathrm{d}t = \sum_h \int p_h\,\mathrm{d}q_h = \sum_h \iint \mathrm{d}p_h\,\mathrm{d}q_h. \tag{6}$$

Hence

$$\overline{2T}/\nu = \sum_h \iint \mathrm{d}p_h\,\mathrm{d}q_h. \tag{7}$$

The double integrations at the right-hand side have the following meaning: When the system performs its periodic motion, its phase-point describes a closed curve† in the $2n$-dimensional q–p space, and its n projections on the two-dimensional surfaces (q_1, p_1), (q_2, p_2)... ...(q_n, p_n) describe n closed curves. $\iint \mathrm{d}p_h\,\mathrm{d}q_h$ is the area of the region enclosed by the hth projection curve.

Remarks. A. The numerical value of $\overline{2T}/\nu$ is not changed if we pass to another system of coordinates for the description of the motion. Hence also the numerical value of the right-hand side of equation (7) is independent of the system of coordinates used.

B. Systems exist possessing the following property: with a *suitable choice of the system of coordinates* not only is the total sum at the right-hand side of (7) an adiabatic invariant, but *each separate integral* $\iint \mathrm{d}p_h\,\mathrm{d}q_h$ is an invariant. Compare the example of § 7.

C. For systems of one degree of freedom we have according to (7):

$$\overline{2T}/\nu = \iint \mathrm{d}q\,\mathrm{d}p = \textit{adiabatic invariant}. \tag{8}$$

* A. Sommerfeld, Sitzungsber. d. bayr. Akad. (1916) pp. 425–500 (§ 7).

† This expression must be altered in some way, if any of the coordinates be angles which increase by 2π in each period.

i.e. for systems of one degree of freedom the area enclosed by the phase-curve in the q–p diagram is an invariant (in this case there exists no other invariant which is independent of the former).

D. A theorem by P. Hertz (1910).* Imagine a system of n degrees of freedom and consider any motion belonging to a set of given values $a_{10}, a_{20} \dots$ of the parameters. The corresponding phase-curve in the $2n$-dimensional q–p space lies wholly on a certain hypersurface of constant energy, $\varepsilon(q, p, a_0) = \varepsilon_0$, which encloses a certain $2n$-dimensional volume:

$$V_0 = \int \dots \int \mathrm{d}q \dots \mathrm{d}q_n. \tag{9}$$

An adiabatic reversible affection $a_0 \to a_1$: firstly, changes the value of the energy (by the amount of the work performed on the system); secondly, alters the form and position of the hypersurfaces $\varepsilon(q, p, a) =$ $=$const. Let the volume enclosed by that surface of constant energy on which lies the phasecurve of the system after the affection be V. Then the theorem of P. Hertz asserts that

$$V = V_0. \tag{10}$$

For systems of one degree of freedom (10) and (8) coincide, for more degrees of freedom this is not the case.

6. Connexion with the formulae of the Theory of Quanta, as proposed by Planck, Debye and others for systems of one degree of freedom

Planck's hypothesis of energy elements (1901) asserts that a harmonically vibrating resonator of frequency ν_0 can contain only the following amounts of energy:†

$$\varepsilon = 0, h\nu_0, 2h\nu_0, \dots. \tag{11}$$

Hence the adiabatic invariant of the resonator may take only the values:

$$\varepsilon/\nu_0 = \overline{2T}/\nu_0 = \iint \mathrm{d}q\, \mathrm{d}p = 0, h, 2h, \dots. \tag{12}$$

Let us consider a resonator with a non-linear equation of motion:

$$\dot{q} = -(\nu_0^2 q + a_1 q^2 + a_2 q^3 \dots). \tag{13}$$

* P. Hertz, Ann. d. Phys. **33** (1910) 225, 537, § 11.

† Comp. note *, § 2.

Its vibrations are not harmonical, and the frequency $\nu \neq \nu_0$ does not depend only on the values of the parameters $a_1, a_2 \ldots$, but also on the exciting force. For the special values of the parameters $a_1 = a_2 = \ldots = 0$, it passes into Planck's resonator. Hence from the adiabatic hypothesis (comp. the formulation in § 3) follows: also for non-harmonically vibrating resonators only those motions are allowed for which

$$\overline{2T}/\nu = \iint \mathrm{d}q \, \mathrm{d}p = 0, h, 2h, \ldots. \tag{14}$$

So by means of the adiabatic hypothesis we have derived Debye's hypothesis on the values of $\iint \mathrm{d}q \, \mathrm{d}p$ for nonharmonical vibrations* from Planck's hypothesis of energy elements.

An electrical doublet with the electrical moment a_1, the moment of inertia a_2, is suspended in such a way that it can turn freely about the z-axis.† An electrical field of intensity a_3 acts parallel to the axis of x. As the coordinate q we choose the position-angle of the doublet. We will begin with very great values of a_1, a_3, and also of a_2; then even for great values of the exciting energy we may consider the vibrations as infinitely small and harmonical – resonator of Planck's type. By diminishing infinitely slowly the values of a_2 and a_3 we can pass in a reversible adiabatic way to vibrations of finite amplitude, and then make the pendulum 'turn over'; if now the moment of inertia a_2 is no more changed, but the directing field a_3 is diminished to zero, we arrive at a molecule which *rotates uniformly, uninfluenced by any force.* For all the motions considered, which are related adiabatically to each other, the adiabatic invariant $\overline{2T}/\nu = \iint \mathrm{d}q \, \mathrm{d}p$ has to retain its original values 0, h, $2h$, ... If for the uniform rotation we identify the frequency ν with the number of rotations of the doublet in unit of time

$$\nu = \pm \dot{q}/2\pi, \tag{15}$$

and observe that

$$2\overline{T} = 2T = p\dot{q}, \tag{16}$$

we must demand that p can take no other values than

$$p = 0, \pm \frac{h}{2\pi}, \pm 2\frac{h}{2\pi}, \ldots. \tag{17}$$

* P. Debye, Quantenhypothese (Göttinger Vorl. Teubner, 1913). Comp. also S. Boguslawski, Phys. Zs. **15** (1914) 569.

† Comp. the treatment and the use of this example in papers B and C. Comp. especially the diagram given in C, § 3.

Remark. The discussion sketched in the preceding lines wants to be developed more sharply, as the adiabatic transformation passes through a singular unperiodical motion, which forms the limit between the oscillatory and the rotatory motions. It is necessary to analyse more precisely the connexion between the adiabatic invariants for both types of motion.

7. Connexion with Sommerfeld's formulae for systems of more than one degree of freedom

We will show that the quantum formulae, which Sommerfeld has given for the motion of a point in a plane about a Newtonian centre of attraction, satisfy the adiabatic hypothesis.

Let $\chi(r, a_1, a_2 \ldots)$ be the potential of a central attractive force. The differential equations of the plane motion of a point, written in polar coordinates, have the form

$$m\ddot{r} - mr\dot{\phi}^2 + \frac{d\chi}{dr} = 0, \tag{18a}$$

$$\frac{d}{dt}(mr^2\dot{\phi}) = 0. \tag{18b}$$

From (18b) we see immediately that the moment of momentum is an invariant against a change of the parameters $a_1, a_2 \ldots$

$$mr^2\dot{\phi} = p_2 = \textit{adiabatic invariant}. \tag{19}$$

Eliminating $\dot{\phi}$ from eq. (18a) with the aid of (19), we get

$$m\ddot{r} = \frac{p_2^2}{mr^3} - \frac{d\chi}{dr}. \tag{20}$$

This equation has the same structure as the differential equation for the motion of a point, which oscillates along a straight line under the influence of a potential

$$\Phi = + \frac{p_2^2}{2mr^2} + \chi(r, a_1, a_2) \tag{21}$$

between two limiting values of $r(r_A > r_B > 0)$. But, according to §§ 4 and 5, this periodical motion of one degree of freedom possesses the adiabatic invariant

$$\overline{2T}_1/\nu_1 = \iint dq_1\, dp_1 = \textit{adiabatic invariant}. \tag{22}$$

Note that also eq. (19) may be written in the same form:

$$\overline{2T_2}/\nu_2 = \iint dq_2\, dp_2 = \textit{adiabatic invariant.} \tag{23}$$

For we have

$$\frac{\overline{2T_2}}{\nu_2} = \frac{p_2\dot{q}_2}{(\dot{q}_2/2\pi)} = 2\pi p_2 = \int_0^{2\pi} p_2\, dq_2 = dq_2\, dp_2.$$

Now Sommerfeld's formulae are*

$$\iint dq_1\, dp_1 = 0, h, 2h \dots nh. \tag{24}$$

$$\iint dq_2\, dp_2 = 0, h \dots n'h. \tag{25}$$

So they satisfy the adiabatic hypothesis (comp. its form in § 3).

Remarks. A. We have seen that the adiabatic invariants (22) and (23) exist not only for periodic motions about a Newtonian or about a quasi-elastic centre of force, but also for the quasi-periodic motions with a more general $\chi(r, a)$. Only in the former case $\nu_1 = \nu_2 = \nu$; so there we may combine the two invariants into the one

$$2\overline{(T_1 + T_2)}/\nu = \overline{2T}/\nu = \textit{adiabatic invariant.}$$

B. It would be of great interest to find the adiabatic invariants for more general quasi-periodic motions, and especially for anisotropic fields of force, instead of the isotropic fields, which are treated above. Then at the same time we should understand better which system of coordinates should be used for the application of Sommerfeld's formulae†.

8. Connexion with the statistical roots of the Second Law of Thermodynamics**

In his statistical mechanical deduction of the Second Law of Thermodycamics, and especially of the equation

$$\frac{\Delta E + A_1 \Delta a_1 + A_2 \Delta a_2 \dots}{\Theta} = k\Delta \log W, \tag{26}$$

Boltzmann based himself upon a certain supposition on the '*probability*

* A. Sommerfeld, Sitzungsber. d. bayr. Akad. (1916) pp. 425–500.

† A. Sommerfeld, l.c., p. 455 at the foot.

** *Cf.* P. Ehrenfest, Phys. Zs. **15** (1914) 657 (paper D).

a priori' of regions in the q–p space of the molecules (μ-space); he considered those regions as '*a priori equally probable*' which correspond with equal volumes $\int \ldots \int dq_1 \ldots dp_n$ in the μ-space. In other words, Boltzmann gave the μ-space everywhere the same '*weight*'

$$G(q, p) = \text{const.} \tag{27}$$

Planck's hypothesis of energy elements and its generalizations destroy this basis; they introduce, as it may be expressed, a weight depending on q, p, and a

$$G(q, p, a); \tag{28}$$

all regions of the μ-space have the '*weight zero*' (are '*forbidden*') with the exception of the discontinuously distributed '*allowed*' regions, the position of which depends on the value of the parameters a.* The latter circumstance is of particular importance.

So we arrive at the following problem: In what manner must the choice of the '*weight-function*' $G(q, p, a)$ – in other words, the choice of the '*allowed*' regions – *be limited, especially in their dependence on the a,* in order that Boltzmann's equation (26) may remain valid?

I have treated this problem first in a special case,† then generally.**

For molecules of one degree of freedom (harmonically and unharmonically vibrating resonators) I could wholly solve the question. The result I arrived at †† may be expressed in the language of this article in the following form:

An ensemble of such-like molecules (resonators) will fulfil Boltzmann's relations between entropy and probability if, and only if, the allowed motions are determined by means of the adiabatically invariant condition ‡

$$\overline{2T}/\nu = \iint dq\, dp = \textit{fixed numerical values } \Omega_1, \Omega_2 \ldots. \tag{29}$$

Planck's hypothesis on the elements of energy for harmonically

* The form and dimensions of the 'allowed' ellipses in the q–p diagram of a Planck's resonator are altered if the inertia and elasticity of the resonator are changed. In an analogous way the 'allowed' ellipses, belonging to the principal modes of vibration of a 'Hohlraum' or of the lattice of a crystal, are altered by a compression.

† Paper A, § 5.

** Paper D.

†† Paper D, § 7, remark.

‡ L.c. § 7, this invariant is noted by i.

vibrating oscillators and its generalization by Debye satisfy this condition; in this case $\Omega_1, \Omega_2 \ldots$ are taken equal to 0, h, $2h$, ...* (comp. § 5, equation 14).

As yet I have not been able to tell if also for molecules of *more than one* degree of freedom the same *necessary* and *sufficient* relations hold between the *adiabatic invariants* on the one hand and the *fulfilment of Boltzmann's theorem* on the other.

Remarks. A. In recent years it has become usual to introduce the relations between probability and entropy (or free energy)

$$S = k \log W \tag{30}$$

$$F = E - Tk \log W \tag{31}$$

simply as a *postulate*. It might seem that in this case the problem discussed in this paragraph becomes superfluous. However, this method of treating only shifts the difficulty to another point, as I have shown.†

B.** A reversible adiabatic compression transforms black body radiation into black body radiation, by which process it is immaterial if a little piece of a black body is present to act as a 'catalyzator' or not. Similarly Maxwell's velocity distribution in a gas consisting of point molecules which exert no forces on each other is transformed into a velocity distribution of the same kind by a reversible adiabatic compression of the gas (in a rough-walled vessel), independently of the occurrence of collisions between the molecules.†† Hence we may ask more generally: Does a *most probable* distribution of states in an

* That Planck's hypothesis on the energy elements is in harmony with the Second Law (and with the adiabatic hypothesis) has come about in the following way: in the deduction of his theory of radiation Planck at a certain moment puts the elements of energy (which were not yet determined before) equal to $h\nu$ in order to make his radiation formula correspond with the displacement law of W. Wien (cf. Planck, 'Vorlesungen über Wärmestrahlung', 1st edition, 1906, p. 153, eq. 226). Compare also the other quantum formulae, paper D, § 6.

† Paper D, Introduction.

** Cf. paper C, § 4.

†† Both examples have the following property in common: The pressure depends only on the *total energy* of the system, it is independent of the distribution of the energy over the different principal modes of vibration or over the molecules. In the cycle compression, catalytic process, dilatation, adiabatic process, the same amount of work is given to the system as is taken from it. For general systems this is no longer the case.

ensemble of molecules (resonators) pass always into a *most probable* distribution when the ensemble is subjected to an adiabatic reversible affection independently of the occurrence of interaction between the molecules during the affection? With the exception of special cases, this question must be answered in the negative.*

9. Difficulties which arise by a passage through singular motions. Aperiodic motions

One of these difficulties has already been mentioned at the end of § 6. A difficulty of a slightly different form arises when we pass in an adiabatic reversible way from the vibrations in an anisotropic quasi-elastic field of force to those in an isotropic field.† If we begin with an anisotropic field, with potential energy**

$$\Phi = \tfrac{1}{2}(\nu_1^2\xi_1^2 + \nu_2^2\xi_2^2), \tag{32}$$

it is usual to treat each of the two principal modes of vibration according to Planck's method; only those motions are allowed for which the energies of the principal modes of vibrations satisfy the equations

$$\varepsilon_1/\nu_1 = n_1 h; \qquad \varepsilon_2/\nu_2 = n_2 h. \tag{33}$$

According to our hypothesis these equations must remain unchanged if ν_1 and ν_2 converge infinitely slowly to the same value. The field becomes isotropic, and the total energy satisfies the equation

$$\varepsilon/\nu = (n_1 + n_2)h. \tag{34}$$

* Without any calculation this may be shown by means of the following example: Imagine an ideal gas with regid ellipsoidal molecules; the walls of the vessel are replaced by a field of force which reflects only the centre of gravity of the molecules (the reflexion is perfectly elastic); if the gas is compressed adiabatically, without collisions between the molecules taking place, the kinetic energy of the translatory motion is increased, but not the energy of the rotatory motion. If collisions between the molecules do take place, this is otherwise. By a short calculation it may be shown by the following example: Point molecules move up and down along a straight line between two fixed points A and B, uninfluenced by any force. An elastic field of force is excited infinitely slowly, so that in the end the molecules perform harmonic vibrations about the centre of the line.

† Comp. a remark made by H. A. Lorentz, Proc. Acad. Amsterdam (1912).

** The mass of the moving particle is supposed to be unity.

At the other hand, an isotropic field of force is a *central* field, hence Sommerfeld's formulae can be applied here. These give:

$$\text{Moment of momentum} = mr^2\dot{\phi} = n(h/2\pi), \tag{35a}$$

$$\text{Total energy} = \varepsilon = (n + n')h\nu. \tag{35b}$$

The motions allowed according to both sets of conditions are *not the same;* in the first place, we cannot see why the moment of momentum (which is not a constant in the anisotropic case, but oscillates between the values $\pm 2h\sqrt{n_1 n_2}$) should converge to one of the distinct values given by (35a) for the isotropic case. These oscillations become slower and slower if ν_1 and ν_2 become more and more equal to each other, hence which value is attained when we have arrived at the isotropic case depends on a *double limiting process*. A second discrepancy, to which Epstein has drawn my attention, is the following; for a circular motion we must have in equation (35b) $n'=0$, in (34), however, $n_1=n_2$, hence in the latter case ε/ν can be equal only to *even multiples of h*, whereas in the former it can be equal to *every integral multiple of h*.

From these considerations it appears that the adiabatic hypothesis wants a special complement in order that in this case (and also in the analogous cases of a passage through singular motions) the double limiting process should lead to a definite value. If such a complement could be found, it would be possible to deduce the quantum formulae for arbitrary central forces from the hypothesis of energy elements for harmonically vibrating resonators.

At this place we must also mention the difficulties which arise if we try to extend the notions of 'reversible adiabatic affection', 'adiabatic invariant', etc., to families of motions which are essentially *unperiodic* – as, for instance, the hyperbolic motions of a point in a Newtonian field of force. In this case, too, the change of the energy and of the moment of momentum of the motion depend on a double limiting process: the course of the whole motion from $t=-\infty$ to $t=+\infty$, and the infinitely slow change of the parameters a_1, a_2....

10. Conclusion

The problems discussed in this paper show, as I hope, that the adiabatic hypothesis and the notion of adiabatic invariants are of importance for the extension of the theory of quanta to still more general classes of motions (§§ 6, 7); furthermore, that they throw some light on the question: What conditions are necessary that Boltzmann's relation between probability and entropy may remain valid (§ 8)?

Hence it would be of great interest to develop a systematic method of finding adiabatic invariants for systems as generally as possible.

The difficulties which arise by the passage through singular motions are yet awaiting their solution; perhaps it will be necessary to seek for some complement of the adiabatic hypothesis. In any case, it seems to me that the validity of Wien's displacement law shows that *reversible adiabatic affections take a prominent place in the theory of quanta.*

Postscriptum. The beautiful researches of Epstein [Ann. d. Phys. **50** (1916) 489, 815], Schwarzschild [Sitzungsber. Berl. Akad. (1916) p. 548], and others which have appeared in the meantime, show the great importance the cases which are integrable by means of Stäckel's method of 'separation of the variables' have for the development of the theory of quanta. Hence the question arises: How far are the different parts into which these authors separate the integral of action according to Stäckel's method adiabatic invariants? In the problem treated by Sommerfeld this is the case, as is shown in § 7.

Related papers

2a P. Ehrenfest, *Zum Boltzmannschen Entropie-Wahrscheinlichkeitstheorem.* I. Phys. Z. **15** (1914) 657.

2b P. Ehrenfest, *Adiabatische Transformationen in der Quantentheorie und ihre Behandlung durch Niels Bohr.* Naturwiss. **11** (1923) 543. In this paper, Ehrenfest gives full references to earlier papers.

Printed April 27, 1918

3

ON THE QUANTUM THEORY OF LINE-SPECTRA

N. BOHR

DEDICATED TO THE MEMORY OF MY VENERATED TEACHER
PROFESSOR C. CHRISTIANSEN
October 9, 1843 † November 28, 1917

INTRODUCTION

In an attempt to develop certain outlines of a theory of line-spectra based on a suitable application of the fundamental ideas introduced by Planck in his theory of temperature-radiation to the theory of the nucleus atom of Sir Ernest Rutherford, the writer has shown that it is possible in this way to obtain a simple interpretation of some of the main laws governing the line-spectra of the elements, and especially to obtain a deduction of the well known Balmer formula for the hydrogen spectrum [1]). The theory in the form given allowed of a detailed discussion only in the case of periodic systems, and obviously was not able to account in detail for the characteristic difference between the hydrogen spectrum and the spectra of other elements, or for the characteristic effects on the hydrogen spectrum of external electric and magnetic fields. Recently, however, a way out of this difficulty has been opened by Sommerfeld [2]) who, by introducing a suitable

Editor's note. This paper is the Introduction and Part I (pp. 1–36) of the paper *On the Quantum Theory of Line-spectra*, which was first published (1918–1922) in the Mémoires de l'Académie Royale des Sciences et des Lettres de Danemark, Copenhague, Section des Sciences, 8me série t. IV, n° 1, fasc. 1–3 (D. Kgl. Danske Vidensk. Selsk. Skrifter, Naturvidensk. og Mathem. Afd. 8. Række, IV.1, 1–3). The Introduction was signed 'Copenhagen, November 1917.' The titles of Parts II and III, which are not reproduced here, are respectively *On the hydrogen spectrum* and *On the spectra of elements of higher atomic number*.

[1]) N. Bohr, Phil. Mag. **26** (1913) 1, 476, 857; **27** (1914) 506; **29** (1915) 332; **30** (1915) 394.

[2]) A. Sommerfeld, Ber. Akad. München. 1915, pp. 425, 459; 1916, p. 131; 1917 p. 83. Ann. d. Phys. **51** (1916) 1.

generalisation of the theory to a simple type of non-periodic motions and by taking the small variation of the mass of the electron with its velocity into account, obtained an explanation of the fine-structure of the hydrogen lines which was found to be in brilliant conformity with the measurements. Already in his first paper on this subject, Sommerfeld pointed out that his theory evidently offered a clue to the interpretation of the more intricate structure of the spectra of other elements. Briefly afterwards Epstein [1]) and Schwarzschild [2]), independent of each other, by adapting Sommerfeld's ideas to the treatment of a more extended class of non-periodic systems obtained a detailed explanation of the characteristic effect of an electric field on the hydrogen spectrum discovered by Stark. Subsequently Sommerfeld [3]) himself and Debye [4]) have on the same lines indicated an interpretation of the effect of a magnetic field on the hydrogen spectrum which, although no complete explanation of the observations was obtained, undoubtedly represents an important step towards a detailed understanding of this phenomenon.

In spite of the great progress involved in these investigations many difficulties of fundamental nature remained unsolved, not only as regards the limited applicability of the methods used in calculating the frequencies of the spectrum of a given system, but especially as regards the question of the polarisation and intensity of the emitted spectral lines. These difficulties are intimately connected with the radical departure from the ordinary ideas of mechanics and electrodynamics involved in the main principles of the quantum theory, and with the fact that it has not been possible hitherto to replace these ideas by others forming an equally consistent and developed structure. Also in this respect, however, great progress has recently been obtained by the work of Einstein [5]) and Ehrenfest [6]). On this state of the theory it might therefore be of interest to make an attempt to discuss the different applications from a uniform point of view, and especially to

[1]) P. Epstein, Phys. Zeitschr. **17** (1916) 148; Ann. d. Phys. **50** (1916) 489; **51** (1916) 168.

[2]) K. Schwarzschild, Ber. Akad. Berlin (1916) 548.

[3]) A. Sommerfeld, Phys. Zeitschr. **17** (1916) 491.

[4]) P. Debye, Nachr. K. Ges. d. Wiss. Göttingen, 1916; Phys. Zeitschr. **17** (1916) 507.

[5]) A. Einstein, Verh. d. D. phys. Ges. **18** (1916) 318; Phys. Zeitschr. **18** (1917) 121.

[6]) P. Ehrenfest, Proc. Acad. Amsterdam **16** (1914) 591; Phys. Zeitschr. **15** (1914) 657; Ann. d. Phys. **51** (1916) 327; Phil. Mag. **33** (1917) 500.

consider the underlying assumptions in their relations to ordinary mechanics and electrodynamics. Such an attempt has been made in the present paper, and it will be shown that it seems possible to throw some light on the outstanding difficulties by trying to trace the analogy between the quantum theory and the ordinary theory of radiation as closely as possible.

The paper is divided into four parts.

Part I contains a brief discussion of the general principles of the theory and deals with the application of the general theory to periodic systems of one degree of freedom and to the class of non-periodic systems referred to above.

Part II contains a detailed discussion of the theory of the hydrogen spectrum in order to illustrate the general considerations.

Part III contains a discussion of the questions arising in connection with the explanation of the spectra of other elements.

Part IV contains a general discussion of the theory of the constitution of atoms and molecules based on the application of the quantum theory to the nucleus atom.

Copenhagen, November 1917.

PART I. ON THE GENERAL THEORY

1. General principles

The quantum theory of line-spectra rests upon the following fundamental assumptions:

I. That an atomic system can, and can only, exist permanently in a certain series of states corresponding to a discontinuous series of values for its energy, and that consequently any change of the energy of the system, including emission and absorption of electromagnetic radiation, must take place by a complete transition between two such states. These states will be denoted as the 'stationary states' of the system.

II. That the radiation absorbed or emitted during a transition between two stationary states is 'unifrequentic' and possesses a frequency ν, given by the relation

$$E' - E'' = h\nu, \tag{1}$$

where h is Planck's constant and where E' and E" are the values of the energy in the two states under consideration

As pointed out by the writer in the papers referred to in the introduction, these assumptions offer an immediate interpretation of the fundamental *principle of combination of spectral lines* deduced from the measurements of the frequencies of the series spectra of the elements. According to the laws discovered by Balmer, Rydberg and Ritz, the frequencies of the lines of the series spectrum of an element can be expressed by a formula of the type:

$$\nu = f_{\tau''}(n'') - f_{\tau'}(n'), \tag{2}$$

where n' and n'' are whole numbers and $f_\tau(n)$ is one among a set of functions of n, characteristic for the element under consideration. On the above assumptions this formula may obviously be interpretated by assuming that the stationary states of an atom of an element form a set of series, and that the energy in the nth state of the τth series, omitting an arbitrary constant, is given by

$$E_\tau(n) = -hf_\tau(n). \tag{3}$$

We thus see that the values for the energy in the stationary states of an atom may be obtained directly from the measurements of the spectrum by means of relation (1). In order, however, to obtain a theoretical connection between these values and the experimental evidence about the constitution of the atom obtained from other sources, it is necessary to introduce further assumptions about the laws which govern the stationary states of a given atomic system and the transitions between these states.

Now on the basis of a vast amount of experimental evidence, we are forced to assume that an atom or molecule consists of a number of electrified particles in motion, and, since the above fundamental assumptions imply that no emission of radiation takes place in the stationary states, we must consequently assume that *the ordinary laws of electrodynamics cannot be applied* to these states without radical alterations. In many cases, however, the effect of that part of the electrodynamical forces which is connected with the emission of radiation will at any moment be very small in comparison with the effect of the simple electrostatic attractions or repulsions of the charged particles corresponding to Coulomb's law. Even if the theory of radiation must be completely altered, it is therefore a natural assumption that it is possible in such cases to obtain a close approximation in the description of the motion in the stationary states, by retaining only the latter forces. In the following we shall therefore, as in all the papers mentioned in the introduction, for the present *calculate the motions of the particles in the stationary states as the motions of mass-points according to ordinary mechanics* including the modifications claimed by the theory of relativity, and we shall later in the discussion of the special applications come back to the question of the degree of approximation which may be obtained in this way.

If next we consider a transition between two stationary states, it is obvious at once from the essential discontinuity, involved in the assumptions I and II, that in general it is impossible even approximately to describe this phenomenon by means of ordinary mechanics or to calculate the frequency of the radiation absorbed or emitted by such a process by means of ordinary electrodynamics. On the other hand, from the fact that it has been possible by means of ordinary mechanics and electrodynamics to account for the phenomenon of temperature-radiation in the limiting region of slow vibrations, we may expect that any theory capable of describing this phenomenon in accordance with observations will form some sort of natural generalisation of the ordinary theory of radiation. Now the theory of temperature-radiation in the form originally given by Planck confessedly lacked internal consistency, since, in the deduction of his radiation formula, assumptions of similar character as I and II were used in connection with assumptions which were in obvious contrast to them. Quite recently, however, Einstein[1]) has succeeded, on the basis of the assumptions I and II, to give a consistent and instructive deduction of Planck's formula by introducing certain supplementary assumptions about the *probability of transition of a system between two stationary states* and about the manner in which this probability depends on the density of radiation of the corresponding frequency in the surrounding space, suggested from analogy with the ordinary theory of radiation. Einstein compares the emission or absorption of radiation of frequency ν corresponding to a transition between two stationary states with the emission or absorption to be expected on ordinary electrodynamics for a system consisting of a particle executing harmonic vibrations of this frequency. In analogy with the fact that on the latter theory such a system will without external excitation emit a radiation of frequency ν,

[1]) A. Einstein, loc. cit. [paper **1** and **1d**].

Einstein assumes in the first place that on the quantum theory there will be a certain probability $A_{n''}^{n'}\mathrm{d}t$ that the system in the stationary state of greater energy, characterised by the letter n', in the time interval $\mathrm{d}t$ will start *spontaneously* to pass to the stationary state of smaller energy, characterised by the letter n''. Moreover, on ordinary electrodynamics the harmonic vibrator will, in addition to the above mentioned independent emission, in the presence of a radiation of frequency ν in the surrounding space, and dependent on the accidental phase-difference between this radiation and the vibrator, emit or absorb radiation-energy. In analogy with this, Einstein assumes secondly that in the presence of a radiation in the surrounding space, the system will on the quantum theory, in addition to the above mentioned probability of spontaneous transition from the state n' to the state n'', possess a certain probability, depending on this radiation, of passing in the time $\mathrm{d}t$ from the state n' to the state n'', as well as from the state n'' to the state n'. These latter probabilities are assumed to be proportional to the intensity of the surrounding radiation and are denoted by $\varrho_\nu B_{n''}^{n'}\mathrm{d}t$ and $\varrho_\nu B_{n'}^{n''}\mathrm{d}t$ respectively, where $\varrho_\nu \mathrm{d}\nu$ denotes the amount of radiation in unit volume of the surrounding space distributed on frequencies between ν and $\nu+\mathrm{d}\nu$, while $B_{n''}^{n'}$ and $B_{n'}^{n''}$ are constants which, like $A_{n''}^{n'}$, depend only on the stationary states under consideration. Einstein does not introduce any detailed assumption as to the values of these constants, no more than to the conditions by which the different stationary states of a given system are determined or to the 'a-priori probability' of these states on which their relative occurrence in a distribution of statistical equilibrium depends. He shows, however, how it is possible from the above general assumptions, by means of Boltzmann's principle on the relation between entropy and probability and Wien's well known displacement-law, to deduce a formula for the temperature radiation which apart from an undetermined constant factor coincides with Planck's, if we only assume that the frequency corresponding to the transition between the two states is determined by (1). It will therefore be seen that by reversing the line of argument, Einstein's theory may be considered as a very direct support of the latter relation.

In the following discussion of the application of the quantum theory to determine the line-spectrum of a given system, it will, just as in the theory of temperature-radiation, not be necessary to introduce detailed assumptions as to the mechanism of transition between two stationary states. We shall show, however, that the conditions which will be used to determine the values of the energy in the stationary states are of such a type that the frequencies calculated by (1), in the limit where the motions in successive stationary states comparatively differ very little from each other, will tend to coincide with the frequencies to be expected on the ordinary theory of radiation from the motion of the system in the stationary states. In order to obtain the necessary relation to the ordinary theory of radiation in the limit of slow vibrations, we are therefore led directly to certain conclusions about the probability of transition between two stationary states in this limit. This leads again to certain general considerations about the connection between the probability of a transition between any two stationary states and the motion of the system in these states, which will be shown to throw light on the question of the polarisation and intensity of the different lines of the spectrum of a given system.

In the above considerations we have by an atomic system tacitly understood a number of electrified particles which move in a field of force which, with the approximation mentioned, possesses a potential depending only on the position of the particles. This may more accurately be denoted as a system under constant external conditions, and the question next arises about the variation in the stationary states which may be expected to take place during a variation of the external conditions, e.g. when exposing the atomic system to some variable external field of force. Now, in general, we must obviously assume that this variation cannot be calculated by ordinary mechanics, no more than the transition between two different stationary states corresponding to constant external conditions. If, however, the variation of the external conditions is very slow, we may from the necessary stability of the stationary states expect that the motion of the system at any given moment during the variation will differ only very little from the motion in a stationary state corresponding to the instantaneous external conditions. If now, moreover, the variation is performed at a constant or very slowly changing rate, the forces to which the particles of the system will be exposed will not differ at any moment from those to which they would be exposed if we imagine that the external forces arise from a number of slowly moving additional particles which together with the original system form a system in a stationary state. From this point of view it seems therefore natural to assume that, with the approximation mentioned, the motion of an atomic system in the stationary states can be calculated by direct application of ordinary mechanics, not only under constant external conditions, but in general also during a slow and uniform variation of these conditions. This assumption, which may be denoted as the principle of the '*mechanical transformability*' of the stationary states, has been introduced in the quantum theory by Ehrenfest[1]) and is, as it will be seen in the following sections, of great importance in the discussion of the conditions to be used to fix the stationary states of an atomic system among the continuous multitude of mechanically possible motions. In this connection it may be pointed out that the principle of the mechanical transformability of the stationary states allows us to overcome a fundamental difficulty which at first sight would seem to be involved in the definition of the energy difference between two stationary states which enters in relation (1). In fact we have assumed that the direct transition between two such states cannot be described by ordinary mechanics, while on the other hand we possess no means of defining an energy difference between two states if there exists no possibility for a continuous mechanical connection between them. It is clear, however, that such a connection is just afforded by Ehrenfest's principle which allows us to transform mechanically the stationary states of a given system into those of another, because for the latter system we may take one in which the forces which act on the particles are very small and where we may assume that the values of the energy in all the stationary states will tend to coincide.

As regards the problem of the statistical distribution of the different stationary states between a great number of atomic systems of the same kind in temperature equilibrium, the number of systems present in the different states may be deduced in the well known way from Boltzmann's fundamental relation between entropy and probability, if we know the values of the energy in these states and the *a-priori probability* to be ascribed to each state in the calculation of the probability of the whole distribution. In contrast to considerations of ordinary statistical mechanics we possess on the quantum theory no direct means of

[1]) P. Ehrenfest, loc. cit. In these papers the principle in question is called the 'adiabatic hypothesis' in accordance with the line of argumentation followed by Ehrenfest in which considerations of thermodynamical problems play an important part. From the point of view taken in the present paper, however, the above notation might in a more direct way indicate the content of the principle and the limits of its applicability.

determining these a-priori probabilities, because we have no detailed information about the mechanism of transition between the different stationary states. If the a-priori probabilities are known for the states of a given atomic system, however, they may be deduced for any other system which can be formed from this by a continuous transformation without passing through one of the singular systems referred to below. In fact, in examining the necessary conditions for the explanation of the second law of thermodynamics Ehrenfest [1]) has deduced a certain general condition as regards the variation of the a-priori probability corresponding to a small change of the external conditions from which it follows, that the a-priori probability of a given stationary state of an atomic system must remain unaltered during a continuous transformation, except in special cases in which the values of the energy in some of the stationary states will tend to coincide during the transformation. In this result we possess, as we shall see, a rational basis for the determination of the a-priori probability of the different stationary states of a given atomic system.

2. Systems of one degree of freedom

As the simplest illustration of the principles discussed in the former section we shall begin by considering systems of a single degree of freedom, in which case it has been possible to establish a general theory of stationary states. This is due to the fact, that *the motion will be simply periodic,* provided the distance between the parts of the system will not increase infinitely with the time, a case which for obvious reasons cannot represent a stationary state in the sense defined above. On account of this, the discussion of the mechanical transformability of the stationary states can, as pointed out by Ehrenfest [2]), for systems of one degree of freedom be based on a mechanical theorem about periodic systems due to Boltzmann and originally applied by this author in a discussion of the bearing of mechanics on the explanation of the laws of thermodynamics. For the sake of the considerations in the following sections it will be convenient here to give the proof in a form which differs slightly from that given by Ehrenfest,

[1]) P. Ehrenfest, Phys. Zeitschr. **15** (1914) 660. The above interpretation of this relation is not stated explicitly by Ehrenfest, but it presents itself directly if the quantum theory is taken in the form corresponding to the fundamental assumption I.

[2]) P. Ehrenfest, loc. cit. Proc. Acad. Amsterdam **16** (1914) 591.

and which takes also regard to the modifications in the ordinary laws of mechanics claimed by the theory of relativity.

Consider for the sake of generality a conservative mechanical system of s degrees of freedom, the motion of which is governed by Hamilton's equations:

$$\frac{dp_k}{dt} = -\frac{\partial E}{\partial q_k}, \qquad \frac{dq_k}{dt} = \frac{\partial E}{\partial p_k}, \qquad (k = 1, \dots s) \tag{4}$$

where E is the total energy considered as a function of the generalised positional coordinates $q_1, \dots q_s$ and the corresponding canonically conjugated momenta $p_1, \dots p_s$. If the velocities are so small that the variation in the mass of the particles due to their velocities can be neglected, the p's are defined in the usual way by

$$p_k = \frac{\partial T}{\partial \dot{q}_k}, \qquad (k = 1, \dots s)$$

where T is the kinetic energy of the system considered as a function of the generalised velocities $\dot{q}_1, \dots \dot{q}_s$ ($\dot{q}_k = dq_k/dt$) and of $q_1, \dots q_s$. If the relativity modifications are taken into account the p's are defined by a similar set of expressions in which the kinetic energy is replaced by $T' = \Sigma m_0 c^2 (1 - \sqrt{1 - v^2/c^2})$, where the summation is to be extended over all the particles of the system, and v is the velocity of one of the particles and m_0 its mass for zero velocity, while c is the velocity of light.

Let us now assume that the system performs a periodic motion with the period σ, and let us form the expression

$$I = \int_0^\sigma \sum_1^s p_k \dot{q}_k \, dt, \tag{5}$$

which is easily seen to be independent of the special choice of coordinates $q_1, \dots q_s$ used to describe the motion of the system. In fact, if the variation of the mass with the velocity is neglected we get

$$I = 2 \int_0^\sigma T \, dt,$$

and if the relativity modifications are included, we get a quite analogous expression in which the kinetic energy is replaced by $T'' = \Sigma \frac{1}{2} m_0 v^2 / \sqrt{1 - v^2/c^2}$.

Consider next some new periodic motion of the system formed by a small variation of the first motion, but which may need the presence of external forces in order to be a mechanically possible motion. For the variation in I we get then

$$\delta I = \int_0^\sigma \sum_1^s (\dot{q}_k\, \delta p_k + p_k\, \delta \dot{q}_k)\, dt + \Big|\sum_1^s p_k \dot{q}_k\, \delta t\Big|_0^\sigma,$$

where the last term refers to the variation of the limit of the integral due to the variation in the period σ. By partial integration of the second term in the bracket under the integral we get next

$$\delta I = \int_0^\sigma \sum_1^s (\dot{q}_k\, \delta p_k - \dot{p}_k\, \delta q_k)\, dt + \Big|\sum_1^s p_k(\dot{q}_k\, \delta t + \delta q_k)\Big|_0^\sigma,$$

where the last term is seen to be zero, because the term in the bracket as well as p_k will be the same in both limits, since the varied motion as well as the original motion is assumed to be periodic. By means of equations (4) we get therefore

$$\delta I = \int_0^\sigma \sum_1^s \left(\frac{\partial E}{\partial p_k}\, \delta p_k + \frac{\partial E}{\partial q_k}\, \delta q_k\right) dt = \int_0^\sigma \delta E\, dt. \tag{6}$$

Let us now assume that the small variation of the motion is produced by a small external field established at a uniform rate during a time interval ϑ, long compared with σ, so that the comparative increase during a period is very small. In this case δE is at any moment equal to the total work done by the external forces on the particles of the system since the beginning of the establishment of the field. Let this moment be $t = -\vartheta$ and let the potential of the external field at $t \geqq 0$ be given by Ω, expressed as a function of the q's. At any given moment $t > 0$ we have then

$$\delta E = -\int_{-\vartheta}^{0} \frac{\vartheta + t}{\vartheta} \sum_1^s \frac{\partial \Omega}{\partial q_k}\, \dot{q}_k\, dt - \int_0^t \sum_1^s \frac{\partial \Omega}{\partial q_k}\, \dot{q}_k\, dt,$$

which gives by partial integration

$$\delta E = \frac{1}{\vartheta} \int_{-\vartheta}^{0} \Omega\, dt - \Omega_t,$$

where the values for the q's to be introduced in Ω in the first term are those corresponding to the motion under the influence of the increasing external field, and the values to be introduced in the second term are those corresponding to the configuration at the time t. Neglecting small quantities of the same order as the square of the external force, however, we may in this expression for δE instead of the values for the q's corresponding to the perturbed motion take those corresponding to the original motion of the system. With this approximation the first term is equal to the mean value of the second taken over a period σ, and we have consequently

$$\int_0^\sigma \delta E \, dt = 0. \tag{7}$$

From (6) and (7) it follows that I will remain constant during the slow establishment of the small external field, if the motion corresponding to a constant value of the field is periodic. If next the external field corresponding to Ω is considered as an inherent part of the system, it will be seen in the same way that I will remain unaltered during the establishment of a new small external field, and so on. Consequently *I will be invariant for any finite transformation of the system which is sufficiently slowly performed,* provided the motion at any moment during the process is periodic and the effect of the variation is calculated on ordinary mechanics.

Before we proceed to the applications of this result we shall mention a simple consequence of (6) for systems for which every orbit is periodic independent of the initial conditions. In that case we may for the varied motion take an undisturbed motion of the system corresponding to slightly different initial conditions. This gives δE constant, and from (6) we get therefore

$$\delta E = \omega \delta I, \tag{8}$$

where $\omega = 1/\sigma$ is the frequency of the motion. This equation forms a simple relation between the variations in E and I for periodic systems, which will be often used in the following.

Returning now to systems of one degree of freedom, we shall take our starting point from Planck's original theory of a *linear harmonic vibrator*. According to this theory the stationary states of a system, consisting of a particle executing linear harmonic vibrations with a constant frequency ω_0 independent of the energy, are given by the well known relation

$$E = nh\omega_0, \tag{9}$$

where n is a positive entire number, h Planck's constant, and E the total energy which is supposed to be zero if the particle is at rest.

From (8) it follows at once, that (9) is equivalent to

$$I = \int_0^\sigma p\dot{q}\, \mathrm{d}t = \int p\, \mathrm{d}q = nh, \tag{10}$$

where the latter integral is to be taken over a complete oscillation of q between its limits. On the principle of the mechanical transformability of the stationary states we shall therefore assume, following Ehrenfest, that (10) holds not only for a Planck's vibrator but for *any periodic system of one degree of freedom* which can be formed in a continuous manner from a linear harmonic vibrator by a gradual variation of the field of force in which the particle moves. This condition is immediately seen to be fulfilled by all such systems in which the motion is of oscillating type i.e. where the moving particle during a period passes twice through any point of its orbit once in each direction. If, however, we confine ourselves to systems of one degree of freedom, it will be seen that systems in which the motion is of rotating type, i.e. where the particle during a period passes only once through every point of its orbit, cannot be formed in a continuous manner from a linear harmonic vibrator without passing through singular states in which the period becomes infinite long and the result becomes ambiguous. We shall not here enter more closely on this difficulty which has been pointed out by Ehrenfest, because it disappears when we consider systems of several degrees of freedom, where we shall see that a simple generalisation of (10) holds for any system for which every motion is periodic.

As regards the application of (9) to statistical problems it was assumed in Planck's theory that the different states of the vibrator corresponding to different values of n are *a-priori equally probable*, and this assumption was strongly supported by the agreement obtained on this basis with the measurements of the specific heat of solids at low temperatures. Now it follows from the considerations of Ehrenfest, mentioned in the former section, that the a-priori probability of a given stationary state is not changed by a continuous transformation, and we shall therefore expect that for any system of one degree of freedom the different states corresponding to different entire values of n in (10) are a-priori equally probable.

As pointed out by Planck in connection with the application of (9), it is simply seen that statistical considerations, based on the assumption of equal probability for the different states given by (10), will show the necessary relation to considerations of ordinary statistical mechanics in the limit where the latter theory has been found to give results in agreement with experiments. Let the configuration and motion of a mechanical system be characterised by s independent variables $q_1, \ldots q_s$ and corresponding momenta $p_1, \ldots p_s$, and let the state of the system be represented in a $2s$-dimensional phase-space by a point with coordinates $q_1, \ldots q_s, p_1, \ldots p_s$. Then, according to ordinary statistical mechanics, the probability for this point to lie within a small element in the phase-space is independent of the position and shape of this element and simply proportional to its volume, defined in the usual way by

$$\delta W = \int \mathrm{d}q_1 \ldots \mathrm{d}q_s \, \mathrm{d}p_1 \ldots \mathrm{d}p_s. \tag{11}$$

In the quantum theory, however, these considerations cannot be directly applied, since the point representing the state of a system cannot be displaced continuously in the $2s$-dimensional phase-space, but can lie only on certain surfaces of lower dimensions in this space. For systems of one degree of freedom the phase-space is a two-dimensional surface, and the points representing the states of some system given by (10) will be situated on closed curves on this surface. Now, in general, the motion will differ considerably for any two states corresponding to successive entire values of n in (10), and a simple general connection between the quantum theory and ordinary statistical mechanics is therefore out of question. In the limit, however, where n is large, the motions in successive states will only differ very little from each other, and it would therefore make little difference whether the points representing the systems are distributed continuously on the phase-surface or situated only on the curves corresponding to (10), provided the number of systems which in the first case are situated between two such curves is equal to the number which in the second case lies on one of these curves. But it will be seen that this condition is just fulfilled in consequence of the above hypothesis of equal a-priori probability of the different stationary states, because the element of phase-surface limited by two successive curves corresponding to (10) is equal to

$$\delta W = \int \mathrm{d}p \, \mathrm{d}q = [\int p \, \mathrm{d}q]_n - [\int p \, \mathrm{d}q]_{n-1} = I_n - I_{n-1} = h, \tag{12}$$

so that on ordinary statistical mechanics the probabilities for the point to lie within any two such elements is the same. We see consequently that the hypothesis of equal probability of the different states given by (10) gives the same result as ordinary statistical mechanics in all such applications in which the states of the great majority of the systems correspond to large values of n. Considerations of this kind have led Debye[1]) to point out that condition (10) might have a general validity for systems of one degree of freedom, already before Ehrenfest, on the basis of his theory of the mechanical transformability of the stationary states, had shown that this condition forms the only rational generalisation of Planck's condition (9).

We shall now discuss the relation between the theory of *spectra of atomic systems of one degree of freedom*, based on (1) and (10), and the ordinary theory of radiation, and we shall see that this relation in several respects shows a close analogy to the relation, just considered, between the statistical applications of (10) and considerations based on ordinary statistical mechanics. Since the values for the frequency ω in two states corresponding to different values of n in (10) in general are different, we see at once that we cannot expect a simple connection between the frequency calculated by (1) of the radiation corresponding to a transition between two stationary states and the motions of the system in these states, except in the limit where n is very large, and where the ratio between the frequencies of the motion in successive stationary states differs very little from unity. Consider now a transition between the state corresponding to $n=n'$ and the state corresponding to $n=n''$, and let us assume that n' and n'' are large numbers and that $n'-n''$ is small compared with n' and n''. In that case we may in (8) for δE put $E'-E''$ and for δI put $I'-I''$, and we get therefore from (1) and (10) for the frequency of the radiation emitted or absorbed during the transition between the two states

$$\nu = \frac{1}{h}(E' - E'') = \frac{\omega}{h}(I' - I'') = (n' - n'')\omega. \tag{13}$$

Now in a stationary state of a periodic system the displacement of the particles in any given direction may always be expressed by means of a Fourier-series as a sum of harmonic vibrations:

$$\xi = \sum C_\tau \cos 2\pi(\tau\omega t + c_\tau), \tag{14}$$

[1]) P. Debye, Wolfskehl-Vortrag. Göttingen (1913).

where the C's and c's are constants and the summation is to be extended over all positive entire values of τ. On the ordinary theory of radiation we should therefore expect the system to emit a spectrum consisting of a series of lines of frequencies equal to $\tau\omega$, but, as it is seen, this is just equal to the series of frequencies which we obtain from (13) by introducing different values for $n'-n''$. As far as the frequencies are concerned we see therefore that in the limit where n is large there exists a close relation between the ordinary theory of radiation and the theory of spectra based on (1) and (10). It may be noticed, however, that, while on the first theory radiations of the different frequencies $\tau\omega$ corresponding to different values of τ are emitted or absorbed at the same time, these frequencies will on the present theory, based on the fundamental assumption I and II, be connected with entirely different processes of emission or absorption corresponding to the transition of the system from a given state to different neighbouring stationary states.

In order to obtain the necessary connection, mentioned in the former section, to the ordinary theory of radiation in the limit of slow vibrations, we must further claim that a relation, as that just proved for the frequencies, will, in the limit of large n, hold also for the intensities of the different lines in the spectrum. Since now on ordinary electrodynamics the intensities of the radiations corresponding to different values of τ are directly determined from the coefficients C_τ in (14), we must therefore expect that for large values of n these coefficients will on the quantum theory determine the *probability of spontaneous transition* from a given stationary state for which $n=n'$ to a neighbouring state for which $n=n''=n'-\tau$. Now, this connection between the amplitudes of the different harmonic vibrations into which the motion can be resolved, characterised by different values of τ, and the probabilities of transition from a given stationary state to the different neighbouring stationary states, characterised by different values of $n'-n''$, may clearly be expected to be of a general nature. Although, of course, we cannot without a detailed theory of the mechanism of transition obtain an exact calculation of the latter probabilities, unless n is large, we may expect that also for small values of n the amplitude of the harmonic vibrations corresponding to a given value of τ will in some way give a measure for the probability of a transition between two states for which $n'-n''$ is equal to τ. Thus in general there will be a certain probability of an atomic system in a stationary state to pass spontaneously to any other state of smaller energy, but if for all motions of a given system the coefficients C in (14) are zero for certain values of τ, we are led to expect that no transition will be possible, for which $n'-n''$ is equal to one of these values.

A simple illustration of these considerations is offered by the linear harmonic vibrator mentioned above in connection with Planck's theory. Since in this case C_τ is equal to zero for any τ different from 1, we shall expect that for this system only such transitions are possible in which n alters by one unit. From (1) and (9) we obtain therefore the simple result that the frequency of any radiation emitted or absorbed by a linear harmonic vibrator is equal to the constant frequency ω_0. This result seems to be supported by observations on the absorption-spectra of diatomic gases, showing that certain strong absorption-lines, which according to general evidence may be ascribed to vibrations of the two atoms in the molecule relative to each other, are not accompanied by lines of the same order of intensity and corresponding to entire multipla of the frequency, such as it should be expected from (1) if the system had any considerable tendency to pass between non-successive states. In this connection it may be noted that the fact, that in the absorption spectra of some diatomic gases faint lines occur corresponding to the double frequency of the main lines,[1] obtains a natural explanation by assuming that for finite amplitudes the vibrations are not exactly harmonic and that therefore the molecules possess a small probability of passing also between non-successive states.

3. Conditionally periodic systems

If we consider systems of several degrees of freedom the motion will be periodic only in singular cases and the general conditions which determine the stationary states cannot therefore be derived by means of the same simple kind of considerations as in the former section. As mentioned in the introduction, however, Sommerfeld and others have recently succeeded, by means of a suitable generalisation of (10), to obtain conditions for an important class of systems of several degrees of freedom, which, in connection with (1), have been found to give results in convincing agreement with experimental results about line-spectra. Subsequently these conditions have been proved

[1]) See E. C. Kemble, Phys. Rev. **8** (1916) 701.

by Ehrenfest and especially by Burgers[1]) to be invariant for slow mechanical transformations.

To the generalisation under consideration we are naturally led, if we first consider such systems for which the motions corresponding to the different degrees of freedom are dynamically independent of each other. This occurs if the expression for the total energy E in Hamilton's equations (4) for a system of s degrees of freedom can be written as a sum $E_1+\ldots+E_s$, where E_k contains q_k and p_k only. An illustration of a system of this kind is presented by a particle moving in a field of force in which the force-components normal to three mutually perpendicular fixed planes are functions of the distances from these planes respectively. Since in such a case the motion corresponding to each degree of freedom in general will be periodic, just as for a system of one degree of freedom, we may obviously expect that the condition (10) is here replaced by a set of s conditions:

$$I_k = \int p_k \, dq_k = n_k h, \qquad (k = 1, \ldots s) \tag{15}$$

where the integrals are taken over a complete period of the different q's respectively, and where $n_1,\ldots n_s$ are entire numbers. It will be seen at once that these conditions are invariant for any slow transformation of the system for which the independency of the motions corresponding to the different coordinates is maintained.

A more general class of systems for which a similar analogy with systems of a single degree of freedom exists and where conditions of the same type as (15) present themselves is obtained in the case where, although the motions corresponding to the different degrees of freedom are not independent of each other, it is possible nevertheless by a suitable choice of coordinates to express each of the momenta p_k as a function of q_k only. A simple system of this kind consists of a particle moving in a plane orbit in a central field of force. Taking the length of the radius-vector from the centre of the field to the particle as q_1, and the angular distance of this radius-vector from a fixed line in the plane of the orbit as q_2, we get at once from (4), since E does not contain q_2, the well known result that during the motion the angular momentum p_2 is constant and that the radial motion, given by the variations of p_1 and q_1 with the time, will be exactly the same as for a system of one degree of freedom. In his fundamental application

1) J. M. Burgers, Versl. Akad. Amsterdam **25** (1917) 849, 918, 1055; **Ann. d.** Phys. **52** (1917) 195; Phil. Mag. **33** (1917) 514.

of the quantum theory to the spectrum of *a non-periodic system* Sommerfeld assumed therefore that the stationary states of the above system are given by two conditions of the form:

$$I_1 = \int p_1 \, dq_1 = n_1 h, \qquad I_2 = \int p_2 \, dq_2 = n_2 h. \tag{16}$$

While the first integral obviously must be taken over a period of the radial motion, there might at first sight seem to be a difficulty in fixing the limits of integration of q_2. This disappears, however, if we notice that an integral of the type under consideration will not be altered by a change of coordinates in which q is replaced by some function of this variable. In fact, if instead of the angular distance of the radius-vector we take for q_2 some continuous periodic function of this angle with period 2π, every point in the plane of the orbit will correspond to one set of coordinates only and the relation between p and q will be exactly of the same type as for a periodic system of one degree of freedom for which the motion is of oscillating type. It follows therefore that the integration in the second of the conditions (16) has to be taken over a complete revolution of the radius-vector, and that consequently this condition is equivalent with the simple condition that the angular momentum of the particle round the centre of the field is equal to an entire multiplum of $h/2\pi$. As pointed out by Ehrenfest, the conditions (16) are invariant for such special transformations of the system for which the central symmetry is maintained. This follows immediately from the fact that the angular momentum in transformations of this type remains invariant, and that the equations of motion for the radial coordinate as long as p_2 remains constant are the same as for a system of one degree of freedom. On the basis of (16), Sommerfeld has, as mentioned in the introduction, obtained a brilliant explanation of the fine structure of the lines in the hydrogen spectrum, due to the change of the mass of the electron with its velocity.[1]) To this theory we shall come back in Part II.

[1]) In this connection it may be remarked that conditions of the same type as (16) were proposed independently by W. Wilson [Phil. Mag. **29** (1915) 795 and **31** (1916) 156] but by him applied only to the simple Keplerian motion described by the electron in the hydrogen atom if the relativity modifications are neglected. Due to the singular position of periodic systems in the quantum theory of systems of several degrees of freedom this application, however, involves, as it will appear from the following discussion, an ambiguity which deprives the result of an immediate physical interpretation. Conditions analogous to (16) have also been established by Planck in his interesting theory of the 'physical

As pointed out by Epstein [1]) and Schwarzschild [2]) the central systems considered by Sommerfeld form a special case of a more general class of systems for which conditions of the same type as (15) may be applied. These are the socalled *conditionally periodic systems*, to which we are led if the equations of motion are discussed by means of the Hamilton-Jacobi partial differential equation [3]). In the expression for the total energy E as a function of the q's and the p's, let the latter quantities be replaced by the partial differential coefficients of some function S with respect to the corresponding q's respectively, and consider the partial differential equation:

$$E\left(q_1, \dots q_s, \frac{\partial S}{\partial q_1}, \dots \frac{\partial S}{\partial q_s}\right) = \alpha_1, \tag{17}$$

obtained by putting this expression equal to an arbitrary constant α_1. If then

$$S = F(q_1, \dots q_s, \alpha_1, \dots \alpha_s) + C,$$

where $\alpha_2,\dots\alpha_s$ and C are arbitrary constants like α_1, is a total integral of (17), we get, as shown by Hamilton and Jacobi, the general solution of the equations of motion (4) by putting

$$\frac{\partial S}{\partial \alpha_1} = t + \beta_1, \qquad \frac{\partial S}{\partial \alpha_k} = \beta_k, \qquad (k = 2, \dots s) \tag{18}$$

and

$$\frac{\partial S}{\partial q_k} = p_k, \qquad (k = 1, \dots s) \tag{19}$$

where t is the time and $\beta_1,\dots\beta_s$ a new set of arbitrary constants. By means of (18) the q's are given as functions of the time t and the $2s$ constants $\alpha_1,\dots\alpha_s$, $\beta_1,\dots\beta_s$ which may be determined for instance from the values of the q's and $\dot{q}$'s at a given moment.

structure of the phase space' of systems of several degrees of freedom [Verh. d. D. Phys. Ges. **17** (1915) 407 and 438; Ann. d. Phys. **50** (1916) 385]. This theory, which has no direct relation to the problem of line-spectra discussed in the present paper, rests upon a profound analysis of the geometrical problem of dividing the multiple-dimensional phase space corresponding to a system of several degrees of freedom into 'cells' in a way analogous to the division of the phase surface of a system of one degree of freedom by the curves given by (10).

1) P. Epstein, loc. cit.

2) K. Schwarzschild, loc. cit.

3) See f. inst. C. V. L. Charlier, Die Mechanik des Himmels, Bd. I, Abt. 2.

Now the class of systems, referred to, is that for which, for a suitable choice of orthogonal coordinates, it is possible to find a total integral of (17) of the form

$$S = \sum_{1}^{s} S_k(q_k, \alpha_1, \dots \alpha_s), \tag{20}$$

where S_k is a function of the s constants $\alpha_1,\dots\alpha_s$ and of q_k only. In this case, in which the equation (17) allows of what is called 'separation of variables', we get from (19) that every p is a function of the α's and of the corresponding q only. If during the motion the coordinates do not become infinite in the course of time or converge to fixed limits, every q will, just as for systems of one degree of freedom, oscillate between two fixed values, different for the different q's and depending on the α's. Like in the case of a system of one degree of freedom, p_k will become zero and change its sign whenever q_k passes through one of these limits. Apart from special cases, the system will during the motion never pass twice through a configuration corresponding to the same set of values for the q's and p's, but it will in the course of time pass within any given, however small, distance from any configuration corresponding to a given set of values $q_1,\dots q_s$, representing a point within a certain closed s-dimensional extension limited by s pairs of $(s-1)$-dimensional surfaces corresponding to constant values of the q's equal to the above mentioned limits of oscillation. A motion of this kind is called 'conditionally periodic'. It will be seen that the character of the motion will depend only on the α's and not on the β's, which latter constants serve only to fix the exact configuration of the system at a given moment, when the α's are known. For special systems it may occur that the orbit will not cover the above mentioned s-dimensional extension everywhere dense, but will, for all values of the α's, be confined to an extension of less dimensions. Such a case we will refer to in the following as a case of 'degeneration'.

Since for a conditionally periodic system which allows of separation in the variables $q_1,\dots q_s$ the p's are functions of the corresponding q's only, we may, just as in the case of independent degrees of freedom or in the case of quasiperiodic motion in a central field, form a set of expressions of the type

$$I_k = \int p_k(q_k, \alpha_1, \dots \alpha_s)\, dq_k, \qquad (k = 1, \dots s) \tag{21}$$

where the integration is taken over a complete oscillation of q_k. As, in general, the orbit will cover everywhere dense an s-dimensional extension limited in the characteristic way mentioned above, it follows that, except in cases of degeneration, a separation of variables will not be possible for two different sets of coordinates $q_1,\ldots q_s$ and $q'_1,\ldots q'_s$, unless $q_1=f_1(q'_1),\ldots q_s=f_s(q'_s)$, and since a change of coordinates of this type will not affect the values of the expressions (21), it will be seen that the values of the I's are completely determined for a given motion of the system. By putting

$$I_k = n_k h, \qquad (k = 1, \ldots s) \tag{22}$$

where $n_1,\ldots n_s$ are positive entire numbers, we obtain therefore *a set of conditions which form a natural generalisation of condition* (10) holding for a system of one degree of freedom.

Since the I's, as given by (21), depend on the constants $\alpha_1,\ldots\alpha_s$ only and not on the β's, the α's may, in general, inversely be determined from the values of the I's. The character of the motion will therefore, in general, be completely determined by the conditions (22), and especially the value for the total energy, which according to (17) is equal to α_1, will be fixed by them. In the cases of degeneration referred to above, however, the conditions (22) involve an ambiguity, since in general for such systems there will exist an infinite number of different sets of coordinates which allow of a separation of variables, and which will lead to different motions in the stationary states, when these conditions are applied. As we shall see below, this ambiguity will not influence the fixation of the total energy in the stationary states, which is the essential factor in the theory of spectra based on (1) and in the applications of the quantum theory to statistical problems.

A well known characteristic example of a conditionally periodic system is afforded by a particle moving under the influence of the attractions from two fixed centres varying as the inverse squares of the distances apart, if the relativity modifications are neglected. As shown by Jacobi this problem can be solved by a separation of variables if so called elliptical coordinates are used, i.e. if for q_1 and q_2 we take two parameters characterising respectively an ellipsoid and a hyperboloid of revolution with the centres as foci and passing through the instantaneous position of the moving particle, and for q_3 we take the angle between the plane through the particle and the centres and a fixed plane through the latter points, or, in closer conformity with the above general description, some continuous periodic function of this angle with period 2π. A limiting case of this problem is afforded by an electron rotating round a positive nucleus and subject to the effect of an additional homogeneous electric field, because this field may be considered as arising from a second nucleus at infinite distance apart from the first. The motion in this case will therefore be conditionally periodic and allow a separation of variables in parabolic coordinates, if the nucleus is taken as focus for both sets of paraboloids of revolution, and their axes are taken parallel to the direction of the electric force. By applying the conditions (22) to this motion Epstein and Schwarzschild have, as mentioned in the introduction, independent of each other, obtained an explanation of the effect of an external electric field on the lines of the hydrogen spectrum, which was found to be in convincing agreement with Stark's measurements. To the results of these calculations we shall return in Part II.

In the above way of representing the general theory we have followed the same procedure as used by Epstein. By introducing the so called 'angle variables' well known from the astronomical theory of perturbations, Schwarzschild has given the theory a very elegant form in which the analogy with systems of one degree of freedom presents itself in a somewhat different manner. The connection between this treatment and that given above has been discussed in detail by Epstein.[1])

As mentioned above the conditions (22), first established from analogy with systems of one degree of freedom, have subsequently been proved generally to be *mechanically invariant for any slow transformation for which the system remains conditionally periodic.* The proof of this invariance has been given quite recently by Burgers [2]) by means of an interesting application of the theory of contact-transformations based on Schwarzschild's introduction of angle variables. We shall not enter here on these calculations but shall only consider some points in connection with the problem of the mechanical transformability of the stationary states which are of importance for the logical consistency of the general theory and for the later applications. In § 2 we saw that in the proof of the mechanical invariance of relation (10) for a periodic system of one degree of freedom, it was essential that the comparative variation of the external conditions during the time of one period could be made small. This may be regarded as an

[1]) P. Epstein, Ann. d. Phys. **51** (1916) 168. See also Note on page 33 of the present paper.

[2]) J. M. Burgers, loc. cit. Versl. Akad. Amsterdam **25** (1917) 1055.

immediate consequence of the nature of the fixation of the stationary states in the quantum theory. In fact the answer to the question whether a given state of a system is stationary, will not depend only on the motion of the particles at a given moment or on the field of force in the immediate neighbourhood of their instantaneous positions, but cannot be given before the particles have passed through a complete cycle of states, and so to speak have got to know the entire field of force of influence on the motion. If thus, in the case of a periodic system of one degree of freedom, the field of force is varied by a given amount, and if its comparative variation within the time of a single period was not small, the particle would obviously have no means to get to know the nature of the variation of the field and to adjust its stationary motion to it, before the new field was already established. For exactly the same reasons it is a necessary condition for the mechanical invariance of the stationary states of a conditionally periodic system, that the alteration of the external conditions during an interval in which the system has passed approximately through all possible configurations within the above mentioned s-dimensional extension in the coordinate-space can be made as small as we like. This condition forms therefore also an essential point in Burgers' proof of the invariance of the conditions (22) for mechanical transformations. Due to this we meet with a characteristic difficulty when during the transformation of the system we pass one of the cases of degeneration mentioned above, where, for every set of values for the α's, the orbit will not cover the s-dimensional extension everywhere dense, but will be confined to an extension of less dimensions. It is clear that, when by a slow transformation of a conditionally periodic system we approach a degenerate system of this kind, the time-interval which the orbit takes to pass close to any possible configuration will tend to be very long and will become infinite when the degenerate system is reached. As a consequence of this *the conditions (22) will generally not remain mechanically invariant when we pass a degenerate system*, what has intimate connection with the above mentioned ambiguity in the determination of the stationary states of such systems by means of (22).

A typical case of a degenerate system, which may serve as an illustration of this point, is formed by a system of several degrees of freedom for which every motion is simply periodic, independent of the initial conditions. In this case, which is of great importance in the physical applications, we have from (5) and (21), for any set of coordinates in which a separation of variables is possible,

$$I = \int_0^\sigma (p_1\dot{q}_1 + \dots + p_s\dot{q}_s)\,\mathrm{d}t = \varkappa_1 I_1 + \dots + \varkappa_s I_s, \tag{23}$$

where the integration is extended over one period of the motion, and where $\varkappa_1,\dots\varkappa_s$ are a set of positive entire numbers without a common divisor. Now we shall expect that every motion, for which it is possible to find a set of coordinates in which it satisfies (22), will be stationary. For any such motion we get from (23)

$$I = (\varkappa_1 n_1 + \dots + \varkappa_s n_s)h = nh, \tag{24}$$

where n is a whole number which may take all positive values if, as in the applications mentioned below, at least one of the $\varkappa$'s is equal to one. Inversely, if the system under consideration allows of separation of variables in an infinite continuous multitude of sets of coordinates, we must conclude that generally every motion which satisfies (24) will be stationary, because in general it will be possible for any such motion to find a set of coordinates in which it satisfies also (22). It will thus be seen that, for a periodic system of several degrees of freedom, condition (24) forms a simple generalisation of condition (10). From relation (8), which holds for two neighbouring motions of any periodic system, it follows further that the energy of the system will be completely determined by the value of I, just as for systems of one degree of freedom.

Consider now a periodic system in some stationary state satisfying (24), and let us assume that an external field is slowly established at a continuous rate and that the motion at any moment during this process allows of a separation of variables in a certain set of coordinates. If we would assume that the effect of the field on the motion of the system at any moment could be calculated directly by means of ordinary mechanics, we would find that the values of the I's with respect to the latter coordinates would remain constant during the process, but this would involve that the values of the n's in (22) would in general not be entire numbers, but would depend entirely on the accidental motion, satisfying (24), originally possessed by the system. That mechanics, however, cannot generally be applied directly to determine the motion of a periodic system under influence of an increasing external field, is just what we should expect according to the singular position of degenerate systems as regards mechanical transformations. In fact, in the presence of a small external field, the motion of a periodic system will undergo slow variations as regards the shape and position of the orbit, and if the perturbed motion is conditionally periodic these variations will be of a periodic nature. Formally, we may therefore compare a periodic system exposed to an external field with a simple mechanical system of one degree of freedom in which the particle performs a slow oscillating motion. Now the frequency of a slow variation of the orbit will be seen to be proportional to the intensity of the external field, and it is therefore obviously impossible to establish the external field at a rate so slow that the comparative change of its intensity during a period of this variation is small. The process which takes place during the increase of the field will thus be analogous to that which takes place if an oscillating particle is subject to the effect of external forces which change considerably during a period. Just as the latter process generally will give rise to emission or absorption of radiation and cannot be described by means of ordinary mechanics, we must expect that the motion of a periodic system of several degrees of freedom under the establishment of the external field cannot be determined by ordinary mechanics, but that the field will give rise to effects of the same kind as those which occur during a transition between two stationary states accompanied by emission or absorption of radiation. Consequently we shall expect that, during the establishment of the field, *the system will in general adjust itself in some unmechanical way* until a stationary state is reached in which the frequency (or frequencies) of the above mentioned slow variation of the orbit has a relation to the additional energy of the system due to the presence of the external field, which is of the same kind as the relation, expressed by (8) and (10), between the energy and frequency of a periodic system of one degree of freedom. As it will be shown in Part II in connection with the physical applications, this condition is just secured if the stationary states in the presence of the field are determined by the conditions (22), and it will be seen that these considerations offer a means of fixing the stationary states of a perturbed periodic system also in cases where no separation of variables can be obtained.

In consequence of the singular position of the degenerate systems in the general theory of stationary states of conditionally periodic systems, we obtain a means of *connecting mechanically two different stationary states of a given system* through a continuous series of stationary states without passing through systems in which the forces are very small and the energies in all the stationary states tend to coincide (comp. page 9). In fact, if we consider a given conditionally periodic system which can be transformed in a continuous way into a system for which every orbit is periodic and for which every state satisfying (24) will also satisfy (22) for a suitable choice of coordinates, it is clear in the first place that it is possible to pass in a mechanical way through a continuous series of stationary states from a state corresponding to a given set of values of the n's in (22) to any other such state for which $\varkappa_1 n_1 + \ldots + \varkappa_s n_s$ possesses the same value. If, moreover, there exists a second periodic system of the same character to which the first periodic system can be transformed continuously, but for which the set of $\varkappa$'s is different, it will be possible in general by a suitable cyclic transformation to pass in a mechanical way between any two stationary states of the given conditionally periodic system satisfying (22).

To obtain an example of such a cyclic transformation let us take the system consisting of an electron which moves round a fixed positive nucleus exerting an attraction varying as the inverse square of the distance. If we neglect the small relativity corrections, every orbit will be periodic independent of the initial conditions and the system will allow of separation of variables in polar coordinates as well as in any set of elliptical coordinates, of the kind mentioned on page 23, if the nucleus is taken as one of the foci. It is simply seen that any orbit which satisfies (24) for a value of $n > 1$, will also satisfy (22) for a suitable choice of elliptical coordinates. By imagining another nucleus of infinite small charge placed at the other focus, the orbit may further be transformed into another which satisfies (24) for the same value of n, but which may have any given value for the eccentricity. Consider now a state of the system satisfying (24), and let us assume that by the above means the orbit is originally so adjusted that in plane polar coordinates it will correspond to $n_1 = m$ and $n_2 = n - m$ in (16). Let then the system undergo a slow continuous transformation during which the field of force acting on the electron remains central, but by which the law of attraction is slowly varied until the force is directly proportional to the distance apart. In the final state, as well as in the original state, the orbit of the electron will be closed, but during the transformation the orbit will not be closed, and the ratio between the mean period of revolution and the period of the radial motion, which in the original motion was equal to one, will during the transformation increase continuously until in the final state it is equal to two. This means that, using polar coordinates, the values of $\varkappa_1$ and $\varkappa_2$ in (22) which for the first state are equal to $\varkappa_1 = \varkappa_2 = 1$, will be for the second state $\varkappa_1 = 2$ and $\varkappa_2 = 1$. Since during the transformation n_1 and n_2 will keep their values, we get therefore in the final state $I = h(2m + (n-m)) = h(n+m)$. Now in the latter state, the system allows a separation of variables not only in polar coordinates but also in any system of rectangular Cartesian coordinates, and by suitable choice of the direction of the axes, we can obtain that any orbit, satisfying (24) for a value of $n > 1$, will also satisfy (22). By an infinite small change of the force components in the directions of the axes, in such a way that the motions in these directions remain independent of each other but possess slightly different periods, it will further be possible to transform the elliptical orbit mechanically into one corresponding to any given ratio between the axes. Let us now assume that in this way the orbit of the electron is transformed into a circular one, so that, returning to plane polar coordinates, we have $n_1 = 0$ and $n_2 = n + m$, and let then by a slow transformation the law of attraction be varied until again it is that of the inverse square. It will be seen that when this state is reached the motion will again satisfy (24), but this time we will have $I = h(n+m)$ instead of $I = nh$ as in the original state. By repeating a cyclic process of this kind we may pass from any stationary state of the system in question which satisfies (24) for a value of $n > 1$ to any other such state without leaving at any moment the region of stationary states.

The theory of the mechanical transformability of the stationary states gives us a means to discuss the question of the *a-priori probability* of the different states of a conditionally periodic system, characterised by different sets of values for the n's in (22). In fact from the considerations, mentioned in § 1, it follows that, if the a-priori probability of the stationary states of a given system is known, it is possible at once to deduce the probabilities for the stationary states of any other system to which the first system can be transformed continuously without passing through a system of degeneration. Now from the analogy with systems of one degree of freedom it seems necessary to assume that, for a system of several degrees of freedom for which the motions corresponding to the different coordinates are dynamically independent of each other, the a-priori probability is the same for all the states corresponding to different sets of n's in (15). According to the above we shall therefore assume that the a-priori probability is the same for all states, given by (22), of any system which can be formed in a continuous way from a system of this kind without passing through systems of degeneration. It will be observed that on this assumption we obtain exactly the same relation to the ordinary theory of statistical mechanics in the limit of large n's as obtained in the case of systems of one degree of freedom. Thus, for a conditionally periodic system, the volume given by (11) of the element of phase-space, including all points $q_1, \ldots q_s, p_1, \ldots p_s$ which represent states for which the value of I_k given by (21) lies between I_k and $I_k + \delta I_k$, is seen at once to be equal to [1])

$$\delta W = \delta I_1 \delta I_2 \dots \delta I_s, \tag{25}$$

if the coordinates are so chosen that the motion corresponding to every degree of freedom is of oscillating type. The volume of the phase-space limited by s pairs of surfaces, corresponding to successive values for the n's in the conditions (22), will therefore be equal to h^s and consequently be the same for every combination of the n's. In the limit where the n's are large numbers and the stationary states corresponding to successive values for the n's differ only very little from each other, we thus obtain the same result on the assumption of equal a-priori probability of all the stationary states, corresponding to different sets of values of $n_1, n_2, \dots n_s$ in (22), as would be obtained by application of ordinary statistical mechanics.

The fact that the last considerations hold for every non-degenerate conditionally periodic system suggests the assumption that in general *the a-priori probability will be the same for all the states determined by (22)*, even if it should not be possible to transform the given system into a system of independent degrees of freedom without passing through degenerate systems. This assumption will be shown to be supported by the consideration of the intensities of the different components of the Stark-effect of the hydrogen lines, mentioned in the next Part. When we consider a degenerate system, however, we cannot assume that the different stationary states are a-priori equally probable. In such a case the stationary states will be characterised by a number of conditions less than the number of degrees of freedom, and the probability of a given state must be determined from the number of different stationary states of some non-degenerate system which will coincide in the given state, if the latter system is continuously transformed into the degenerate system under consideration.

In order to illustrate this, let us take the simple case of a degenerate system formed by an electrified particle moving in a plane orbit in a central field, the stationary states of which are given by the two conditions (16). In this case the plane of the orbit is undetermined, and it follows already from a comparison with ordinary statistical mechanics, that the a-priori probability of the states characterized by different combinations of n_1 and n_2 in (16) cannot be the same. Thus the volume of the phase-space, corresponding to states for which

[1]) Comp. A. Sommerfeld, Ber. Akad. München, 1917, p. 83.

I_1 lies between and I_1 and $I_1+\delta I_1$ and for which I_2 lies between I_2 and $I_2+\delta I_2$, is found by a simple calculation * to be equal to $\delta W = 2I_2\delta I_1\delta I_2$, if the motion is described by ordinary polar coordinates. For large values of n_1 and n_2, we must therefore expect that the a-priori probability of a stationary state corresponding to a given combination (n_1, n_2) is proportional to n_2. The question of the a-priori probability of states corresponding to small values of the n's has been discussed by Sommerfeld in connection with the problem of the intensities of the different components in the fine structure of the hydrogen lines (see Part II). From considerations about the volume of the extensions in the phase-space, which might be considered as associated with the states characterised by different combinations (n_1, n_2), Sommerfeld proposes several different expressions for the a-priori probability of such states. Due to the necessary arbitrariness involved in the choice of these extensions, however, we cannot in this way obtain a rational determination of the a-priori probability of states corresponding to small values of n_1 and n_2. On the other hand, this probability may be deduced by regarding the motion of the system under consideration as the degeneration of a motion characterised by three numbers n_1, n_2 and n_3, as in the general applications of the conditions (22) to a system of three degrees of freedom. Such a motion may be obtained for instance by imagining the system placed in a small homogeneous magnetic field. In certain respects this case falls outside the general theory of conditionally periodic systems discussed in this section, but, as we shall see in Part II, it can be simply shown that the presence of the magnetic field imposes the further condition on the motion in the stationary states that the angular momentum round the axis of the field is equal to $n'h/2\pi$, where n' is a positive entire number equal to or less than n_2, and which for the system considered in the spectral problems must be assumed to be different from zero. When regard is taken to the two opposite directions in which the particle may rotate round the axis of the field, we see therefore that for this system a state corresponding to a given combination of n_1 and n_2 in the presence of the field can be established in $2n_2$ different ways. The a-priori probability of the different states of the system may consequently for all combinations of n_1 and n_2 be assumed to be proportional to n_2.

* See A. Sommerfeld, loc. cit.

The assumption just mentioned that the angular momentum round the axis of the field cannot be equal to zero is deduced from considerations of systems for which the motion corresponding to special combinations of the n's in (22) would become physically impossible due to some singularity in its character. In such cases we must assume that no stationary states exist corresponding to the combinations $(n_1, n_2, \ldots n_s)$ under consideration, and on the above principle of the invariance of the a-priori probability for continuous transformations we shall accordingly expect that the a-priori probability of any other state, which can be transformed continuously into one of these states without passing through cases of degeneration, will also be equal to zero.

Let us now proceed to consider the *spectrum of a conditionally periodic system*, calculated from the values of the energy in the stationary states by means of relation (1). If $E(n_1, \ldots n_s)$ is the total energy of a stationary state determined by (22) and if ν is the frequency of the line corresponding to the transition between two stationary states characterised by $n_k = n_k'$ and $n_k = n_k''$ respectively, we have

$$\nu = \frac{1}{h}\left[E(n_1', \ldots n_s') - E(n_1'', \ldots n_s'')\right]. \tag{26}$$

In general, this spectrum will be entirely different from the spectrum to be expected on the ordinary theory of electrodynamics from the motion of the system. Just as for a system of one degree of freedom we shall see, however, that in the limit where the motions in neighbouring stationary states differ very little from each other, there exists a close relation between the spectrum calculated on the quantum theory and that to be expected on ordinary electrodynamics. As in § 2 we shall further see, that this connection leads to certain general considerations about the probability of transition between any two stationary states and about the nature of the accompanying radiation which are found to be supported by observations. In order to discuss this question we shall first deduce a general expression for the energy difference between two neighbouring states of a conditionally periodic system, which can be simply obtained by a calculation analogous to that used in § 2 in the deduction of the relation (8).

Consider some motion of a conditionally periodic system which allows of separation of variables in a certain set of coordinates $q_1, \ldots q_s$, and let us assume that at the time $t = \vartheta$ the configuration of the system will to a close approximation be the same as at the time $t=0$. By taking ϑ large enough we can make this approximation as close as we like. If next we consider some conditionally periodic motion, obtained by a small variation of the first motion, and which allows of separation of variables in a set of coordinates $q_1',\ldots q_s'$ which may differ slightly from the set $q_1,\ldots q_s$, we get by means of Hamilton's equations (4), using the coordinates $q_1',\ldots q_s'$,

$$\int_0^\vartheta \delta E \, dt = \int_0^\vartheta \sum_1^s \left(\frac{\partial E}{\partial p_k'} \delta p_k' + \frac{\partial E}{\partial q_k'} \delta q_k'\right) dt = \int_0^\vartheta \sum_1^s (\dot{q}_k' \, \delta p_k' - \dot{p}_k' \, \delta q_k') \, dt.$$

By partial integration of the second term in the bracket this gives:

$$\int_0^\vartheta \delta E \, dt = \int_0^\vartheta \sum_1^s \delta(p_k' \dot{q}_k') \, dt - \left|\sum_1^s p_k' \, \delta q_k'\right|_{t=0}^{t=\vartheta}. \tag{27}$$

Now we have for the unvaried motion

$$\int_0^\vartheta \sum_1^s p_k' \dot{q}_k' \, dt = \int_0^\vartheta \sum_1^s p_k \dot{q}_k \, dt = \sum_1^s N_k I_k,$$

where I_k is defined by (21) and where N_k is the number of oscillations performed by q_k in the time interval ϑ. For the varied motion we have on the other hand:

$$\int_0^\vartheta \sum_1^s p_k' \dot{q}_k' \, dt = \int_{t=0}^{t=\vartheta} \sum_1^s p_k' \, dq_k' = \sum_1^s N_k I_k' + \left|\sum_1^s p_k' \, \delta q_k'\right|_{t=0}^{t=\vartheta},$$

where the I''s correspond to the conditionally periodic motion in the coordinates $q_1',\ldots q_s'$, and the $\delta q'$'s which enter in the last term are the same as those in (27). Writing $I_k' - I_k = \delta I_k$, we get therefore from the latter equation

$$\int_0^\vartheta \delta E \, dt = \sum_1^s N_k \, \delta I_k. \tag{28}$$

In the special case where the varied motion is an undisturbed motion belonging to the same system as the unvaried motion we get, since δE will be constant,

$$\delta E = \sum_1^s \omega_k \, \delta I_k, \tag{29}$$

where $\omega_k = N_k/\vartheta$ is the mean frequency of oscillation of q_k between its limits, taken over a long time interval of the same order of magnitude as ϑ. This equation forms a simple generalisation of (8), and in the general case in which a separation of variables will be possible only for one system of coordinates leading to a complete definition of the I's it might have been deduced directly from the analytical theory of the periodicity properties of the motion of a conditionally periodic system, based on the introduction of angle variables.[1]) From

[1]) See Charlier, Die Mechanik des Himmels, Bd. I Abt. 2, and especially P. Epstein, Ann. d. Phys. **51** (1916) 178. By means of the well known theorem of Jacobi about the change of variables in the canonical equations of Hamilton, the connection between the notion of angle-variables and the quantities I, discussed by Epstein in the latter paper, may be briefly exposed in the following elegant manner which has been kindly pointed out to me by Mr. H. A. Kramers. Consider the function $S(q_1, \ldots q_s, I_1, \ldots I_s)$ obtained from (20) by introducing for the α's their expressions in terms of the I's given by the equations (21). This function will be a many valued function of the q's which increases by I_k if q_k describes one oscillation between its limits and comes back to its original value while the other q's remain constant. If we therefore introduce a new set of variables $w_1, \ldots w_s$ defined by

$$w_k = \frac{\partial S}{\partial I_k}, \qquad (k = 1, \ldots s) \tag{1*}$$

it will be seen that w_k increases by one unit while the other w's will come back to their original values if q_k describes one oscillation between its limits and the other q's remain constant. Inversely it will therefore be seen that the q's, and also the p's which were given by

$$p_k = \frac{\partial S}{\partial q_k}, \qquad (k = 1, \ldots s), \tag{2*}$$

when considered as functions of the I's and w's will be periodic functions of every of the w's with period 1. According to Fourier's theorem any of the q's may therefore be represented by an s-double trigonometric series of the form

$$q = \Sigma A_{\tau_1, \ldots \tau_s} \cos 2\pi(\tau_1 w_1 + \ldots \tau_s w_s + \alpha_{\tau_1, \ldots \tau_s}), \tag{3*}$$

where the A's and α's are constants depending on the I's and where the summation is to be extended over all entire values of $\tau_1, \ldots \tau_s$. On account of this property of the w's, the quantities $2\pi w_1, \ldots 2\pi w_s$ are denoted as 'angle variables'. Now from (1*) and (2*) it follows according to the above mentioned theorem of Jacobi (see for instance Jacobi, Vorlesungen über Dynamik § 37) that the variations with the time of the I's and w's will be given by

$$\frac{dI_k}{dt} = -\frac{\partial E}{\partial w_k}, \qquad \frac{dw_k}{dt} = \frac{\partial E}{\partial I_k}, \qquad (k = 1, \ldots s) \tag{4*}$$

where the energy E is considered as a function of the I's and w's. Since E, however, is determined by the I's only we get from (4*), besides the evident

(29) it follows moreover that, if the system allows of a separation of variables in an infinite continuous multitude of sets of coordinates, the total energy will be the same for all motions corresponding to the same values of the I's, independent of the special set of coordinates used to calculate these quantities. As mentioned above and as we have already shown in the case of purely periodic systems by means of (8), the total energy is therefore also in cases of degeneration completely determined by the conditions (22).

Consider now a transition between two stationary states determined by (22) by putting $n_k = n'_k$ and $n_k = n''_k$ respectively, and let us assume that $n'_1, \ldots n'_s, n''_1, \ldots n''_s$ are large numbers, and that the differences

result that the I's are constant during the motion, that the w's will vary linearly with the time and can be represented by

$$w_k = \omega_k t + \delta_k, \qquad \omega_k = \frac{\partial E}{\partial I_k}, \qquad (k = 1, \ldots s) \tag{5*}$$

where δ_k is a constant, and where ω_k is easily seen to be equal to the mean frequency of oscillation of q_k. From (5*) eq. (28) follows at once, and it will further be seen that by introducing (5*) in (3*) we get the result that every of the q's, and consequently also any one-valued function of the q's, can be represented by an expression of the type (31).

In this connection it may be mentioned that the method of Schwarzschild of fixing the stationary states of a conditionally periodic system, mentioned on page 117, consists in seeking for a given system a set of canonically conjugated variables $Q_1, \ldots Q_s, P_1, \ldots P_s$ in such a way that the positional coordinates of the system $q_1, \ldots q_s$ and their conjugated momenta $p_1, \ldots p_s$, when considered as functions of the Q's and P's, are periodic in every of the Q's with period 2π, while the energy of the system depends only on the P's. In analogy with the condition which fixes the angular momentum in Sommerfeld's theory of central systems Schwarzschild next puts every of the P's equal to an entire multiplum of $h/2\pi$. In contrast to the theory of stationary states of conditionally periodic systems based on the possibility of separation of variables and the fixation of the I's by (22), this method does not lead to an absolute fixation of the stationary states, because, as pointed out by Schwarzschild himself, the above definition of the P's leaves an arbitrary constant undetermined in every of these quantities. In many cases, however, these constants may be simply determined from considerations of mechanical transformability of the stationary states, and as pointed out by Burgers [loc. cit. Versl. Akad. Amsterdam **25** (1917) 1055] Schwarzschild's method possesses on the other hand the essential advantage of being applicable to certain classes of systems in which the displacements of the particles may be represented by trigonometric series of the type (31), but for which the equations of motion cannot be solved by separation of variables in any fixed set of coordinates. An interesting application of this to the spectrum of rotating molecules, given by Burgers, will be mentioned in Part IV.

$n'_k - n''_k$ are small compared with these numbers. Since the motions of the system in these states will differ relatively very little from each other we may calculate the difference of the energy by means of (29), and we get therefore, by means of (1), for the frequency of the radiation corresponding to the transition between the two states

$$\nu = \frac{1}{h}(E' - E'') = \frac{1}{h}\sum_1^s \omega_k(I'_k - I''_k) = \sum_1^s \omega_k(n'_k - n''_k), \tag{30}$$

which is seen to be a direct generalisation of the expression (13) in § 2.

Now, in complete analogy to what is the case for periodic systems of one degree of freedom, it is proved in the analytical theory of the motion of conditionally periodic systems mentioned above that for the latter systems the coordinates $q_1, \ldots q_s$, and consequently also the displacements of the particles in any given direction, may be expressed as a function of the time by an s-double infinite Fourier series of the form:

$$\xi = \sum C_{\tau_1, \ldots \tau_s} \cos 2\pi\{(\tau_1\omega_1 + \ldots \tau_s\omega_s)t + c_{\tau_1, \ldots \tau_s}\}, \tag{31}$$

where the summation is to be extended over all positive and negative entire values of the τ's, and where the ω's are the above mentioned mean frequencies of oscillation for the different q's. The constants $C_{\tau_1, \ldots \tau_s}$ depend only on the α's in the equations (18) or, what is the same, on the I's, while the constants $c_{\tau_1, \ldots \tau_s}$ depend on the α's as well as on the β's. In general the quantities $\tau_1\omega_1 + \ldots \tau_s\omega_s$ will be different for any two different sets of values for the τ's, and in the course of time the orbit will cover everywhere dense a certain s-dimensional extension. In a case of degeneration, however, where the orbit will be confined to an extension of less dimensions, there will exist for all values of the α's one or more relations of the type $m_1\omega_1 + \ldots m_s\omega_s = 0$ where the m's are entire numbers and by the introduction of which the expression (31) can be reduced to a Fourier series which is less than s-double infinite. Thus in the special case of a system of which every orbit is periodic we have $\omega_1/\varkappa_1 = \ldots = \omega_s/\varkappa_s = \omega$, where the $\varkappa$'s are the numbers which enter in eq. (23), and the Fourier series for the displacements in the different directions will in this case consist only of terms of the simple form $C_\tau \cos 2\pi\{\tau\omega t + c_\tau\}$, just as for a system of one degree of freedom.

On the ordinary theory of radiation, we should expect from (31) that the spectrum emitted by the system in a given state would consist of an s-double infinite series of lines of frequencies equal to $\tau_1\omega_1+\ldots+\tau_s\omega_s$. In general, this spectrum would be completely different from that given by (26). This follows already from the fact that the ω's will depend on the values for the constants $\alpha_1,\ldots\alpha_s$ and will vary in a continuous way for the continuous multitude of mechanically possible states corresponding to different sets of values for these constants. Thus in general the ω's will be quite different for two different stationary states corresponding to different sets of n's in (22), and we cannot expect any close relation between the spectrum calculated on the quantum theory and that to be expected on the ordinary theory of mechanics and electrodynamics. In the limit, however, where the n's in (22) are large numbers, the ratio between the ω's for two stationary states, corresponding to $n_k=n'_k$ and $n_k=n''_k$ respectively, will tend to unity if the differences $n'_k-n''_k$ are small compared with the n's, and as seen from (30) the spectrum calculated by (1) and (22) will in this limit just tend to coincide with that to be expected on the ordinary theory of radiation from the motion of the system.

As far as the frequencies are concerned, we thus see that for conditionally periodic systems there exists a connection between the quantum theory and the ordinary theory of radiation of exactly the same character as that shown in § 2 to exist in the simple case of periodic systems of one degree of freedom. Now on ordinary electrodynamics the coefficients $C_{\tau_1,\ldots\tau_s}$ in the expression (31) for the displacements of the particles in the different directions would in the well known way determine the intensity and polarisation of the emitted radiation of the corresponding frequency $\tau_1\omega_1+\ldots\tau_s\omega_s$. As for systems of one degree of freedom we must therefore conclude that, in the limit of large values for the n's, the probability of spontaneous transition between two stationary states of a conditionally periodic system, as well as the polarisation of the accompanying radiation, can be determined directly from the values of the coefficient $C_{\tau_1,\ldots\tau_s}$ in (31) corresponding to a set of τ's given by $\tau_k=n'_k-n''_k$, if $n'_1,\ldots n'_s$ and $n''_1,\ldots n''_s$ are the numbers which characterise the two stationary states.

Without a detailed theory of the mechanism of transition between the stationary states we cannot, of course, in general obtain an exact determination of the *probability of spontaneous transition* between two such states, unless the n's are large numbers. Just as in the case of systems of one degree of freedom, however, we are naturally led from the above considerations to assume that, also for values of the

n's which are not large, there must exist an intimate connection between the probability of a given transition and the values of the corresponding Fourier coefficient in the expressions for the displacements of the particles in the two stationary states. This allows us at once to draw certain important conclusions. Thus, from the fact that in general negative as well as positive values for the τ's appear in (31), it follows that we must expect that in general not only such transitions will be possible in which all the n's decrease, but that also transitions will be possible for which some of the n's increase while others decrease. This conclusion, which is supported by observations on the fine structure of the hydrogen lines as well as on the Stark effect, is contrary to the suggestion, put forward by Sommerfeld with reference to the essential positive character of the I's, that every of the n's must remain constant or decrease under a transition. Another direct consequence of the above considerations is obtained if we consider a system for which, for all values of the constants $\alpha_1, \ldots \alpha_s$, the coefficient $C_{\tau_1, \ldots \tau_s}$ corresponding to a certain set $\tau_1^0, \ldots \tau_s^0$ of values for the τ's is equal to zero in the expressions for the displacements of the particles in every direction. In this case we shall naturally expect that no transition will be possible for which the relation $n'_k - n''_k = \tau_k^0$ is satisfied for every k. In the case where $C_{\tau_1^0, \ldots \tau_s^0}$ is equal to zero in the expressions for the displacement in a certain direction only, we shall expect that all transitions, for which $n'_k - n''_k = \tau_k^0$ for every k, will be accompanied by a radiation which is polarised in a plane perpendicular to this direction.

A simple illustration of the last considerations is afforded by the system mentioned in the beginning of this section, and which consists of a particle executing motions in three perpendicular directions which are independent of each other. In that case all the Fourier coefficients in the expressions for the displacements in any direction will disappear if more than one of the τ's are different from zero. Consequently we must assume that only such transitions are possible for which only one of the n's varies at the same time, and that the radiation corresponding to such a transition will be linearly polarised in the direction of the displacement of the corresponding coordinate. In the special case where the motions in the three directions are simply harmonic, we shall moreover conclude that none of the n's can vary by more than a single unit, in analogy with the considerations in the former section about a linear harmonic vibrator.

Another example which has more direct physical importance, since it includes all the special applications of the quantum theory to spectral problems mentioned in the introduction, is formed by a conditionally periodic system possessing an axis of symmetry. In all these applications a separation of variables is obtained in a set of three coordinates q_1, q_2 and q_3, of which the first two serve to fix the position of the particle in a plane through the axis of the system, while the last is equal to the angular distance between this plane and a fixed plane through the same axis. Due to the symmetry, the expression for the total energy in Hamilton's equations will not contain the angular distance q_3 but only the angular momentum p_3 round the axis. The latter quantity will consequently remain constant during the motion, and the variations of q_1 and q_2 will be exactly the same as in a conditionally periodic system of two degrees of freedom only. If the position of the particle is described in a set of cylindrical coordinates z, ϱ, ϑ, where z is the displacement in the direction of the axis, ϱ the distance of the particle from this axis and ϑ is equal to the angular distance q_3, we have therefore

$$z = \Sigma\, C_{\tau_1, \tau_2} \cos 2\pi\{(\tau_1\omega_1 + \tau_2\omega_2)t + c_{\tau_1, \tau_2}\}$$

and

$$\varrho = \Sigma\, C'_{\tau_1, \tau_2} \cos 2\pi\{(\tau_1\omega_1 + \tau_2\omega_2)t + c'_{\tau_1, \tau_2}\}, \qquad (32)$$

where the summation is to be extended over all positive and negative entire values of τ_1 and τ_2, and where ω_1 and ω_2 are the mean frequencies of oscillation of the coordinates q_1 and q_2. For the rate of variation of ϑ with the time we have further

$$\frac{d\vartheta}{dt} = \dot{q}_3 = \frac{\partial E}{\partial p_3} = f(q_1, q_2, p_1, p_2, p_3) =$$

$$= \pm\, \Sigma\, C''_{\tau_1, \tau_2} \cos 2\pi\{(\tau_1\omega_1 + \tau_2\omega_2)t + c''_{\tau_1, \tau_2}\},$$

where the two signs correspond to a rotation of the particle in the direction of increasing and decreasing q_3 respectively, and are introduced to separate the two types of symmetrical motions corresponding to these directions. This gives

$$\pm\, \vartheta = 2\pi\omega_3 t + \Sigma\, C'''_{\tau_1, \tau_2} \cos 2\pi\{(\tau_1\omega_1 + \tau_2\omega_2)t + c'''_{\tau_1, \tau_2}\}, \qquad (33)$$

where the positive constant $\omega_3 = C''_{0,0}/2\pi$ is the mean frequency of rotation round the axis of symmetry of the system. Considering now the displacement of the particle in rectangular coordinates x, y and z, and taking as above the axis of symmetry as z-axis, we get from (32) and (33) after a simple contraction of terms

$$x = \varrho \cos \vartheta = \quad \sum D_{\tau_1, \tau_2} \cos 2\pi\{(\tau_1\omega_1 + \tau_2\omega_2 + \omega_3)t + d_{\tau_1, \tau_2}\}$$

and (34)

$$y = \varrho \sin \vartheta = \pm \sum D_{\tau_1, \tau_2} \sin 2\pi\{(\tau_1\omega_1 + \tau_2\omega_2 + \omega_3)t + d_{\tau_1, \tau_2}\},$$

where the D's and d's are new constants, and the summation is again to be extended over all positive and negative values of τ_1 and τ_2.

From (32) and (34) we see that the motion in the present case may be considered as composed of a number of linear harmonic vibrations parallel to the axis of symmetry and of frequencies equal to the absolute values of $(\tau_1\omega_1+\tau_2\omega_2)$, together with a number of circular harmonic motions round this axis of frequencies equal to the absolute values of $(\tau_1\omega_1+\tau_2\omega_2+\omega_3)$ and possessing the same direction of rotation as that of the moving particle or the opposite if the latter expression is positive or negative respectively. According to ordinary electrodynamics the radiation from the system would therefore consist of a number of components of frequency $|\tau_1\omega_1+\tau_2\omega_2|$ polarised parallel to the axis of symmetry, and a number of components of frequencies $|\tau_1\omega_1+\tau_2\omega_2+\omega_3|$ and of circular polarisation round this axis (when viewed in the direction of the axis). On the present theory we shall consequently expect that in this case only two kinds of transitions between the stationary states given by (22) will be possible. In both of these n_1 and n_2 may vary by an arbitrary number of units, but in the first kind of transition, which will give rise to a radiation polarised parallel to the axis of the system, n_3 will remain unchanged, while in the second kind of transition n_3 will decrease or increase by one unit and the emitted radiation will be circularly polarised round the axis in the same direction as or the opposite of that of the rotation of the particle respectively.

In the next Part we shall see that these conclusions are supported in an instructive manner by the experiments on the effects of electric and magnetic fields on the hydrogen spectrum. In connection with the discussion of the general theory, however, it may be of interest to show that the formal analogy between the ordinary theory of radiation and the theory based on (1) and (22), in case of systems possessing an axis of symmetry, can be traced not only with respect to frequency relations but also by considerations of *conservation of angular momentum.* For a conditionally periodic system possessing an axis of symmetry the angular momentum round this axis is, with the above choice of coordinates, according to (22) equal to $I_3/2\pi = n_3h/2\pi$. If therefore, as assumed above for a transition corresponding to an emission of linearly polarised light, n_3 is unaltered, it means that the angular momentum of the system remains unchanged, while if n_3 alters by one unit, as assumed for a transition corresponding to an emission of circularly polarised light, the angular momentum will be altered by $h/2\pi$. Now it is easily seen that the ratio between this amount of angular momentum and the amount of energy $h\nu$ emitted during the transition is just equal to the ratio between the amount of angular momentum and energy possessed by the radiation which according to ordinary electrodynamics would be emitted by an electron rotating in a circular orbit in a central field of force. In fact, if α is the radius of the orbit, ν the frequency of revolution and F the force of reaction due to the electromagnetic field of the radiation, the amount of energy and of angular momentum round an axis through the centre of the field perpendicular to the plane of the orbit, lost by the electron in unit of time as a consequence of the radiation, would be equal to $2\pi\nu\alpha F$ and αF respectively. Due to the principles of conservation of energy and of angular momentum holding in ordinary electrodynamics, we should therefore expect that the ratio between the energy and the angular momentum of the emitted radiation would be $2\pi\nu$, [1] but this is seen to be equal to the ratio between the energy $h\nu$ and the angular momentum $h/2\pi$ lost by the system considered above during a transition for which we have assumed that the radiation is circularly polarised. This agreement would seem not only to support the validity of the above considerations but also to offer a direct support, independent of the equations (22), of the assumption that, *for an atomic system possessing an axis of symmetry, the total angular momentum round this axis is equal to an entire multiple of* $h/2\pi$.

A further illustration of the above considerations of the relation between the quantum theory and the ordinary theory of radiation is obtained if we consider a conditionally periodic system subject to the *influence of a small perturbing field of force.* Let us assume that the original system allows of separation of variables in a certain set of

[1]) Comp. K. Schaposchnikow, Phys. Zeitschr. **15** (1914) 454.

coordinates $q_1, \ldots q_s$, so that the stationary states are determined by (22). From the necessary stability of the stationary states we must conclude that the perturbed system will possess a set of stationary states which only differ slightly from those of the original system. In general, however, it will not be possible for the perturbed system to obtain a separation of variables in any set of coordinates, but if the perturbing force is sufficiently small the perturbed motion will again be of conditionally periodic type and may be regarded as a superposition of a number of harmonic vibrations just as the original motion. The displacements of the particles in the stationary states of the perturbed system will therefore be given by an expression of the same type as (31) where the fundamental frequencies ω_k and the amplitudes $C_{\tau_1, \ldots \tau_s}$ may differ from those corresponding to the stationary states of the original system by small quantities proportional to the intensity of the perturbing forces. If now for the original motion the coefficients $C_{\tau_1, \ldots \tau_s}$ corresponding to certain combinations of the τ's are equal to zero for all values of the constants $\alpha_1, \ldots \alpha_s$, these coefficients will therefore for the perturbed motion, in general, possess small values proportional to the perturbing forces. From the above considerations we shall therefore expect that, in addition to the main probabilities of such transitions between stationary states which are possible for the original system, there will for the perturbed system exist small probabilities of new transitions corresponding to the above mentioned combinations of the τ's. Consequently we shall expect that the effect of the perturbing field on the spectrum of the system will consist partly in a small displacement of the original lines, partly in the appearance of new lines of small intensity.

A simple example of this is afforded by a system consisting of a particle moving in a plane and executing harmonic vibrations in two perpendicular directions with frequencies ω_1 and ω_2. If the system is undisturbed all coefficients C_{τ_1, τ_2} will be zero, except $C_{1,0}$ and $C_{0,1}$. When, however, the system is perturbed, for instance by an arbitrary small central force, there will in the Fourier expressions for the displacements of the particle, in addition to the main terms corresponding to the fundamental frequencies ω_1 and ω_2, appear a number of small terms corresponding to frequencies given by $\tau_1\omega_1 + \tau_2\omega_2$ where τ_1 and τ_2 are entire numbers which may be positive as well as negative. On the present theory we shall therefore expect that in the presence of the perturbing force there will appear small probabilities for new transitions which will give rise to radiations analogous to the socalled harmonics and combination tones in acoustics, just as it should be expected on the ordinary theory of radiation where a direct connection between the emitted radiation and the motion of the system is assumed. Another example of more direct physical application is afforded by the effect of an external homogeneous electric field in producing new spectral lines. In this case the potential of the perturbing force is a linear function of the coordinates of the particles and, whatever is the nature of the original system, it follows directly from the general theory of perturbations that the frequency of any additional term in the expression for the perturbed motion, which is of the same order of magnitude as the external force, must correspond to the sum or difference of two frequencies of the harmonic vibrations into which the original motion can be resolved. With applications of these considerations we will meet in Part II in connection with the discussion of Sommerfeld's theory of the fine structure of the hydrogen lines and in Part III in connection with the problem of the appearance of new series in the spectra of other elements under the influence of intense external electric fields.

As mentioned we cannot without a more detailed theory of the mechanism of transition between stationary states obtain quantitative information as regards the general question of the intensities of the different lines of the spectrum of a conditionally periodic system given by (26), except in the limit where the n's are large numbers, or in such special cases where for all values of the constants $\alpha_1, \ldots \alpha_s$ certain coefficients $C_{\tau_1, \ldots \tau_s}$ in (31) are equal to zero. From considerations of analogy, however, we must expect that it will be possible also in the general case to obtain an *estimate of the intensities* of the different lines in the spectrum by comparing the intensity of a given line, corresponding to a transition between two stationary states characterised by the numbers $n'_1, \ldots n'_s$ and $n''_1, \ldots n''_s$ respectively, with the intensities of the radiations of frequencies $\omega_1(n'_1 - n''_1) + \ldots + \omega_s(n'_s - n''_s)$ to be expected on ordinary electrodynamics from the motions in these states; although of course this estimate becomes more uncertain the smaller the values for the n's are. As it will be seen from the applications mentioned in the following Parts this is supported in a general way by comparison with the observations.

Related papers

3a N. Bohr, *On the Constitution of Atoms and Molecules*. Phil. Mag. [6] **26** (1913) 1; (1913) 476; (1913) 857.

3b N. Bohr, *On the Effect of Electric and Magnetic Fields on Spectral Lines*. Phil. Mag. [6] **27** (1914) 506.

3c N. Bohr, *On the Series Spectrum of Hydrogen and the Structure of the Atom*. Phil. Mag. [6] **29** (1915) 332.

3d N. Bohr, *On the Quantum Theory of Radiation and the Structure of the Atom*. Phil. Mag. [6] **30** (1915) 394.

3e H. A. Kramers, *Intensities of Spectral Lines*. Ph.D. Thesis. Kgl. Danske Vid. Selsk. 8e række VI, **3** (1919) 327.

3f N. Bohr, *Über die Serienspektra der Elemente*. Z. Phys. **2** (1920) 423. Received July 21, 1920.

3g H. A. Kramers, *Über den Einfluss eines elektrischen Feldes auf die Feinstruktur der Wasserstofflinien*. Z. Phys. **3** (1920) 199. Received October 1, 1920.

3h N. Bohr, *Über die Anwendung der Quantentheorie auf die Atomstruktur I: Die Grundpostulate der Quantentheorie*. Z. Phys. **13** (1922) 117. Received Nov. 15, 1922. English translation: *On the application ... I. The fundamental postulates of quantum theory*. Publications Cambridge Phil. Soc. 1923.

Received February 8, 1921

4

THE QUANTUM-THEORETICAL INTERPRETATION OF THE NUMBER OF DISPERSION ELECTRONS

R. LADENBURG

According to classical electron theory, the absorption of isolated spectral lines is characterised above all by the number $\mathfrak{N}$ of dispersion electrons per unit volume, the 'dispersion constant', apart from the frequency ν and the damping coefficient ν' in Voigt's [1]) notation. In quantum theory, on the other hand, the absorption is produced by a transition of the molecules from a state i to a state k, and the strength of the absorption is determined by the probability of such transitions $i \rightarrow k$. This result follows from Einstein's well-known considerations [2]) and has recently been used by Füchtbauer [3]) in calculations connected with his absorption measurements on alkali vapours. Einstein's above-mentioned theory (derivation of Planck's radiation formula for Bohr atoms) now leads to an important relation between this probability factor and the probability of the spontaneous (reverse) transition from state k to state i. It will be shown that it is this latter probability, multiplied by the ratio of the statistical weights of the two quantum states occurring in Einstein's relation, which takes the place of the dispersion constant $\mathfrak{N}$, so that it is directly obtainable from absorption measurements, as well as from emission measurements and those of anomalous dispersion and magnetic rotation at and near the spectral lines. For these phenomena are all essentially determined by the same dispersion constant $\mathfrak{N}$. On the basis of existing measurements of the anomalous dispersion, etc., for different lines of a given series, one can therefore reach important conclusions on the probability of different

Editor's note. This paper was published as Zs. f. Physik **4** (1921) 451–468. It was signed 'Breslau, Physikal. Institut der Universität, 5th February, 1921.'

[1]) W. Voigt, Magneto-Elektrooptik (B. G. Teubner, Leipzig, 1908) p. 103 ff.

[2]) A. Einstein, Phys. Zs. **18** (1917) 121. [This volume, paper **1**.]

[3]) Chr. Füchtbauer, Phys. Zs. **21** (1920) 322.

spontaneous transitions and its connection with the Bohr Theory; this is the subject of the present paper.[1])

The desired relation between the dispersion constant and the probability is obtained directly by calculating the emitted and absorbed energy of N molecules in thermal equilibrium, from classical electron theory on the one hand, and quantum theory on the other.

Let us assume that there are N molecules per unit volume, and $\mathfrak{N}$ oscillatory electrons having natural frequency ν_0. Then, according to classical electron theory, the energy radiated by them per second is given by

$$J_{\mathrm{El}} = \frac{\overline{U}\mathfrak{N}}{\tau},$$

where

$$\tau = \frac{3mc^3}{8\pi^2 e^2 \nu_0^2} \tag{1a}$$

is the decay time of the oscillating electron, damped only by emission, and $\overline{U}$ is its mean energy. If the molecules are in equilibrium at radiation temperature T, and if the electrons are regarded as three-dimensional oscillators with three degrees of freedom, then the following well-known relation between $\overline{U}$ and the radiation density u_0 exists, according to Planck [2]),

$$\overline{U} = \frac{3c^3}{8\pi\nu_0^2} u_0,$$

and

$$J_{\mathrm{El}} = \frac{3c^3}{8\pi\nu_0^2\tau} \mathfrak{N} u_0, \tag{2}$$

i.e. from (1a)

$$= \frac{\pi e^2}{m} \mathfrak{N} u_0.$$

This is equal to the energy A absorbed per second in the radiation field. The same result is obtained if, in addition to radiation damping, other damping factors are assumed to act (e.g. collisions in the Lorentz

[1]) I would like, here too, to express my sincere thanks to Miss H. Kohn and Mr. F. Reiche for much useful advice.

[2]) For the factor 3, cf. F. Reiche, Ann. d. Phys. **58** (1919) 693.

Theory) or if the Doppler effect is taken into account[1]). But it has to be presupposed here that the layer thickness under consideration is sufficiently small, so that the emitted or absorbed energy can still be regarded as proportional to it.

In the quantum-theoretical treatment of Bohr and Einstein, the emission of radiation by molecules is produced in two ways: first, by means of spontaneous transitions from state k to state i, and secondly by the interactions with the radiation field present in the space under consideration. The first type of transition is characterised by the probability coefficient $a_{k\to i}$, the second by $b_{k\to i}$. Both coefficients are assumed to be temperature-independent. For each transition, an energy $h\nu_{ik}$ is emitted, so that the total energy emitted per second is given by

$$J_Q = h\nu_{ik} N_k(a_{k\to i} + b_{k\to i} u_{ik}), \tag{3a}$$

where N_k is the number of molecules (out of a total N) in state k; this is related to the number N_i of molecules in state i which are in thermal equilibrium at temperature T, by means of the equation

$$\frac{N_k}{N_i} = \frac{g_k[\exp(-E_k/\mathfrak{k}T)]}{g_i[\exp(-E_i/\mathfrak{k}T)]}. \tag{4}$$

Here g_i is a temperature-independent quantity which characterises the state i and is called its statistical weight[2]); E_i is the energy in state i. One can correspondingly calculate, according to Einstein, the absorbed energy[3])

$$A_Q = h\nu_{ik} N_i b_{i\to k} u_{ik}, \tag{3b}$$

where $b_{i\to k}$ measures the probability for transitions $i \to k$. According to Einstein (loc. cit.) the following relations have to hold[4]),

$$g_i b_{ik} = g_k b_{ki} \tag{5a}$$

and

$$a_{ki} = b_{ki} \frac{8\pi h\nu_{ik}^3}{c^3}, \tag{5b}$$

[1]) Cf. R. Ladenburg, Verh. d. D. Phys. Ges. **16** (1914) 765. Detailed (but unpublished) calculations were carried out in this connection by F. Reiche in the years 1913 to 1915.

[2]) Cf. M. Planck, Verh. d. D. Phys. Ges. **17** (1915) 407.

[3]) Cf. also Chr. Füchtbauer, Phys. Zs. **21** (1920) 322.

[4]) We are introducing here the abbreviated notation a_{ik} for $a_{i\to k}$, b_{ik} for $b_{i\to k}$, etc. The dimensions of a are sec^{-1}, those of b, $cm^3\,sec^{-2}\,erg^{-1}$.

i.e.

$$b_{ik} = a_{ki} \frac{g_k}{g_i} \frac{c^3}{8\pi h \nu_{ik}^3}, \tag{5c}$$

so that in thermal equilibrium u tends to infinity with T, and so that in the limit of small frequencies the Rayleigh radiation law is satisfied. Furthermore, it follows from the basic assumption of Bohr's Theory [1]) that

$$E_k - E_i = h\nu_{ik}.$$

Equating (3a) and (3b) and using relations (4) and (5), we obtain Planck's radiation formula. At the same time, we get for the quantised emission or absorption energy the expression

$$J_Q = A_Q = N_i \frac{g_k}{g_i} a_{ki} \frac{c^3}{8\pi\nu_{ik}^2} u_{ik}. \tag{6}$$

The relations (5) must also be valid outside the black body cavity, at least as long as collision effects can be neglected [2]), since a and b are temperature-independent properties of the molecules. In any event, for the case of thermal equilibrium, collisions cannot affect equations (5) or (6). In addition, it should be noted that equation (4) also tells us how the (generally small) percentage of molecules in state k provides for just the required number of transitions to produce the black body radiation [3]), and how thermal radiation can also be produced in gases in the absence of chemical reactions or other such processes, purely by the mutual irradiation of different space regions.

Eq. (2) can be looked upon as the definition for the experimentally determinable quantity $\mathfrak{N}$, which has of course no definite meaning in

[1]) Therefore the same basic assumptions are needed for Einstein's derivation as for Bohr's Theory, and no others; cf. R. Ladenburg, Jahrb. d. Rad. u. El. **17** (1921) 101ff. (Report on Planck's action constant h.)

[2]) Collisions can, however, produce a change in the values of the a_{ik} in the presence of strong electric fields, in which case the selection rules are violated and transitions are produced which would otherwise be forbidden.

[3]) E.g. for $\nu = 0.5 \times 10^{15}$ (corresponding to the D-lines) and temperatures $T = 300°$, 2000°, 3000°, 4000°, the exponent becomes

$$\frac{E_k - E_i}{\mathfrak{k}T} = \frac{h\nu}{\mathfrak{k}T} = 80,\ 12,\ 8,\ 6,$$

and the ratio

$$\frac{N_k g_k}{N_i g_i} \sim 3 \times 10^{-35},\ 6 \times 10^{-6},\ 3.3 \times 10^{-4},\ 2.5 \times 10^{-3}.$$

quantum theory. If we identify eq. (2) with eq. (6), we obtain the quantum-theoretical interpretation for the dispersion constant $\mathfrak{N}$, i.e.[1])

$$\mathfrak{N} \equiv \frac{Am}{\pi e^2 u_0} = N_i h\nu_{ik} b_{ik} \frac{m}{\pi e^2} = N_i \frac{g_k}{g_i} a_{ki} \frac{mc^3}{8\pi^2 e^2 \nu_{ik}^2}. \tag{7}$$

The expression $mc^3/8\pi^2e^2\nu_{ik}^2$ is seen to be $\frac{1}{3}$ of the decay time of a classical linear oscillator, cf. eq. (1a).

So far, only the single transition $i \rightleftarrows k$ has been considered. But it is almost always possible for a molecule to suffer transitions from state k to different states i, h, g etc. (see figs. 1 and 2). In such cases the fundamental equations (5) must, according to Einstein, hold separately for each of the transitions $i \rightleftarrows k$, $h \rightleftarrows k$, etc., because statistical equilibrium must exist for each such elementary process. If a_k is the probability that the state k decays at all, the following equation holds for the spontaneous decay of the N_k molecules

$$-\Delta N_k = N_k^0 a_k \Delta t. \tag{8a}$$

The reciprocal of a_k is therefore also called the mean life, in analogy to the phenomena of radioactive decay.[2]) Since the different possible transitions represent mutually exclusive events in the sense of probability theory, we have

$$a_k = a_{ki} + a_{kh} + a_{kg} + \dots . \tag{8b}$$

We shall attempt to deduce from absorption measurements, etc., the magnitude of the various transition probabilities, which can at the same time be calculated approximately from Bohr's correspondence principle in very simple cases.

In electron theory the number $\mathfrak{N}$ is basic not only for the absorption and emission of spectral lines, but also for the phenomena of anomalous dispersion and magnetic rotation in the neighbourhood of the spectral lines. It is therefore to be expected that these latter phenomena, too, will essentially be determined by the various transition probabilities. Admittedly there exists as yet no quantum theory of anomalous dispersion. But one should be able to explain these phenomena by

[1]) Here, ν_0 and u_0 have been identified with ν_{ik} and u_{ik} resp. The first part of the equation thus obtained can also be taken directly from Füchtbauer's calculations (loc. cit.); his factor P is equal to b_{ik}/c.

[2]) Cf. O. Stern and M. Volmer, Phys. Zs. **20** (1919) 183.

assuming the molecules to suffer a change in electric moment (produced when the incident radiation brings about a transition from i to state k) and thus to interact with the radiation field. The electric fields of the neighbouring molecules or other causes (perhaps, e.g., the electric field of the incident light itself) will produce modified states $i', i'', \ldots k', k'', \ldots$ in addition to i and k (Stark effect). The transitions between them are closely connected with the emission and absorption of frequencies $\nu \gtrless \nu_{ik}$. In addition to absorption this also leads to a change in the velocity with which radiation of other frequencies ν is propagated. Such a change will be all the more pronounced, the closer ν is to ν_{ik}, since the majority of atoms absorb frequency ν_{ik}, i.e. we have anomalous dispersion. As in Planck's dispersion theory, the absorption here certainly consists of a transformation of the incident radiation into diffuse radiation which is dispersed in all directions by the molecules. The rôle of the radiation damping ($1/\tau$) is taken over by the decay constant a_k. In the absence of other sources of damping, its magnitude thus determines the width of the absorbed and emitted spectral lines. That the decay constant is equal to the classical radiation damping is easy to demonstrate on the basis of the correspondence principle [1]), or e.g. for the case of a resonator with a single quantum $h\nu$ [2]).

Without going into the quantum-theoretical calculation of anomalous dispersion and magnetic rotation, we shall nevertheless, in what follows, apply eq. (7) to the $\mathfrak{R}$-values measured in this way. To obviate possible objections to this procedure, I have initiated at this Institute magnetic rotation experiments on sodium vapour *in vacuo*. These measurements are to resolve the question whether the $\mathfrak{R}$-values determined in this way agree with those obtained from absorption measurements. Reasonable agreement has already been obtained from Gouy's measurements of the absolute intensity and from Senftleben's magnetic rotation measurements on the D-lines in sodium flames (see pp. 153–154).

[1]) Cf. Stern and Volmer, Phys. Zs. **20** (1919) 183.

[2]) The same result will be derived below from the experimental data, without direct appeal to the correspondence principle, for the case of the resonance state of the alkalis.

$\mathfrak{N}$-value ratios for different series terms

The existing measurements on different spectral lines of a series always show a decrease in the $\mathfrak{N}$-values with increasing term number, which can now be interpreted in the light of Bohr's theory of series spectra [1].

(a) *The Balmer series in hydrogen.* The values for the Balmer series are of particular interest, since the relevant electron orbits may be considered as accurately known. Hydrogen does not absorb the Balmer lines in its normal state, as is well known, but only when it is under strong (electrical) excitation [2]. Evidently the normal molecules need first to be converted into atoms with orbits of quantum number 2, which are the limit orbits of the Balmer series according to Bohr. In the same state, hydrogen exhibits [3] small anomalous dispersion and magnetic rotation, which has so far only been established for the first and second terms (H_α and H_β). The ratio of the $\mathfrak{N}$-values could only be determined approximately, since the effects for H_β are still very much smaller than those for H_α. From the dispersion experiments, the author [2] obtained the rough value 4.5 for $\mathfrak{N}_\alpha/\mathfrak{N}_\beta$ (lying between 6 and 3); from experiments on magnetic rotation it could only be deduced that $\mathfrak{N}_\alpha/\mathfrak{N}_\beta > 2.3$, which is compatible with the first value. From relation (7),

$$\mathfrak{N} = N_i \frac{g_k}{g_i} a_{ki} \frac{mc^3}{8\pi^2 e^2 \nu_{ik}^2}.$$

It therefore follows that

$$\frac{b_{ik}^\alpha}{b_{ik}^\beta} = \frac{g_\alpha a_\alpha}{g_\beta a_\beta} \approx 4.5 \left(\frac{\lambda_\beta}{\lambda_\alpha}\right)^2 \approx 2.5,$$

since N_i has the same value for both lines.

As is well known, the hydrogen lines are made up of different components which always merged into a single broadened line in the

[1] On the classical explanation of these phenomena, see Cl. Schaefer, Ann. d. Phys. **41** (1913) 868.

[2] G. D. Liveing and J. Dewar, Proc. Roy. Soc. **35** (1883) 74; A. Pflüger, Ann. d. Phys. **24** (1907) 515; R. Ladenburg, Verh. d. D. Phys. Ges. **10** (1908) 550.

[3] R. Ladenburg und St. Loria, Verh. d. D. Phys. Ges. **10** (1908) 858; R. Ladenburg, Phys. Zs. **10** (1909) 497; **12** (1911) 10; Ann. d. Phys. **38** (1912) 249.

experiments [1]). Using corresponding average values, one can put

$$g_n = \tfrac{1}{2}n(n+1),$$

where n is the principal quantum number of the state [2]) and find

$$g_\alpha / g_\beta = \tfrac{3}{5},$$

or

$$a_\alpha / a_\beta \approx 4,$$

i.e. the probability for transitions H_α is about four times as great as that for H_β. It follows from the Bohr–Sommerfeld theory that transitions from state 3 into state 2 are about four times as probable as those for state 4 into state 2 (see fig. 1, where the series terms are represented in the usual way as energy levels [3])). This is evidently related to the fact that state 4 can decay in many more different ways than state 3. The latter only allows transitions to 2 or 1 (Balmer and Lyman series), whereas the former can also have transitions to 3 with extremely small changes in energy (corresponding to the first term of the infra-red Paschen series). One could thus estimate the relative probabilities for the different transitions of a state and compare them with results obtained from Bohr's correspondence principle [2]); this could be done provided one could establish that the total decay probability for state 4 (the decay constant a_4) was either equal to, or smaller than, a_3.

The equality could be deduced from Wien's [4]) experiments, the inequality from J. Stark's [5]) experiments on canal rays. Both research

[1]) From Bohr's theory, and the validity of the selection rules, it follows that each line of the Balmer series is produced in three different ways, as is shown in fig. 1, whose energy differences are very close (e.g. H_α is produced by the transitions 2.5s–3p, 2p–3.5s, 2p–3d). The weights of the individual states are known here, and the relative transition probabilities can be calculated approximately from the correspondence principle.

[2]) A. Sommerfeld, Atombau und Spektrallinien (Braunschweig, 1921) 415.

[3]) Cf. N. Bohr, Zs. f. Phys. **2** (1920) 434; A. Sommerfeld, Phys. Zs. **21** (1920) 619; W. Grotrian, Phys. Zs. **21** (1920) 638. To save space, the level for the first term has been drawn too high. With the scale to which the figure is drawn, the differences in the s-, p-, d-, b- levels corresponding to the fine structure of the hydrogen lines could of course not be shown.

[4]) W. Wien, Ann. d. Phys. **60** (1919) 592.

[5]) J. Stark, Ann. d. Phys. **49** (1916) 731.

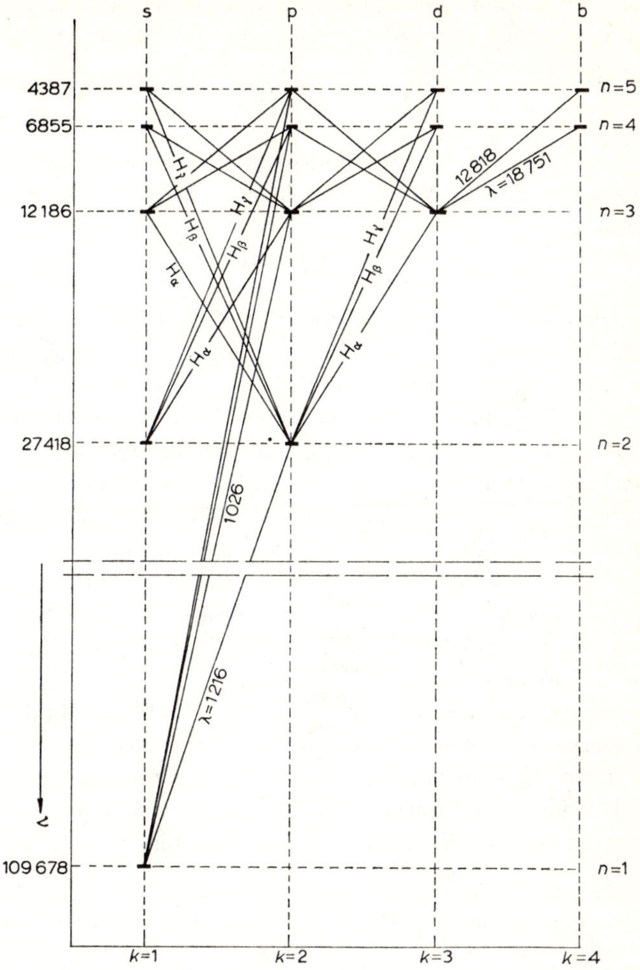

Fig· 1. Hydrogen.

workers observed a measurable decay in time of the Balmer lines emitted by radiating hydrogen atoms in motion. But whereas Stark came to the conclusion that the radiation time for H_γ and H_δ was appreciably greater than that for H_α and H_β, Wien concluded that, within the experimental error (barely 10%), the radiation times for H_α, H_β and H_γ were of the same magnitude, and nearly equal to the classical decay time for H_α (for its definition, see p. 140). According to quantum theory, the radiation intensity of initially excited free atoms is given by

$$J_Q = N_k h\nu_{ik} a_{ki},$$

from eq. (3a). The decrease in intensity with time originates from the decrease in the number N_k of atoms in state k, which is given by the exponential law

$$N_k^t = N_k^0 \, e^{-akt}$$

from eq. (8a), p. 143, and also agrees with Wien's measurements. The alternative conclusions to be drawn are then: (i) the decay constant a_n decreases with increasing quantum number n (Stark)[1]; (ii) it has the same value for the different states (Wien; Seeliger[2]) however believes that one can obtain a change in the radiation time also from Wien's measurements, in the spirit of Stark's experiments, if one takes the data from each separate exposure instead of using Wien's average values). When the decay constant for states with larger quantum numbers is smaller than for those with smaller quantum numbers, one should expect, *a fortiori*, that the probability for the specific transition H_β is smaller than that for H_α – as we had deduced for the case of anomalous dispersion for eq. (7).

(b) *The principal alkali series.* Wood[3]) was probably the first to note in his observations on sodium vapour in a tube, that the anomalous dispersion for the D-lines far outweighs that for the next lines of the principal series (3303 and 2853 Å). Bevan[4]), who made extensive investigations on alkali vapours, found the following ratios of $\mathfrak{N}$-

[1]) Thus the value found for H_β, e.g., not only depends on the frequency associated with this line (4→2), but also on the frequencies for transitions 4→1 and 4→3, since the decay constant a_4 of the initial state is equal to the sum of the probabilities of the individual transitions, according to eq. (8b) ($a_4 = a_{41} + a_{42} + a_{43}$).

[2]) R. Seeliger, Jahrb. d. Radioakt. u. Elektron. **16** (1920) 415. Seeliger there discusses different models for the time behaviour of the electron jumps. We shall assume, following Einstein, that the time for the transition $k \to i$ is immeasurably small compared with the time which the atom spends in state k. As opposed to Seeliger, we make no assumption here about the decay time of the emitted radiation.

[3]) R. W. Wood, Phil. Mag. (6) **8** (1904) 293.

[4]) P. V. Bevan, Proc. Roy. Soc. **84** (1910) 209; **85** (1911) 58.

values: for the first three sodium lines[1] $1:\frac{1}{81}:\frac{1}{480}$; for the first four potassium lines $1:\frac{1}{192}:\frac{1}{2060}:\frac{1}{6620}$ and from other experiments the ratios $1:\frac{1}{161}:\frac{1}{1340}$. For the first six rubidium lines, Bevan's measurements give the corresponding ratios $1:\frac{1}{104}:\frac{1}{575}:\frac{1}{1660}:\frac{1}{1850}:\frac{1}{3000}$. Here too, the decrease in the value of $\mathfrak{N}$ is by far the most pronounced for the transition from the first to the second term and then diminishes rapidly. Apart from that, Bevan considers the measurements for the fifth and sixth terms to be very inaccurate, so that no significance should be attached to the irregularity in the rate of decrease there. It also follows from the absorption measurements of Füchtbauer and Hofmann[2]) that the $\mathfrak{N}$-value for the second caesium line (4554 Å) is about six times that for the third (3877 Å).

There is therefore no doubt that the $\mathfrak{N}$-values decrease very rapidly with the term number, particularly so for the first terms.

TABLE 1

	$m=2$	$m=3$	$m=4$	$m=5$	$m=6$	$m=7$
Na	1	$\frac{1}{26}$	$\frac{1}{110}$	—	—	—
K	1	$\frac{1}{50}$	$\frac{1}{225}$	$\frac{1}{1100}$	—	—
Rb	1	$\frac{1}{31}$	$\frac{1}{102}$	$\frac{1}{305}$	$\frac{1}{318}$	$\frac{1}{490}$
Cs	—	$(\frac{1}{35})$	$\frac{1}{150}$	—	—	—

Table 1 shows roughly the behaviour of the product of statistical weight and decay constant for the principal alkali series with increasing term number, using eq. (7) and taking the decrease in wave length into account.

The differences in corresponding values for the different metals are consistent with the relatively large experimental errors, so that these values are possibly all equal. Since the statistical weight cannot decrease with increasing quantum number, it follows that the proba-

1) Bevan's values of the constant m in Sellmeier's formula

$$n^2 - 1 = \Sigma \frac{m_i \lambda^2}{\lambda^2 - \lambda_i^2}$$

have to be divided by λ_i^2 to obtain our $\mathfrak{N}$-values. Here, the values given are always those for the more intense line of the doublet (shorter wave length). About the separation of the doublet lines, see (c).

2) Chr. Füchtbauer and W. Hofmann, Ann. d. Phys. **43** (1914) 96.

bility of corresponding transitions for the higher terms decreases as much as is shown in the table, or even more so.

To gain further insight, we look again at the representation of the individual terms (see fig. 2, drawn for Na [1])); for the alkali series, the s-, p-, d-, b- terms of equal principal quantum number (e.g. 3.5s, 3p, 3d) differ markedly from each other: their values decrease in that order. From state 2p only a transition to 1.5s is possible, and this distinguishes it from all the others. From 3p, apart from the transition to 1.5s (3302.₅ Å), transitions to 2.5s and 3d are also possible. These correspond to a very much smaller energy difference (λ=2.2084 μ and 9.085 μ, as observed by Paschen [2]) in an arc discharge). The transition 3p→2p has not so far been observed. It would contradict Bohr's correspondence principle (in Sommerfeld's terminology: constancy of the azimuthal quantum number). From 4p, apart from the transition to 1.5s (third term in principal series, λ=2853 Å), transitions to 2.5s, 3.5s, 3d, 4d are possible, which again correspond to much smaller energy differences, i.e. longer wavelengths. Transitions to 3p, 2p, and 4b violate the correspondence principle. The higher the term number, therefore, the greater the number of individual states into which transitions are possible. In particular, the difference between the 2p- and 3p-states (which form the starting points for the first and second terms in the principal series resp.) is the greatest, since only one kind of transition is possible for 2p, whereas three distinct transitions are possible from 3p. Thus there corresponds to each individual type of transition a probability which decreases with increasing term number – unless it were balanced by a corresponding increase in the decay constant (the probability for the sum of all transitions from a given state), which is extremely unlikely. This decrease must be largest from the first to the second term of the series, just as is demonstrated by the dispersion experiments.

That Bohr's concepts about the origin of the individual series spectral lines are correct, is demonstrated by Strutt's [3]) experiments. He illuminated sodium vapour with the second term of the principal series, λ3303, and observed that the fluorescence consisted apparently only to a small extent of light of this wavelength and that in addition

[1]) Of the two p-terms, only p_2 has been drawn in.

[2]) F. Paschen, Ann. d. Phys. **27** (1908) 537; **33** (1910) 717; see also B. Dunz, Diss. Tübingen 1911.

[3]) J. Strutt, Proc. Roy. Soc. (A) **96** (1920) 272.

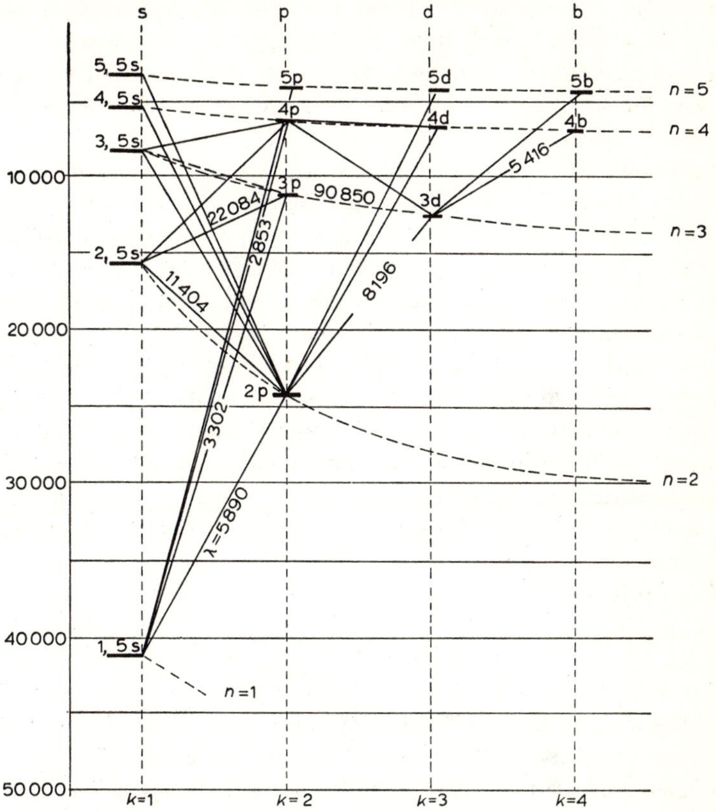

Fig. 2. Sodium.

the D-lines appeared very clearly. In other words, indirect transitions via 2.5s and 3d happen frequently [1]). Bohr [2]) even deduced from Strutt's experiments that the transitions 3p→1.5s (λ3303) are less frequent than those from 3p to 2.5s, as would be expected from the correspondence principle. The same result would follow from the above-mentioned dispersion experiments on different alkalis: one

[1]) One would certainly expect from the series picture to see the lines 2.208, 9.085, 1.1404, 0.8196 μ, but these have not as yet been observed in fluorescence experiments, as far as I know. Strutt did not look for these lines in the above experiment, but tried (without success) to obtain the line 3p–2p, which is forbidden by the selection rules for this experiment.

[2]) N. Bohr, Zs. f. Phys. **2** (1920) 438.

would have to assume that the decay constant a_{3p} is equal to, or only slightly less than, a_{2p}, in analogy to Stark's and Wien's experiments on the hydrogen series. For the probability for the transitions 3p→1.5s is, on average, only 1/35 of that for the transition 2p→1.s5.

(c) *The doublets of the alkali principal series.* For the $\mathfrak{N}$-value ratios for the doublets of the alkali principal series, there are intensity and absorption measurements available as well determinations of dispersion and magnetic rotation. In particular, there is a large number of investigations on the two D-lines. Combining the results of the various measurements [cf. table in R. Ladenburg, Zs. f. Phys. 4 (1921) 469], we obtain the integral value 2.0 for the ratio of the $\mathfrak{N}$-values, within the experimental error. This value is independent of the temperature (varied between 174° and 1800°C), vapour pressure (investigated over a range $1:10^5$) and pressure (between 10^{-6} and 3 atm). For the quantum-theoretical interpretation it can be assumed that the decay constants of the initial states corresponding to the two D-lines are equal [1]).

It then follows from eq. (7)

$$\mathfrak{N} = N_i \frac{g_k}{g_i} a_{ki} \frac{mc^3}{8\pi^2 e^2 \nu_{ik}^2},$$

that the integral property of the ratio of the $\mathfrak{N}$-values is due to the fact that the ratio of the statistical weights g is an integer, since the frequencies of the two lines only differ by 1‰. In other words, state $2p_2$ is twice as frequent as state $2p_1$ [2]). This is in agreement with the result that the Zeeman effect produces more components for the D_2- than for the D_1-line [3]).

For the higher-order series terms, the ratios are again more complicated [cf. the regular increase in the $\mathfrak{N}$-value ratios of the doublets as suspected by Füchtbauer, Phys. Zs. 21 (1920) 322].

[1]) Since for these initial states $2p_2$ and $2p_1$ only the transition to 1.5s is allowed, their probabilities are equal to the decay constants a_{2p_2} and a_{2p_1}, respectively.

[2]) On the other hand we know that Wood succeeded in exciting each state separately, by illuminating sodium vapour with the individual D-lines at low pressures.

[3]) For electron-theoretical reasons it follows from these observations [cf. W. Voigt, Ann. d. Phys. **41** (1913) 403] that the number of degrees of freedom determining the D_2-line is double that for the D_1-line.

The absolute magnitude of the $\mathfrak{N}$-values

For the quantum-theoretical interpretation of the absolute magnitude of the alkali principal series $\mathfrak{N}$-values one has to consider in the main the first series term, since only one type of transition is possible in this case. Thus the probability of these transitions is equal to the decay constant a_{2p} of the initial state 2p. Of the available experimental data only a few can be used. Some time ago I took Gouy's results on flames [1]), and in particular his measurements of the absolute brightness of the D-lines and of the sodium content of the flames, and estimated [2]) that the value of $\mathfrak{N}$ for both D-lines together was approximately half as great as the number of sodium atoms present in the flame. Gouy chose as unit the very low luminous intensity, which had the value 0.1 for the 'line absorption' of the D-lines [3]), and determined the absolute magnitude of this intensity by comparing it with that of a black body radiating at the same temperature. By means of eq. (2)

$$J_{\mathrm{El}} = \frac{\pi e^2}{m} \mathfrak{N} u_0$$

one can thus compute $\mathfrak{N}$ from Gouy's results [4]),

$$\mathfrak{N} = 5.75 \times 10^{10}.$$

For the same intensity, the observed number of sodium atoms (assuming total dissociation) was $N = 1.4 \times 10^{11}$, so that $\mathfrak{N}_{\mathrm{D_2+D_1}} = 0.41 N$ [5]).

A similar value is obtained from magnetic rotation measurements, which were carried out by Senftleben in his Breslau Dissertation at my instigation [6]). He found for sodium flames, whose line absorption was

1) G. L. Gouy, C.R. **88** (1879) 420; **154** (1912) 1764; cf. in particular his new summarising report, Annales de physique (Paris) (9) **13** (1920) 188.

2) R. Ladenburg, Verh. d. D. Phys. Ges. **16** (1914) 765. This is based on eq. (2) of the present paper, which follows from electron theory.

3) This is the absorption by a sodium flame of light from a second, identical, flame. Cf. R. Ladenburg and F. Reiche, Ann. d. Phys. **42** (1913) 181.

4) In the paper quoted in [2]) I found $\mathfrak{N} \cdot L = 3.5 \times 10^{11}$. For a layer thickness $L = 2.5$ cm the above result follows.

5) Since the dissociation of NaCl into atoms is presumably not complete, the ratio $\mathfrak{N}/N$ becomes slightly smaller.

6) H. Senftleben, Diss. Breslau 1915; Ann. d. Phys. **47** (1915) 949.

0.1 [1]), that

$$\varrho_{D_2+D_1} = 0.071 \times 10^{21} = 4\pi \mathfrak{N} e^2/m,$$

i.e.

$$\mathfrak{N}_{D_2+D_1} = 2.23 \times 10^{10},$$

or about half Gouy's value [2]). Unfortunately, Senftleben did not measure the intensity and sodium content of his flames.

With Gouy's value for N,

$$\mathfrak{N}_{D_2+D_1} = 1.16N.$$

A similar result was obtained by Füchtbauer and Hofman (loc. cit.) in their absorption measurements on sodium vapour in a tube at 174°C. Admittedly, the vapour pressure could only be extrapolated from existing measurements at very much higher temperatures. Their result is

$$\mathfrak{N}_{D_2} = 0.14N,$$

and since

$$\mathfrak{N}_{D_1} = \tfrac{1}{2}\mathfrak{N}_{D_2},$$

it follows that

$$\mathfrak{N}_{D_2+D_1} = 0.21N.$$

Similarly, Füchtbauer and Hofman deduced from their measurements on the second term of the caesium principal series that the $\mathfrak{N}$-value of the corresponding first term is not appreciably smaller than the number of atoms; admittedly, they had to use Bevan's rough measurements of the $\mathfrak{N}$-value ratio for the first and second terms for rubidium (see p. 149).

It is therefore obvious that this question has to be investigated more closely. From experimental results available today it follows that, on average, the value of $\mathfrak{N}_2$ of the first term of the sodium principal series is about equal to 1/5 the number of atoms. In this case,

1) According to electron theory [cf. Ladenburg and Reiche, Ann. d. Phys. **42** (1913) 181; also C.R. **158** (1914) 1788] the line absorption only depends on $\mathfrak{N}$ and possibly on the damping coefficient, and is therefore fairly independent of the particular type of flame used; apart from this, Senftleben modelled both the form of his burners as well as his flames accurately on those used by Gouy.

2) In some measure one can see in this agreement a proof for the equality of the results obtained for $\mathfrak{N}$ from emission and magnetic rotation (see p. 144).

eq. (7) takes the form

$$\mathfrak{N} = N_{1.5s} \frac{g_{2p}}{g_{1.5s}} a_{2p} \frac{\tau}{3}.$$

Here $N_{1.5s}$ can undoubtedly be taken to mean the number of all atoms, for all practical purposes: at the temperatures in question, the number of the other states is negligibly small [1]). If we therefore put

$$\mathfrak{N}_{D_2}/N_{1.5s} = \tfrac{1}{5},$$

it follows [2]) that

$$\frac{1}{a_{2p_2}} = \frac{g_{2p_2}}{g_{1.5s}} \frac{5\tau}{3}.$$

The ratio $g_{2p_2}/g_{1.5s}$ is as yet unknown. It is probably smaller, but certainly not appreciably greater, than 1. The meaning of our provisional result, when put into the language of quantum theory, is therefore that the mean life time of sodium atoms in the initial state of the first term of the principal series ($2p_2$) is of the same magnitude as the classical decay time τ, i.e. $\sim 10^{-8}$ sec. Stern and Volmer [3]) reach the same conclusion for the case of a monochromatic resonator, possessing only a single quantum $h\nu$. And indeed the atom in the 2p-state resembles a monochromatic resonator, insofar as it is only capable of emitting a single frequency. It should be noted that Wien (loc. cit.) found accurate agreement of the mean life (in our interpretation) with decay time [4]) in his measurements on the radiation time of the hydrogen line H_α in canal rays. For higher series terms, the probabilities for the individual allowed transitions are smaller than for the first term; nevertheless the decay constant, which is

[1]) Cf. footnote [3]), p. 142. To a sufficiently good approximation, the temperature can be set equal to the black body temperature of the radiation passing through the vapour. Thus the number of atoms excited from their ground state into the 2p-state is comparatively small even for an arc lamp as light source, and under normal optical conditions.

[2]) Similarly, $b_{1.5s\rightarrow 2p_2} = 4.84 \times 10^{19}$.

[3]) *Loc. cit.*, see also pp. 143–144.

[4]) Here the decay time is calculated for the frequency of the H_α-line. But one is dealing in this case with an initial state of principal quantum number 3, so that transitions to both 2 (H_α) and 1 are possible. The comparison with a monochromatic resonator is therefore no longer appropriate (cf. also footnote [1]), p. 148).

the sum of the various probabilities, and therefore also the mean life, can have the same value as for the first term [1]).

As for the absolute value of $\mathfrak{N}$, obtained from measurements of the anomalous dispersion of hydrogen for H_α, the number of molecules is indeed known accurately for this case. But N_i in eq. (7) now represents the number of excited atoms in state 2 (see fig. 1) which have to be produced from the molecules by the electric current in the first place. In other words, the observed ratio of $\mathfrak{N}$ to the number of molecules is that of the number of excited atoms in state 2 to the number of molecules. The value obtained for this from experiments with moderate condenser discharge is of order 10^{-4} and increased in proportion to the current amplitude [$V\sqrt{(C/L)}$, where V = voltage, C = capacity, L = self-inductance] which is to be expected from the above interpretation. This shows what percentage of the molecules must be in an excited state so that the various phenomena of absorption, anomalous dispersion, etc., can be observed for the lines originating from this state.

[1]) From recently published experimental results of Füchtbauer and Joos for the mercury line $\lambda 2537$ [Phys. Zs. **21** (1920) 635] it follows that

$$b_{ik} = cP = 2.4 \times 10^{18},$$

and thus from eq. (5)

$$\frac{g_k}{g_i} a_{ki} = 2.4 \times 10^7 = \frac{0.07}{\tau_{2537}},$$

i.e. b_{ik} is appreciably smaller than $1/\tau$. In the same way, Wood's investigation of the anomalous dispersion at $\lambda 2537$ [Phil. Mag. (6) **25** (1913) 433] leads to a ratio $\mathfrak{N}/N$ of about 1/30 [as calculated in my report in Beibl. **38** (1914) 240]. Therefore, according to eq. (7),

$$\frac{g_k}{g_i} a_{ki} = \frac{0.1}{\tau_{2537}}.$$

On the other hand we obtain from the above results for the D_2-line

$$\frac{g_k}{g_i} a_{ki} = 3.8 \times 10^7 = \frac{0.6}{\tau_{5890}}.$$

As far as we know at present, the only transition possible from the initial $2p_2$-state of the Hg atom is that to 1.5s, with emission of the $\lambda 2537$ line. But the existence of the three states $2p_1$, $2p_2$ and $2p_3$, and the fact that Hg is divalent, shows that atomic surface conditions are quite different here from those for the alkalis.

Summary of results

1. From measurements of emission, absorption, dispersion and magnetic rotation the number $\mathfrak{N}$ of dispersion electrons is obtained. According to quantum theory this is given by the expression

$$N_i \, a_{ki} \frac{g_k}{g_i} \frac{mc^3}{8\pi^2 e^2 \nu_{ik}^2}$$

where N_i is the number of atoms in state i, a_{ik} the probability for spontaneous transitions $k \rightarrow i$ which produce the frequency ν_{ik}, and g_k and g_i are the statistical weights of the relevant states.

2. The approximate agreement of $\mathfrak{N}$, for the first member of the alkali principal series, with the number of atoms shows that the mean life of the atom in the initial state 2p (i.e. $1/a_{2\mathrm{p}}$) is approximately equal to $3mc^3/8\pi^2 e^2 \nu^2$, i.e. that it is equal to the decay time of classical resonators with corresponding frequency (1.5s–2p). In that state the atoms can only emit this one frequency, in analogy to the case of monochromatic resonators.

3. The decrease of the value of $\mathfrak{N}$ with increasing series term number implies the decrease in the probability a_{ik}. This decrease is related to the fact that the number of allowed individual transitions increases with increasing term number.

4. The ratio $\mathfrak{N}_{D_2}/\mathfrak{N}_{D_1} = 2$ for the two D-lines implies that the state $2p_2$ is twice as probable as state $2p_1$, if one assumes the same mean life for both initial states.

Related papers

4a R. Ladenburg und F. Reiche, *Absorption, Zerstreuung und Dispersion in der Bohrschen Atomtheorie*. Naturwiss. **11** (1923) 584.

Dated January 1924

5

THE QUANTUM THEORY OF RADIATION

N. BOHR, H. A. KRAMERS AND J. C. SLATER

Introduction

In the attempts to give a theoretical interpretation of the mechanism of interaction between radiation and matter, two apparently contradictory aspects of this mechanism have been disclosed. On the one hand, the phenomena of interference, on which the action of all optical instruments essentially depends, claim an aspect of continuity of the same character as that involved in the wave theory of light, especially developed on the basis of the laws of classical electrodynamics. On the other hand, the exchange of energy and momentum between matter and radiation, on which the observation of optical phenomena ultimately depends, claims essentially discontinuous features. These have even led to the introduction of the theory of light-quanta, which in its most extreme form denies the wave constitution of light. At the present state of science it does not seem possible to avoid the formal character of the quantum theory which is shown by the fact that the interpretation of atomic phenomena does not involve a description of the mechanism of the discontinuous processes, which in the quantum theory of spectra are designated as transitions between stationary states of the atom. On the correspondence principle it seems nevertheless possible, as it will be attempted to show in this paper, to arrive at a consistent description of optical phenomena by connecting the discontinuous effects occurring in atoms with the continuous radiation field in a somewhat different manner from what is usually done. The essentially new assumption introduced in § 2 that the atom, even before a process of transition between two stationary states takes

Editor's note. This paper was published as Phil. Mag. [6] **47** (1924) 785–802. It was signed 'Institute for Theoretical Physics, Copenhagen, January 1924.' A German version appeared in Zs. f. Phys. **24** (1924) 69–87.

place, is capable of communication with distant atoms through a virtual radiation field, is due to Slater.* Originally his endeavour was in this way to obtain a harmony between the physical pictures of the electrodynamical theory of light and the theory of light-quanta by coupling transitions of emission and absorption of communicating atoms together in pairs. It was pointed out by Kramers, however, that instead of suggesting an intimate coupling between these processes, the idea just mentioned leads rather to the assumption of a greater independence between transition processes in distant atoms than hitherto perceived. The present paper is the result of a mutual discussion between the authors concerning the possible importance of these assumptions for the elaboration of the quantum theory, and may in various respects be considered as a supplement to the first part of a recent treatise by Bohr, dealing with the principles of the quantum theory, in which several of the problems dealt with here are treated more fully.†

1. Principles of the quantum theory

The electromagnetic theory of light not only gives a wonderfully adequate picture of the propagation of radiation through free space, but has also to a wide extent shown itself adapted for the interpretation of the phenomena connected with the interaction of radiation and matter. Thus a general description of the phenomena of emission, absorption, refraction, scattering, and dispersion of light may be obtained on the assumption that the atoms contain electrified particles which can perform harmonic oscillations round positions of stable equilibrium, and which will exchange energy and momentum with the radiation fields according to the classical laws of electrodynamics. On the other hand, it is well known that these phenomena exhibit a number of features which are contradictory to the consequences of the classical electrodynamical theory. The first phenomenon where such contradictions were firmly established was the law of tempera-

* J. C. Slater, Nature, March 1st, 1924, p. 307.

† N. Bohr, Zs. f. Phys. **13** (1923) 117. An English translation has recently appeared in the Publications of the Cambridge Philosophical Society under the title: 'On the application of the Quantum Theory to atomic structure. I: The fundamental postulates of the quantum theory.' This treatise, which also contains more detailed references to the literature, will in the following be quoted as 'P.Q.T.'

ture radiation. Starting from the classical conception of emission and absorption of radiation by a particle performing harmonic oscillations, Planck found that, in order to obtain agreement with the experiments on temperature radiation, it was necessary to introduce the auxiliary assumption that in a statistical distribution only certain states of the oscillating particles have to be taken into account. For these distinguished states the energy was found to be equal to a multiple of the quantum $h\omega$, where ω is the natural frequency of the oscillator and h is a universal constant. Independent of radiation phenomena, this result obtained, as Einstein pointed out, a direct support from experiments on the specific heat of solids. At the same time, this author put forward his well-known theory of 'light-quanta', according to which radiation should not be propagated through space as continuous trains of waves in the classical theory of light, but as entities, each of which contains the energy $h\nu$, concentrated in a minute volume, where h is Planck's constant and ν the quantity which in the classical picture is described as the number of waves passing in unit time. Although the great heuristic value of this hypothesis is shown by the confirmation of Einstein's predictions concerning the photoelectric phenomenon, still the theory of light-quanta can obviously not be considered as a satisfactory solution of the problem of light propagation. This is clear even from the fact that the radiation 'frequency' ν appearing in the theory is defined by experiments on interference phenomena which apparently demand for their interpretation a wave constitution of light.

In spite of the fundamental difficulties involved in the ideas of the quantum theory, it has nevertheless been possible to a certain extent to apply these conceptions, combined with information on the structure of the atom derived from other sources, to the interpretation of the results of investigations of the emission and absorption spectra of the elements. This interpretation is based on the fundamental postulate: that an atom possesses a number of distinguished states, the so-called 'stationary states', which are supposed to possess a remarkable stability, for which no interpretation can be derived from the concepts of classical electrodynamics. This stability comes to light in the circumstance that any change of the state of the atom must consist of a complete process of transition from one of these stationary states to another. The postulate obtains a connection with optical phenomena through the further assumption that when a transition between two stationary states is accompanied by emission of radiation, this consists of a train of harmonic waves, whose frequency is given by the relation

$$h\nu = E_1 - E_2, \tag{1}$$

where E_1 and E_2 are the values of the energy of the atom in the initial and in the final state of the process respectively. Inversely it is assumed that the reversed process of transition can take place by illumination with light with this same frequency. The applicability of these assumptions to the interpretation of the spectra of the elements is essentially due to the fact that it has been found possible in many cases to fix the energy in the stationary states of an isolated atom by means of simple rules referring to motions which with a high approximation obey the ordinary laws of electrodynamics (P.Q.T., Ch. I, § 2). The concepts of this theory, however, do not allow us to describe the details of the mechanism underlying the process of transition between the various stationary states.

At the present state of science it seems necessary, as regards the occurrence of transition processes, to content ourselves with considerations of probability. Such considerations have been introduced by Einstein,* who has shown how a remarkably simple deduction of Planck's law of temperature radiation can be obtained by assuming that an atom in a given stationary state may possess a certain probability of a 'spontaneous' transition in unit time to a stationary state of smaller energy content, and that in addition an atom, by illumination with external radiation of suitable frequency, may acquire a certain probability of performing an 'induced' transition to another stationary state with higher or smaller energy content. In connexion with the conditions of thermal equilibrium between radiation and matter, Einstein further arrived at the conclusion that the exchange of energy by the transition process is accompanied by an exchange of momentum of the amount $h\nu/c$, just as would be the case if the transition were accompanied by the starting or stopping of a small entity moving with the velocity of light c and containing the energy $h\nu$. He concluded that the direction of this momentum for the induced transitions is the same as the direction of propagation of the illuminating light-waves, but that for the spontaneous transitions the direction of the impulse is distributed according to probability laws. These results, which were considered as an argument for ascribing a

* A. Einstein, Phys. Zs. **18** (1917) 121.

certain physical reality to the theory of light-quanta, have recently found an important application in explaining the remarkable phenomena of the change of wave-length of radiation scattered by free electrons brought to light by A. H. Compton's * investigation on X-ray scattering. The application of probability considerations to the problem of temperature equilibrium between free electrons and radiation suggested by this discovery has recently been successfully treated by Pauli,† and the formal analogy of his results with the laws governing transition processes between stationary states of atoms has been emphasized by Einstein and Ehrenfest.**

In spite of the fundamental departure of the quantum theory of atomic processes from a picture based on the ordinary concepts of electrodynamics, the former must in a certain sense ultimately appear as a natural generalization of the latter. This is evident from the condition that in the limit, where we consider processes which depend on the statistical behaviour of a large number of atoms, and which involve stationary states where the difference between neighbouring stationary states is comparatively little, the classical theory leads to conclusions in agreement with the experiments. In the case of emission and absorption of spectral lines, the connexion between the two theories has led to the establishment of the 'correspondence principle', which postulates a general conjugation of each of the various possible transitions between stationary states with one of the harmonic oscillation components in which the electrical moment of the atom, considered as a function of the time, can be resolved (P.Q.T., Ch. II, § 2). This principle has afforded a basis for an estimation of probabilities of transition, and thereby for bringing the problem of intensities and polarization of spectral lines in close connexion with the motion of the electrons in the atom.

The correspondence principle has led to comparing the reaction of an atom on a field of radiation with the reaction on such a field which, according to the classical theory of electrodynamics, should be expected from a set of 'virtual' harmonic oscillators with frequencies equal to those determined by the equation (1) for the various possible transitions between stationary states (P.Q.T., Ch. III, § 3). Such a

* A. H. Compton, Phys. Rev. **21** (1923) 483. See also P. Debye, Phys. Zs. **24** (1923) 161.

† W. Pauli, Zs. f. Phys. **18** (1923) 272.

** A. Einstein and P. Ehrenfest, Zs. f. Phys. **14** (1923) 301.

picture has been used by Ladenburg * in an attempt to connect the experimental results on dispersion quantitatively with considerations on the probability of transitions between stationary states. Also in the phenomenon of interaction between free electrons and radiation, the possibility of applying similar considerations is suggested by the analogy, emphasized by Compton, between the change of wave-length of the scattered rays and the classical Doppler effect of radiation from a moving source.

Although the correspondence principle makes it possible through the estimation of probabilities of transition to draw conclusions about the mean time which an atom remains in a given stationary state, great difficulties have been involved in the problem of the time-interval in which emission of radiation connected with the transition takes place. In fact, together with other well-known paradoxes of the quantum theory, the latter difficulty has strengthened the doubt, expressed from various sides,† whether the detailed interpretation of the interaction between matter and radiation can be given at all in terms of a causal description in space and time of the kind hitherto used for the interpretation of natural phenomena (P.Q.T., Ch. III, § 1). Without in any way removing the formal character of the theory, it nevertheless appears, as mentioned in the introduction, that a definite advance as regards the interpretation of the observable radiation phenomena may be made by connecting these phenomena with the stationary states and the transitions between them in a way somewhat different from that hitherto followed.

2. Radiation and transition processes

We will assume that a given atom in a certain stationary state will communicate continually with other atoms through a time-spatial mechanism which is virtually equivalent with the field of radiation which on the classical theory would originate from the virtual harmonic oscillators corresponding with the various possible transitions to other stationary states. Further, we will assume that the occurrence of transition processes for the given atom itself, as well as for the other atoms with which it is in mutual communication, is connected with

* R. Ladenburg, Zs. f. Phys. **4** (1921) 451. See also R. Ladenburg and P. Reiche, Naturwiss. **11** (1923) 584.

† Such a view has perhaps for the first time been clearly expressed by O. W. Richardson, 'The Electron Theory of Matter,' 2nd ed. (Cambridge 1916) p. 507.

this mechanism by probability laws which are analogous to those which in Einstein's theory hold for the induced transitions between stationary states when illuminated by radiation. On the one hand, the transitions which in this theory are designated as spontaneous are, on our view, considered as induced by the virtual field of radiation which is connected with the virtual harmonic oscillators conjugated with the motion of the atom itself. On the other hand, the induced transitions of Einstein's theory occur in consequence of the virtual radiation in the surrounding space due to other atoms.

While these assumptions do not involve any change in the connexion between the structure of the atom and the frequency, intensity, and polarization of the spectral lines derived by means of the relation (1) and of the correspondence principle, they lead to a picture as regards the time-spatial occurrence of the various transition processes on which the observations of the optical phenomena ultimately depend which in an essential respect differs from the usual concepts. In fact, the occurrence of a certain transition in a given atom will depend on the initial stationary state of this atom itself and on the states of the atoms with which it is in communication through the virtual radiation field, but not on the occurrence of transition processes in the latter atoms.

On the one hand it will be seen that our view, in the limit where successive stationary states differ only little from each other, leads to a connexion between the virtual radiation field and the motion of the particles in the atom which gradually merges into that claimed by the classical radiation theory. In fact neither the motion nor the constitution of the radiation field will in this limit undergo essential changes through the transitions between stationary states. As regards the occurrence of transitions, which is the essential feature of the quantum theory, we abandon on the other hand any attempt at a causal connexion between the transitions in distant atoms, and especially a direct application of the principles of conservation of energy and momentum, so characteristic for the classical theories. The application of these principles to the interaction between individual atomic systems is, on our view, limited to interactions which take place when the atoms are so close that the forces which would be connected with the radiation field on the classical theory are small compared with the conservative parts of the fields of force originating from the electric charges in the atom. Interactions of this type, which may be termed 'collisions', offer, as is well known, remarkable illustrations of the stability of stationary states postulated in the quantum theory. In fact, an analysis of the experimental results based on the theory of conservation of energy and momentum is in agreement with the view that the colliding atoms before as well as after the process will always find themselves in stationary states (P.Q.T., Ch. I, § 4)*. By interaction between atoms at greater distances from each other, where according to the classical theory of radiation there would be no question of simultaneous mutual action, we shall assume an independence of the individual transition processes, which stands in striking contrast to the classical claim of conservation of energy and momentum. Thus we assume that an induced transition in an atom is not directly caused by a transition in a distant atom for which the energy difference between the initial and the final stationary state is the same. On the contrary, an atom which has contributed to the induction of a certain transition in a distant atom through the virtual radiation field conjugated with the virtual harmonic oscillator corresponding with one of the possible transitions to other stationary states, may nevertheless itself ultimately perform another of these transitions.

At present there is unfortunately no experimental evidence at hand which allows to test these ideas, but it may be emphasized that the degree of independence of the transition processes assumed here would

* These considerations hold obviously only in so far as the radiation connected with the collisions can be neglected. Although in many cases the energy of this radiation is very small, its occurrence might be of essential importance. This has been emphasized by Franck in connexion with the explanation of Ramsauer's important results regarding collisions between atoms and slow electrons [Ann. d. Phys. Leipzig **64** (1922) 513; **66** (1922) 546], from which it seems to follow that in certain cases the electron can pass freely through the atom, without being influenced by its presence. In fact, if in these 'collisions' a change in the motion of the electron actually took place, the classical theory would involve so large a radiation, that a rational conjugation of the radiation with the possible transition processes, as claimed by the correspondence principle, could hardly be established [compare F. Hund, Zs. f. Phys. **13** (1923) 241]. On the view presented in this paper, such an explanation might on the one hand be regarded as the more natural, since the origin of radiation is not directly sought in the occurrence of transitions but in the motion of the electron. On the other hand, it must be remembered that we are here dealing with a case where, on account of the large magnitude of the classical reaction of radiation, the theory does not allow a sharp distinction between stationary motion and transition processes.

seem the only consistent way of describing the interaction between radiation and atoms by a theory involving probability considerations. This independence reduces not only conservation of energy to a statistical law, but also conservation of momentum. Just as we assume that any transition process induced by radiation is accompanied by a change of energy of the atom of the amount $h\nu$, we shall assume, following Einstein, that any such process is also accompanied by a change of momentum of the atom of an amount $h\nu/c$. If the transition is induced by virtual radiation fields from distant atoms, the direction of this momentum is the same as that of the wave propagation in this virtual field. In case of a transition by its own virtual radiation, we shall naturally assume that the change of momentum is distributed according to probability laws in such a way that changes of momentum due to the transitions in other atoms are statistically compensated for any direction in space.

The cause of the observed statistical conservation of energy and momentum we shall not seek in any departure from the electrodynamic theory of light as regards the laws of propagation of radiation in free space, but in the peculiarities of the interaction between the virtual field of radiation and the illuminated atoms. In fact, we shall assume that these atoms will act as secondary sources of virtual wave radiation which interferes with the incident radiation. If the frequency of the incident waves coincides closely with the frequency of one of the virtual harmonic oscillators corresponding to the various possible transitions, the amplitudes of the secondary waves will be especially large, and these waves will possess such phase relations with the incident waves that they will diminish or augment the intensity of the virtual radiation field, and thereby weaken or strengthen its power of inducing transitions in other atoms. Whether it is a diminishing or an augmentation of the intensity which takes place, will depend on whether the virtual harmonic oscillator, which is called into play by the incident radiation, corresponds with a transition by which the energy of the atom is increased or diminished respectively. It will be seen that this view is closely related to the ideas which led Einstein to introduce probabilities of two kinds of induced transitions between stationary states corresponding with an increase or decrease of the energy of the atom respectively. In spite of the time-spatial separation of the processes of absorption and emission of radiation characteristic for the quantum theory, we may nevertheless expect, on our view, a far-reaching analogy with the classical theory of electrodynamics as regards the interaction of the virtual radiation field and the virtual harmonic oscillators conjugated with the motion of the atom. It seems actually possible, guided by this analogy, to establish a consistent and fairly complete description of the general optical phenomena accompanying the propagation of light through a material medium, which accounts at the same time for the close connexion of the phenomena with the spectra of the atoms of the medium.

3. Capacity of interference of spectral lines

Before we enter more closely on the general problem of the reaction of atoms on a virtual radiation field, responsible for the phenomena accompanying the propagation of light through material media, we shall here briefly consider the properties of the field originating from a single atom, as far as they are connected with the capacity of interference of light from one and the same source. The constitution of this field must obviously not be sought in the peculiarities of the transition processes themselves, the duration of which we shall assume at any rate not to be large compared with the period of the corresponding harmonic component in the motion of the atom. These processes will, on our view, simply mark the termination of the time-interval in which the atom will be able to communicate with other atoms through the corresponding virtual oscillator. An upper limit of the capacity of interference, however, will clearly be given by the mean time interval in which the atom remains in the stationary state representing the initial state of the transition under consideration. The estimation of the time of duration of states based on the correspondence principle has obtained a general confirmation from the well-known beautiful experiments on the duration of the luminosity of high speed atoms emerging from a luminescent discharge into a high vacuum. (Compare P.Q.T., Ch. II, § 4.) On the present point of view these experiments obtain a very simple interpretation. In fact it will be seen that on this view the variation of the luminosity along the path of the atoms will not depend on the peculiarity of the transitions, but only on the relative number of atoms in the various stationary states in the different parts of the path. If all the emerging atoms have the same speed and are initially in the same state, we must thus expect that for any spectral line conjugated with a transition from this state the luminosity will decrease exponentially along the path at one and the same rate. At present the experimental material at hand is hardly sufficient to test these considerations.

When we ask for the capacity of interference of spectral lines, determined by optical apparatus, the mean time of duration of the stationary states will certainly constitute an upper limit for this capacity, but it must be remembered that the sharpness of a given spectral line which is due to the statistical result of the action from a large number of atoms will depend not only on the lengths of the individual wave trains terminated by the transition processes, but clearly also on any uncertainty in the definition of the frequency of these waves. In view of the way this frequency through relation (1) is related to the energy in the stationary states, it is of interest to note that the above-mentioned upper limit of capacity of interference may be brought in close connexion with the limit of definition of the motion and of the energy in the stationary states. In fact, the postulate of the stability of stationary states imposes an *a priori* limit to the accuracy with which the motion in these states can be described by means of classical electrodynamics, a limit which on our picture is directly involved in the assumption that the virtual radiation field is not accompanied by a continuous change in the motion of the atom, but only acts by its induction of transitions involving finite changes of the energy and the momentum of the atom (P.Q.T., Ch. II, § 4). In the limiting region where the motions in the two stationary states involved in the transition process differ only comparatively little from each other, the upper limit of capacity of interference of the individual wave trains coincides with the limit of definition of the frequency of the radiation determined by (1), if the influence of the lack of definition of the energy in the two states is treated as independent errors. In the general case where the motions in these states may differ considerably from each other, the upper limit of the capacity of interference of the wave trains is closely related with the definition of the motion in the stationary state which forms the starting point of the transition process. Also here we may, however, expect that the observable sharpness of the spectral lines will be determined according to relation (1) by adding the effect of any possible lack of definition of the energy in the stationary state terminating the transition process to the effect of the lack of definition in the starting state in a similar way as independent errors. Just this influence of the lack of definition of both stationary states on the sharpness of a spectral line makes it possible to ensure the reciprocity which will exist between the constitution of a line when appearing in an emission and in an absorption spectrum, and which is claimed by the condition for thermal equilibrium expressed by Kirchhoff's law. In this connexion it may be remembered how the apparent deviations from this law exhibited by the remarkable difference often shown by the structure of the emission and the absorption spectra of an element as regards the number of lines present are directly accounted for on the quantum theory when account is taken of the difference in the statistical distributions of the atoms over the various stationary states under different external conditions.

A problem closely related to the sharpness of spectral lines originating from atoms under constant external conditions, is the problem of the spectrum to be expected from atoms under the influence of external forces which vary considerably within a time-interval of the same order of magnitude as the mean duration of the stationary states. Such a problem is met with in certain of the experiments by Stark on the influence of electric fields on spectral lines. In these experiments the emitting atoms move with large velocities, and the time-intervals in which they pass between two points where the intensity of the electric field differs very much, are only a small fraction of the mean time of duration of the stationary states connected with the investigated spectral lines. Nevertheless Stark found that, except for a Doppler effect of the usual kind, the radiation from the moving atoms was influenced by the electric field at any point of the path in the same way as the radiation from resting atoms subject to the constant action of the field at this point. While, as emphasized by various authors,* the interpretation of this result obviously presents difficulty on the usual quantum theory description of the connexion between radiation and transition processes, it is clear that Stark's results are in conformity with the picture adopted in this paper. In fact, during the passing of the atoms through the field, the motion in the stationary states changes in a continuous way, and in consequence also the virtual harmonic oscillators corresponding with the possible transitions. The effect of the virtual radiation field originating from the moving atoms will therefore not be different from that which would occur if the atoms along their whole path had moved in

* Compare K. Försterling, Zs. f. Phys. **10** (1922) 387; A. J. Dempster, Astrophys. Journ. **57** (1923) 193.

a field of constant intensity, at any rate if – as in Stark's experiments – the radiation originating from the other parts of their paths is prevented from reaching those parts of the apparatus on which the observation of the phenomenon depends. In a problem of this kind it will also be seen how a far reaching reciprocity in the observable phenomena of emission and absorption is ensured on account of the symmetry exhibited by our picture as regards the coupling of the radiation field with the transition processes in the one or in the other direction.

4. Quantum theory of spectra and optical phenomena

Although on the quantum theory the observation of the optical phenomena ultimately depends on discontinuous transition processes, an adequate interpretation of these phenomena must, as already emphasized in the introduction, nevertheless involve an element of continuity similar to that exhibited by the classical electrodynamical theory of the propagation of light through material media. On this theory the phenomena of reflexion, refraction, and dispersion are attributed to a scattering of light by the atom due to the forced vibration in the individual electric particles, set up by the electromagnetic forces of the radiation field. The postulate of the stability of stationary states might at first sight seem to involve a fundamental difficulty on this point. The contrast, however, was to a certain extent bridged over by the correspondence principle, which, as mentioned in § 1, led to comparing the reaction of an atom on a radiation field with the scattering which, according to the classical theory, would arise from a set of virtual harmonic oscillators conjugated with the various possible transitions. It must still be remembered that the analogy between the classical theory and the quantum theory as formulated through the correspondence principle is of an essentially formal character, which is especially illustrated by the fact that on the quantum theory the absorption and emission of radiation are coupled to different processes of transition, and thereby to different virtual oscillators. Just this point, however, which is so essential for the interpretation of the experimental results on emission and absorption spectra, seems to afford a guidance as regards the way in which the scattering phenomena are related with the activity of the virtual oscillators concerning emission and absorption of radiation. In a later paper it is hoped to show how on the present view a quantitative theory of dispersion resembling Ladenburg's theory can be established.* Here we shall confine ourselves to emphasizing once more the continuous character of the optical phenomena, which seemingly does not permit an interpretation based on a simple causal connexion with transition processes in the propagating medium.

An instructive example of these considerations is offered by the experiments on absorption spectra. In fact, the pronounced absorption by monatomic vapours for light of frequencies coinciding with certain lines in the emission spectra of the atoms strictly cannot be said, as often done for brevity, to be caused by the transition processes which take place in the atoms of the vapour induced by wave trains in the incident radiation possessing the frequencies of the absorption lines. The appearance of these lines in the spectroscope is due to the decrease of the intensity of the incident waves in consequence of the peculiarities of the secondary spherical wavelets set up by each of the illuminated atoms, while the induced transitions appear only as an accompanying effect by which a statistical conservation of energy is ensured. The presence of the secondary coherent wave-trains is at the same time responsible for the anomalous dispersion connected with the absorption lines, and is especially clearly shown by the phenomenon, discovered by Wood†, of selective reflexion from the wall of a vessel containing metallic vapour under sufficiently high pressure. The occurrence of included transitions between stationary states is on the other hand directly observed in the fluorescent radiation, which for an essential part originates from the presence of a small number of atoms which through the illumination have been transferred to a stationary state of higher energy. As is well known, the fluorescent radiation can be suppressed through the admixture of foreign gases. As regards the part played by atoms in the higher stationary states this phenomenon is explained by collisions which cause a considerable increase of the probability of the atoms to return into their normal state. At the same time any part of the fluorescent radiation due to the coherent wavelets will, through the admixture of foreign gases, just as the phenomena of absorption, dispersion, and reflexion, undergo such changes as can be brought in connexion with

* *Note added during the proof.* The outline of such a theory is briefly described by Kramers in a letter to 'Nature' published in April, 1925.

† R. W. Wood, Phil. Mag. **23** (1915) 689.

a broadening of the spectral lines*. It will be seen that a view on absorption phenomena differing essentially from that just described can hardly be maintained, if it can be shown that the selective absorption of spectral lines is a phenomenon qualitatively independent of the intensity of the source of radiation, in a similar way to what has already been found to be the case for the usual phenomena of reflexion and refraction, whose transitions in the medium do not occur to a similar extent (compare P.Q.T., Ch. III, § 3).

Another interesting example is offered by the theory of the scattering of light by free electrons. As has been shown by Compton by means of reflexion of X-rays from crystals, this scattering is accompanied by a change of frequency, different in different directions, and corresponding with the constitution of the radiation which on the classical theory would be emitted by an imaginary moving source. As mentioned, Compton has reached a formal interpretation of this effect on the theory of light-quanta by assuming that the electron may take up a quantum of the incident light and simultaneously re-emit a light-quantum in some other direction. By this process the electron acquires a velocity in a certain direction, which is determined, just as the frequency of the re-emitted light, by the laws of conservation of energy and momentum, an energy of $h\nu$ and a momentum $h\nu/c$ being ascribed to each light-quantum. In contrast to this picture, the scattering of the radiation by the electrons is, on our view, considered as a continuous phenomenon to which each of the illuminated electrons contributes through the emission of coherent secondary wavelets. Thereby the incident virtual radiation gives rise to a reaction from each electron, similar to that to be expected on the classical theory from an electron moving with a velocity coinciding with that of the above-mentioned imaginary source and performing forced oscillations under the influence of the radiation field. That in this case the virtual oscillator moves with a velocity different from that of the illuminated electrons themselves is certainly a feature strikingly unfamiliar to the classical conceptions. In view of the fundamental departures from the classical space-time description, involved in the very idea of virtual oscillators, it seems at the present state of science hardly justifiable to reject a formal interpretation as that under consideration as inadequate. On the contrary, such an interpretation seems unavoidable in order to account for the effects observed, the description of which

* See for instance, Chr. Füchtbauer and G. Joos, Phys. Zs. **23** (1922) 73.

involves the wave-concept of radiation in an essential way. At the same time, however, we shall assume, just as in Compton's theory, that the illuminated electron possesses a certain probability of taking up in unit time a finite amount of momentum in any given direction. By this effect, which in the quantum theory takes the place of the continuous transfer of momentum to the electrons which on the classical theory would accompany a scattering of radiation of the type described, a statistical conservation of momentum is secured in a way quite analogous to the statistical conservation of energy in the phenomena of absorption of light discussed above. In fact, the laws of probability for the exchange of momentum by interaction of free electrons and radiation derived by Pauli are essentially analogous to the laws governing transition processes between well-defined states of an atomic system. Especially the considerations of Einstein and Ehrenfest, referred to in § 1, are suited to bring out this analogy.

A problem similar to that of the scattering of light by free electrons is presented by the scattering of light by an atom, even in the case where the frequency of the radiation is not large enough to induce transitions by which an electron is wholly removed from the atom. In fact, in order to secure statistical conservation of momentum, we must, as emphasized by various authors,* assume the occurrence of transition processes by which the momentum of the scattering atom changes by finite amounts without, however, the relative motion of the particles of the atom being changed, as in transition processes of the usual type considered in the spectral theory. It will also be seen, on our picture, that transition processes of the type mentioned will be closely connected with the scattering phenomena, in a way analogous with the connexion of the spectral phenomena with the transition processes by which the internal motion of the atom undergoes a change. Due to the large mass of the atomic nucleus the velocity change which the atom undergoes by these transitions is so small, that it will not have a perceptible effect on the energy of the atom and the frequency of the scattered radiation. Nevertheless, it is of principal importance that the transference of momentum is a discontinuous process, while the scattering itself is an essentially continuous phenomenon, in which all the illuminated atoms take part, independent of the intensity of the incident light. The discontinuous changes in momentum of the atoms, however, are the cause of the observable reactions on the atoms

* W. Pauli, Zs. f. Phys. **18** (1923) 272; A. Smekal, Naturwissensch. **11** (1923) 875.

described as radiation pressure. This view fulfils clearly the conditions for thermal equilibrium between a (virtual) radiation field and a reflecting surface, derived by Einstein * and considered as an argument for the light-quantum theory. At the same time it needs hardly be emphasized that it is also consistent with the apparent continuity exhibited by actual observations on radiation pressure. In fact, if we consider a solid, a change of $h\nu/c$ in its total momentum will be totally imperceptible, and for visible light even vanishingly small compared with the irregular changes of this momentum of a body in thermal equilibrium with the surroundings. In the discussion of the actual experiments it may, however, be noted at the same time, that the frequency of the occurrence of such processes may often be so large that the problem arises whether the time involved in the transitions themselves can be neglected, or, in other words, whether the limit has been reached inside which the formulation of the principles of the quantum theory can be maintained (compare P.Q.T., Ch. II, § 5).

The last considerations may illustrate how our picture of optical phenomena offers a natural connexion with the ordinary continuous description of macroscopic phenomena for the interpretation of which Maxwell's theory has shown itself so wonderfully adapted. The advantage in this respect of the present formulation of the principles of the quantum theory over the usual representation of this theory will perhaps be still more clearly illustrated if we consider the phenomenon of emission of electromagnetic waves, say from an antenna as used in wireless telegraphy. In this case no adequate description of the phenomenon is offered on the picture of emission of radiation during separate successive transition processes between imaginary stationary states of the antenna. In fact, when the smallness of the energy changes by the transitions, and the magnitude of the energy radiation from the antenna per unit time, are taken into account, it will be seen that the duration of the individual transition processes can only be an exceedingly small fraction of the period of oscillation of the electricity in the antenna, so that there would be no justification in describing the result of one of these processes as the emission of a train of waves of this period. On the present view, however, we will describe the action of the oscillation of the electricity in the antenna as producing a (virtual) radiation field which through probability laws again induces changes in the motion of the electrons which may

* A. Einstein, Phys. Zs. **10** (1909) 817.

be regarded as continuous. In fact, even if a distinction between different energy steps *hν* could be kept upright, the size of these steps would be quite negligible compared with the energy associated with the antenna. It will in this connexion be observed that the emphasizing of the 'virtual' character of the radiation field, which at the present state of science seems so essential for an adequate description of atomic phenomena, automatically loses its importance in a limiting case like that just considered, where the field, as regards its observable interaction with matter, is endowed with all the attributes of an electromagnetic field in classical electrodynamics.

Related papers

5a J. C. Slater, *Radiation and Atoms*. Nature **113** (1924) 307.

5b H. Geiger und W. Bothe, *Über das Wesen des Comptoneffektes; ein experimenteller Beitrag zur Theorie der Strahlung*. Z. Phys. **32** (1925) 639, received April 25, 1925.

Dated March 25, 1924

6

THE LAW OF DISPERSION AND BOHR'S THEORY OF SPECTRA

H. A. KRAMERS

It is well known that a consistent description of the phenomena of dispersion, reflection, and scattering of electromagnetic waves by material media can be given on the fundamental assumption that an atom, when exposed to radiation, becomes a source of secondary spherical wavelets, which are coherent with the incident waves. If we imagine that the incident radiation consists of a train of polarised harmonic waves of frequency ν, the electric vector of which at the point in space where the atom is situated at rest can be represented by

$$\mathfrak{E} = Ev \cos 2\pi\nu t, \tag{1}$$

where E is the amplitude and v is a unit vector; the secondary wavelets can be described as originating from a varying electrical doublet, the strength of which is given by

$$\mathfrak{P} = Pw \cos (2\pi\nu t - \varphi), \tag{2}$$

where P is the amplitude and w also a unit vector, while φ represents the phase difference between the secondary and primary waves. The quantities P, w, and φ depend on v, ν, and on the peculiarities of the atom; moreover, the amplitude P will be proportional to the amplitude E of the incident waves.

If we consider an atom containing an electron of charge $-e$ and mass m, which is isotropically bound to a position of equilibrium, we find on the classical theory that the vectors v and w will coincide and the following expression for P is found to hold for frequencies which differ sensibly from the natural frequency ν_1 of the electron.

$$P = E\frac{e^2}{m}\frac{1}{4\pi^2(\nu_1^2 - \nu^2)}. \tag{3}$$

Editor's note. This paper was published as Nature **113** (1924) 673–676. It was signed 'Institute for Theoretical Physics, Copenhagen, March 25.'

In the region where this formula holds the phase difference φ is very small and of such a magnitude as to ensure energy-balance. For substances exhibiting absorption lines at the frequencies $\nu_1, \nu_2 \ldots \nu_i$, a formula of the type

$$P = E \sum_i f_i \frac{e^2}{m} \frac{1}{4\pi^2(\nu_i^2 - \nu^2)}, \tag{4}$$

where the quantities f_i are constants, has actually been found to represent the results of experiment with considerable accuracy. Especially the experiments of Wood and Bevan on the dispersion of light in monatomic vapours of alkali metals have confirmed formula (4) and allowed a determination of the constants f_i which are conjugated to the different absorption lines of the vapour.

In Bohr's theory of spectra, the picture of electrons which are elastically bound inside the atom is abandoned, and for it is substituted a picture according to which an atom exhibiting an absorption line of frequency ν is capable of performing under the influence of the illumination a transition from the state under consideration to a stationary state the energy content of which is $h\nu$ greater. On Bohr's principle of correspondence, the possibility for such transitions is considered as being directly connected with the periodicity properties of the motion of the atom, in such a way that every possible transition between two stationary states is conjugated with a certain harmonic oscillating component in the motion.

In a paper by Bohr, Slater, and the writer, which gives a more detailed discussion of an idea briefly described by Dr. Slater in a recent letter to *Nature* (Vol. **113**, p. 307), and is to appear shortly in the *Philosophical Magazine*, it will be shown that, with the correspondence principle as a guide, it seems possible to arrive at an adequate description of the activity of the atoms regarding their interaction with radiation. On this theory the picture described above of the mechanism underlying dispersion and scattering phenomena is essentially preserved, and the important question arises concerning the quantitative laws connecting the quantities P, w, and φ appearing in (2), which characterise the reaction of an atom in a given state against external radiation, with the peculiarities of the transitions which the atom may perform to other stationary states. The present state of the quantum theory does not allow a rigorous deduction of these laws. It is, however, possible to establish a very simple expression for P, which fulfils the condition, claimed by the correspondence principle, that, in the region where successive stationary states of an atom differ only comparatively little from each other, the interaction between the atom and the field of radiation tends to coincide with the interaction to be expected on the classical theory of electrons.

Consider an atom in a stationary state which by absorption of radiation of frequencies $\nu_1^{\mathrm{a}}, \nu_2^{\mathrm{a}},\ldots$ may perform transitions to states of higher energy, and by emission of radiation of frequencies $\nu_1^{\mathrm{e}}, \nu_2^{\mathrm{e}},\ldots$ may perform spontaneous transitions to states of lower energy. We will, following Einstein, denote the probability of the isolated atom performing in unit time one of the latter transitions by $A_1^{\mathrm{e}}, A_2^{\mathrm{e}},\ldots$, whereas the analogous probability coefficients for the spontaneous transitions of which the state under consideration represents the final state are denoted by $A_1^{\mathrm{a}}, A_2^{\mathrm{a}},\ldots$. For the sake of simplicity we will further assume that the statistical weights of all the states involved are the same, and that the atom is so oriented in space that the electrical vector in the spontaneous radiation conjugated with the different transitions under consideration is always parallel to the electrical vector of the incident waves. The expression for P alluded to above takes, then, the following form:

$$P = E\sum_i A_i^{\mathrm{a}}\tau_i^{\mathrm{a}}\frac{e^2}{m}\,\frac{1}{4\pi^2(\nu_i^{\mathrm{a}2}-\nu^2)} - E\sum_j A_j^{\mathrm{e}}\tau_j^{\mathrm{e}}\frac{e^2}{m}\,\frac{1}{4\pi^2(\nu_j^{\mathrm{e}2}-\nu^2)}, \qquad (5)$$

where $\tau_i^{\mathrm{a}}=3mc^3/8\pi^2e^2\nu_i^{\mathrm{a}2}$ and $\tau_j^{\mathrm{e}}=3mc^3/8\pi^2e^2\nu_j^{\mathrm{e}2}$ represent the time in which on the classical theory the energy of a particle of charge e and mass m performing linear harmonic oscillations of frequency ν is reduced to $1/\varepsilon$ of its original value, where ε is the base of the natural logarithms. In analogy with the region of applicability of formula (3), this formula only applies in the regions for ν which lie outside the absorption and the emission lines, where the phase angle φ is negligibly small.

The reaction of the atom against the incident radiation can thus formally be compared with the action of a set of virtual harmonic oscillators inside the atom, conjugated with the different possible transitions to other stationary states. These oscillators might be thought of as electrical particles with such charge e^* and mass m^* that the classical formula (3) would give the right result directly, but if we do so, we meet with the remarkable circumstance that, while for the 'absorption oscillators' $e^{*2}/m^*=A_j^{\mathrm{a}}\tau_j^{\mathrm{a}}e^2/m$ is a positive quantity, the corresponding expression for the 'emission oscillators' $e^{*2}/m^* = -A_i^e\tau_i^e e^2/m$ becomes negative. Denoting the quantity $A\tau$ which thus can be conjugated with a given transition and has the dimensions of a number, by f, one might introduce the following terminology: in the final state of the transition the atom acts as a 'positive virtual oscillator' of relative strength $+f$; in the initial state it acts as a negative virtual oscillator of strength $-f$. However unfamiliar this 'negative dispersion' might appear from the point of view of the classical theory, it may be noted that it exhibits a close analogy with the 'negative absorption' which was introduced by Einstein, in order to account for the law of temperature radiation on the basis of the quantum theory.

Led by considerations of the close connexion between dispersion and selective absorption, Ladenburg has proposed a formula equivalent to ours if the second term on the right side is omitted. In the case where the dispersing atoms are present in the normal states and only positive oscillators come into play, his formula is thus equivalent to ours. In the general case of a stationary state where the atom can perform spontaneous transitions to states with lower energy, negative virtual oscillators also come into play, corresponding to the second term in our formula.

As shown by Ladenburg, there is considerable experimental evidence in favour of the connexion between selective absorption and dispersion as indicated by the formula when applied to atoms in their normal state. The experiments at hand scarcely allow testing the complete formula in a more general case. It may be remembered, however, that the presence of the second term in (5) is necessary if the classical theory can be applied in the limiting region where the motions in successive stationary states differ only by small amounts from each other.

Received June 13, 1924

7

QUANTUM MECHANICS

M. BORN

This paper contains an attempt to make a first step towards a quantum theory of coupling, which takes into account some of the more important properties of atoms (stability, resonance for transition frequencies, correspondence principle) and which arises in a natural way from the classical laws. This theory includes Kramers' dispersion formula and reveals a close affinity to Heisenberg's formulation of the rules for the anomalous Zeeman effect.

Introduction

The breakdown of the quantum theory in all cases which deal with the motion of several electrons (e.g. helium), has been accounted for on several occasions in the past by the existence of an oscillating field which acts on each of the electrons and whose frequency is of the same order of magnitude as that of a light wave. Since one knows that, in certain circumstances, atoms react to light waves completely 'non-mechanically' (i.e. they are excited to quantum jumps), it is not to be expected either that the interaction between the electrons of one and the same atom should comply with the laws of classical mechanics; this disposes of any attempt to calculate the stationary orbits by using a classical perturbation theory complemented by quantum rules. For as long as one does not know the laws for the interaction of light with atoms, i.e. the connection of dispersion with atomic structure and quantum jumps, one is left all the more in the dark about the laws of interaction between several electrons of the same atom.

Recently, however, considerable progress has been made by Bohr,

Editor's note. This paper was published as Zs. f. Phys. **26** (1924) 379–395.

Kramers and Slater [1]) on just this matter of the connection between radiation and atomic structure. In my opinion, this progress consists above all in the fact that classical optics comes largely into its own again. How fruitful these ideas are, is also shown by Kramers' [2]) success in setting up a dispersion formula and in proving that it satisfies all the conditions of quantum theory, in particular the correspondence principle.

In this situation, one might consider whether it would not be possible to extend Kramers' ideas, which he applied so successfully to the interaction between radiation field and radiating electron, to the case of the interaction between several electrons of an atom. Closer study of Kramers' dispersion formula leads one to investigate whether the method of quantisation used by him is not based on some general property of perturbed mechanical systems. The present paper is an attempt to carry out this idea.[3])

What we shall do, is to bring the classical laws for the perturbation of a mechanical system, caused by internal couplings or external fields, into one and the same form, which would very strongly suggest the formal passage from classical mechanics to a 'quantum mechanics'. For this, the quantum rules as such will be retained essentially unchanged; as multiples of the action quantum h there will appear the action integrals of the unperturbed system, [4]) which is assumed to be separable and non-degenerate. On the other hand, mechanics itself will undergo a change, in the sense of a transition from differential to difference equations, as already exhibited by Bohr's frequency conditions. In the simple case of non-degenerate systems, there seems then to be no room left for arbitrariness.

The combination of this new 'quantum mechanics' with the old quantum rules now leads to interaction laws, which will first of all have to be tested to see whether they contain Kramers' dispersion formula. This is in fact the case, and thus a basis for other investigations is established.

[1]) N. Bohr, H. A. Kramers and J. C. Slater, Zs. f. Phys. **24** (1924) 69.

[2]) H. A. Kramers, Nature **113** (1924) 673.

[3]) By a happy coincidence I was able to discuss the contents of this paper with Mr. Niels Bohr, which contributed greatly to a clarification of the concepts. I am also greatly indebted to Mr. W. Heisenberg for much advice and help with the calculations.

[4]) It is not necessary to define action integrals for a perturbed system with its intricate, and in general no longer multiperiodic, solutions.

One would think here in the first place of the theory of the anomalous Zeeman effect, which Heisenberg put into a form whose affinity with our postulates becomes immediately apparent. But it should of course not be expected that Heisenberg's quantum prescriptions are covered by this first stage of a quantum mechanics of coupled systems, since the multiplets and Zeeman effects are characterised by complicated degeneracies. For the time being, one is therefore forced to base the multiplet formulae on a quasi-classical model; the formal analogy of these formulae with the rules of quantum mechanics, which will be established here for the simplest cases, shows clearly that these rules describe the essence of the coupling processes.

Our attempt at formulating a quantum mechanics of coupled systems seems to contain many of the features which are needed for a description of the properties of the atom: stability, resonance for transition frequencies, fulfilment of the correspondence principle, etc. Whether it will in fact stand the test, can only be shown by quantitative calculations of simple systems. For this one would still have to overcome considerable difficulties, which stem from the variety of possible degeneracies.

1. Classical perturbation theory for systems which are subject to external forces

We consider a mechanical system whose equations of motion can be solved by separation of variables; let w_k^0, J_k^0 ($k=1,\ldots f$) be the appropriate angle and action variables and

$$H_0(w_1^0, w_2^0, \ldots w_f^0, J_1^0, \ldots J_f^0) \tag{1}$$

the Hamiltonian. We assume at first that we are not dealing with a (proper) degeneracy, i.e. that the frequencies

$$\nu_k = \frac{\partial H_0}{\partial J_k^0} \tag{2}$$

are all different from zero.

We now consider the effect of a perturbation which may be due to internal (previously neglected) interactions or to some external periodic force or to both.

In the former case, let us suppose that the perturbation function can be expanded in a Fourier series of the form

$$\sum_{\tau_1 \ldots \tau_f} C_{\tau_1 \ldots \tau_f} \, e^{2\pi i(w_1{}^0\tau_1 + \ldots w_f{}^0\tau_f)}. \tag{3}$$

The external force will be supposed to have frequency ν_0 and to be represented by the Fourier series

$$\sum_{\tau_0} C_{\tau_0} e^{2\pi i w_0{}^0 \tau_0}, \tag{4}$$

where we have put

$$w_0^0 = \nu_0 t. \tag{5}$$

We assume here that the C_{τ_0} in turn are themselves periodic functions of the $w_1^0, \ldots w_f^0$, which can be expanded in the form (3). We assume further that ν_0 is not commensurable with any of the natural frequencies $\nu_1, \ldots \nu_f$ of the system, i.e. for any set of integers $\tau_0, \tau_1, \ldots \tau_f$ the quantity

$$(\tau\nu) = \tau_0\nu_0 + \tau_1\nu_1 + \ldots + \tau_f\nu_f \neq 0. \tag{5a}$$

The general Hamiltonian which we want to consider is therefore of the form

$$H = H_0 + \lambda H_1, \tag{6}$$

where

$$H_1 = \sum_{\tau_0, \tau_1, \ldots \tau_f} C_{\tau_0, \tau_1, \ldots \tau_f} e^{2\pi i (w_0{}^0 \tau_0 + w_1{}^0 \tau_1 + \ldots w_f{}^0 \tau_f)}. \tag{7}$$

In abbreviated notation, this is written

$$H_1 = \sum_{\tau} C_\tau e^{2\pi i (w^0 \tau)} \qquad (\tau = \tau_0, \tau_1, \ldots \tau_f). \tag{7a}$$

Here, the $C_\tau = C_{\tau_0 \ldots \tau_f}$ are functions of the $J_1, \ldots J_f$. In order that H_1 should be real, the following conditions must be satisfied

$$C_{-\tau_0 \ldots -\tau_f} = \tilde{C}_{\tau_0 \ldots \tau_f},$$

or in short

$$C_{-\tau} = \tilde{C}_\tau, \tag{7b}$$

where the symbol $\sim$ denotes the complex conjugate.

The case where the external field is absent is obtained from the general case by putting $\nu_0 = 0$; one can then simply omit the (lower) index 0 everywhere.

Because of (5), the function H contains the time explicitly. Nevertheless we know that the canonical equations

$$\dot{w}_k^0 = \frac{\partial H}{\partial J_k^0}, \qquad \dot{J}_k^0 = -\frac{\partial H}{\partial w_k^0} \tag{8}$$

are valid.

To solve them, we shall try to introduce new variables w_k, J_k ($k=1,\dots f$) by means of a canonical transformation with generating function $S(w_0^0, w_1^0, \dots w_f^0, J_1, \dots J_f)$

$$J_k^0 = \frac{\partial S}{\partial w_k^0}, \qquad w_k = \frac{\partial S}{\partial J_k}, \qquad H + \frac{\partial S}{\partial t} = W \qquad (k = 1, \dots f), \tag{9}$$

such that W depends[1] only on the J_k; it then follows from the transformed equations of motion,

$$\dot{w}_k = \frac{\partial W}{\partial J_k}, \qquad \dot{J}_k = -\frac{\partial W}{\partial w_k} = 0, \tag{10}$$

that the J_k are constant and the w_k are linear functions of time.

Except when $\nu_0=0$, W can certainly not represent the energy of the system, since that energy cannot be constant for the case of external forces which vary with time; we shall see that it stands for the mean energy of the system in the presence of the external field (including the interaction energy with such a field).

We choose S in the form of a power series,

$$S = \sum_{k=1}^{f} w_k^0 J_k + \lambda S_1 + \lambda^2 S_2 + \dots, \tag{11}$$

and insert it into the equation

$$H + \nu_0 \frac{\partial S}{\partial w_0^0} = W_0 + \lambda W_1 + \lambda^2 W_2 + \dots, \tag{12}$$

where W_0 (the constant energy of the unperturbed system), $W_1, W_2, \dots$ are functions of the J_k alone. One then obtains the equations for the successive approximations

$$\sum_{k=0}^{f} \nu_k \frac{\partial S_1}{\partial w_k^0} + H_1 = W_1, \tag{13a}$$

$$\sum_{k=0}^{f} \nu_k \frac{\partial S_2}{\partial w_k^0} + \frac{1}{2} \sum_{k,l=1}^{f} \frac{\partial^2 H_0}{\partial J_k \partial J_l} \frac{\partial S_1}{\partial w_k^0} \frac{\partial S_1}{\partial w_l^0} + \sum_{k=1}^{f} \frac{\partial H_1}{\partial J_k} \frac{\partial S_1}{\partial w_k^0} = W_2, \tag{13b}$$

. . . .

[1]) Since H depends on t explicitly, one will obviously also have to choose S so that it depends on t (i.e. on w_0^0), and to introduce a new Hamiltonian W in place of H.

These equations can be solved in the usual way. First of all, one averages (13a) over all the w_k^0 (k=0, 1,...f) and obtains

$$\bar{H}_1 = C_{0,0,\dots 0} = C_0 = W_1. \tag{14}$$

Eq. (13a) can then be integrated to give

$$S_1 = -\frac{1}{2\pi i}\sum_\tau{}' \frac{C_\tau}{(\nu\tau)} e^{2\pi i(w\tau)}, \tag{15}$$

where the prime on the summation sign signifies that the term $\tau_0=0$, $\tau_1=0,\dots,\tau_f=0$ is to be omitted. Eq. (13b) can now be written as follows:

$$\sum_k \nu_k \frac{\partial S_2}{\partial w_k^0} + \tfrac{1}{2}\sum_{kl}\frac{\partial \nu_l}{\partial J_k}\sum_\tau{}'\sum_{\tau'}{}' \frac{\tau_k\tau_l' C_\tau C_{\tau'}}{(\nu\tau)(\nu\tau')} e^{2\pi i(w^0,\,\tau+\tau')}$$
$$+\sum_k\sum_\tau\sum_{\tau'}{}' \frac{\partial C_\tau}{\partial J_k}\frac{\tau_k' C_{\tau'}}{(\nu\tau')} e^{2\pi i(w^0,\,\tau+\tau')} = W_2.$$

Forming averages, this gives

$$\tfrac{1}{2}\sum_{kl}\frac{\partial \nu_l}{\partial J_k}\sum_\tau{}' \tau_k\tau_l \frac{C_\tau C_{-\tau}}{(\nu\tau)^2} - \sum_k\sum_\tau{}' \frac{\partial C_\tau}{\partial J_k}\frac{\tau_k C_{-\tau}}{(\nu\tau)} = W_2.$$

This can be written as

$$W_2 = -\tfrac{1}{2}\sum_\tau{}'\sum_k \tau_k \frac{\partial}{\partial J_k}\left(\frac{|C_\tau|^2}{(\nu\tau)}\right), \tag{16}$$

or because of (5a),

$$W_2 = -\sum_{(\nu\tau)>0}\sum_k \tau_k \frac{\partial}{\partial J_k}\left(\frac{|C_\tau|^2}{(\nu\tau)}\right). \tag{16a}$$

We shall content ourselves with this approximation.

Formulae (14) and (16) represent the effect of the internal interactions and of the external forces. If only the former are present, we simply have to put $\nu_0=0$, as previously remarked, and are then led back to known perturbation formulae. What is important for us, is that these formulae remain unchanged [except for the addition of the term $\tau_0\nu_0$ in $(\tau\nu)$] if an external periodic field acts.

2. Classical dispersion theory

As an example we shall consider the effect of a monochromatic, linearly polarised light wave on the unperturbed system. Let the electric vector oscillate parallel to the X-axis:

$$\mathfrak{E}_x = E \cos 2\pi\nu_0 t = E\,\tfrac{1}{2}(e^{2\pi i w_0{}^0} + e^{-2\pi i w_0{}^0}). \tag{17}$$

The work done on the system is $\mathfrak{p}^0\mathfrak{E} = \mathfrak{p}^0_x\mathfrak{E}_x$, where $\mathfrak{p}^0$ is the electric moment of the (unperturbed) system. For $\mathfrak{p}^0_x$ we use an expansion of the form

$$\mathfrak{p}^0_x = \sum_\tau A_\tau\, e^{2\pi i (w^0\tau)}, \qquad (\tau = \tau_1, \tau_2, \ldots \tau_f), \tag{18}$$

where we must have

$$A_{-\tau} = \tilde{A}_\tau. \tag{18a}$$

The perturbation function is therefore given by

$$H_1 = \mathfrak{p}^0_x\mathfrak{E}_x = \tfrac{1}{2}E \sum_\tau (A_\tau\, e^{2\pi i[(w\tau)+w_0]}$$
$$+ A_{-\tau}\, e^{-2\pi i[(w\tau)-w_0]}), \qquad (\tau = \tau_1 \ldots \tau_f). \tag{19}$$

This is of the form (7a) ($\lambda = 1$), with

$$\begin{aligned} C_{1,\tau_1\ldots\tau_f} &= \tfrac{1}{2}E\, A_{\tau_1\ldots\tau_f} &&= \tfrac{1}{2}E\, A_\tau, \\ C_{-1,-\tau_1\ldots\tau_f} &= \tfrac{1}{2}E\, A_{-\tau_1\ldots -\tau_f} &&= \tfrac{1}{2}E\, A_{-\tau}. \end{aligned} \tag{19a}$$

For $\tau_0 \neq \pm 1$, C_τ vanishes.

From (14) and (16) we now have

$$W_1 = 0,$$
$$W_2 = -\frac{E^2}{4}\,\tfrac{1}{2}\sum_\tau \sum_k \tau_k \frac{\partial}{\partial J_k}\left(\frac{|A_\tau|^2}{(\nu\tau)+\nu_0} + \frac{|A_{-\tau}|^2}{(\nu\tau)-\nu_0}\right). \tag{20}$$

The second expression can also be written

$$W_2 = -\frac{E^2}{4}\sum_{(\nu\tau)>0}\sum_k \tau_k \frac{\partial}{\partial J_k}\left(\frac{2|A_\tau|^2(\nu\tau)}{(\nu\tau)^2 - \nu_0^2}\right), \qquad (\tau = \tau_1 \ldots \tau_f). \tag{21}$$

In order to assess the influence of the radiation field on the motions of the system, we now want to calculate its effect on the electric moment.

Using (9) in (18), we can put

$$w^0_k = w_k - \lambda \frac{\partial S_1}{\partial J_k}, \qquad J^0_k = J_k + \lambda \frac{\partial S_1}{\partial w^0_k}.$$

Then

$$\mathfrak{p}_x = \mathfrak{p}_x^0 + \lambda \mathfrak{p}_x^{(1)}, \tag{22}$$

where

$$\mathfrak{p}_x^{(1)} = \sum_k \left(\frac{\partial \mathfrak{p}_x^0}{\partial J_k^0} \frac{\partial S_1}{\partial w_k^0} - \frac{\partial \mathfrak{p}_x^0}{\partial w_k^0} \frac{\partial S_1}{\partial J_k} \right). \tag{22a}$$

It now follows from (15) and (19a) that

$$S_1 = -\frac{E}{2} \cdot \frac{1}{2\pi i} \sum_\tau \left(\frac{A_\tau}{(\nu\tau) + \nu_0} e^{2\pi i[(w^0\tau) + w_0{}^0]} - \frac{A_{-\tau}}{(\nu\tau) - \nu_0} e^{-2\pi i[(w^0\tau) + w_0{}^0]} \right). \tag{23}$$

Inserting (18) and (23) in (22a), a straightforward calculation gives the following result for the term in $\mathfrak{p}_x^{(1)}$ which are proportional to $\cos 2\pi\nu_0 t$:

$$\mathfrak{p}_x^{(1)} = -E \cos (2\pi\nu_0 t) \sum_k \sum_{(\nu\tau)>0} \tau_k \frac{\partial}{\partial J_k} \left(\frac{2|A_\tau|^2(\nu\tau)}{(\nu\tau)^2 - \nu_0^2} \right) + \ldots . \tag{24}$$

Hence the average value becomes

$$\tfrac{1}{2}\overline{\mathfrak{p}_x^{(1)}\mathfrak{E}_x} = W_2. \tag{25}$$

This shows the connection between W_2 and the mean energy of the system in the radiation field. The factor $\frac{1}{2}$ in formula (25) requires some elucidation. For this purpose we consider, as the simplest model of a polarisable system, a resonator with quasi-elastic energy $\frac{1}{2}ax^2$, acted on by an electric field. The total energy in the field is

$$W = \tfrac{1}{2}ax^2 + Ex.$$

The equilibrium condition reads

$$x_0 = -E/a,$$

so that the equilibrium energy is

$$W_0 = -\frac{1}{2a} E^2 = \tfrac{1}{2}x_0 E.$$

One can see immediately that the factor $\frac{1}{2}$ in (25), just as here, is due to the production of a distortion which compensates for part of the work done by the external field.

In their dependence on the light frequency ν_0, formulae (21) and (24) display the typical structure of resonance formulae, such as they always occur in classical dispersion theory; what is more, the resonance values are the natural frequencies and all their upper harmonics [1]).

3. Passage to quantum theory

The question now arises, how to effect the transition from the classical formulae to quantum theory.

For this it will be profitable to make use of the intuitive ideas, introduced by Bohr, Kramers and Slater in the paper quoted above, about the connection between frequencies and quantum jumps; but our line of reasoning will be independent of the critically important and still disputed conceptual framework of that theory, such as the statistical interpretation of energy and momentum transfer.

According to Bohr, each stationary state carries with it a number of 'virtual' resonators, whose frequencies correspond to transitions to other stationary states by the frequency condition

$$h\nu = |W_1 - W_2|.$$

If we now concentrate our attention on the quantum jumps which arise from a given stationary state,

$$n_1, n_2, \ldots n_l \ldots n_f,$$

they fall into two categories, depending on whether the final energy level is lower or higher. The former kind, of which there is a finite number, are associated with the emission of light; to these correspond an equal number of 'emission resonators' of the state n_l with frequencies

$$\nu(n, n') = \frac{1}{h} [W(n) - W(n')].$$

The second class of jumps, of which there is generally an infinite number, occurs in the case of light absorption; to them correspond the 'absorption resonators' of the state n_l with frequencies

$$\nu(n', n) = \frac{1}{h} [W(n') - W(n)].$$

To each 'virtual resonator' (transition), there 'corresponds' a higher harmonic of the stationary state n_l, which has to be calculated

1) A treatment of dispersion theory by perturbation methods, which is somewhat different from the above, has been given by P. Epstein, Zs. f. Phys. **9** (1922) 92.

from the classical laws for the mechanical model. Its frequency is $(\nu\tau)$, where $\tau_l = |n_l - n_l'|$ and the ν_l depend on the quantum number of the stationary state under discussion.

The following quantitative connection exists between the classical frequency $(\nu\tau)$ and the quantum-theoretical absorption frequency $\nu(n', n)$. Let us imagine that the transition $n_k \to n_{k'} = n_k + \tau_k$ is performed in a 'linear' way; i.e. let us set for the action integrals

$$J_k = h(n_k + \mu\tau_k), \qquad 0 \leqq \mu \leqq 1. \tag{26}$$

Then we obtain on the one hand,

$$(\nu\tau) = \sum_k \nu_k\tau_k = \sum_k \frac{\partial H_0}{\partial J_k}\tau_k = \frac{1}{h}\sum_k \frac{\partial H_0}{\partial J_k}\frac{\mathrm{d}J_k}{\mathrm{d}\mu} = \frac{1}{h}\frac{\mathrm{d}H_0}{\mathrm{d}\mu}, \tag{27}$$

and on the other,

$$\nu(n', n) = \frac{1}{h}\left[H_0(n+\tau) - H_0(n)\right]; \tag{28}$$

therefore

$$\nu(n+\tau, n) = \int_0^1 (\nu\tau)\,\mathrm{d}\mu. \tag{29}$$

The actual (quantum-theoretical) frequency of the resonator is the 'linear' average of the corresponding (classical) frequency. Alternatively, one can say that the ways in which $\nu(n+\tau, n)$ and $(\nu\tau)$ are obtained from H_1 stand in the same relationship as differential coefficients stand to difference-quotients.

Let us now consider the interaction process. In our model this was described by the perturbation function λH_1. With the present approach, which looks upon the virtual resonators as the real primary thing and upon the classical calculation methods only as a means to a rational tracking down of the true laws of quantum mechanics, we arrive at the following interpretation:

The interaction consists of a mutual influence (*irradiation*) exerted by the virtual resonators on each other. To find the interaction law, let us consider the corresponding law for the higher harmonics in the motion of the model. We shall therefore have to seek a description of the perturbation energy in which the latter appears as the sum of contributions from the higher harmonics.

But this is exactly what our basic formula (16) does for us. Furthermore, it is of the same form as the frequency $(\nu\tau)$, which is characterised by the operator

$$\sum_k \tau_k \frac{\partial}{\partial J_k} = \frac{1}{h} \frac{\mathrm{d}}{\mathrm{d}\mu},$$

by (27). We are therefore as good as forced to adopt the rule that we have to replace a classically calculated quantity, wherever it is of the form

$$\sum_\tau \tau_k \frac{\partial \Phi}{\partial J_k} = \frac{1}{h} \frac{\mathrm{d}\Phi}{\mathrm{d}\mu},$$

by the linear average or difference quotient

$$\int_0^1 \sum_\tau \tau_k \frac{\partial \Phi}{\partial J_k} \mathrm{d}\mu = \frac{1}{h} [\Phi(n+\tau) - \Phi(n)]. \tag{30}$$

Apart from expressions of this kind, however, the basic formula (16a) also contains the quantities $|C_\tau|^2 = C_\tau C_{-\tau}$, which are a measure of the vibration energies of the higher harmonics in the classical model. Naturally these, too, will have to be replaced by 'corresponding' expressions for the virtual resonators. In this case it would not seem reasonable to look for quantities corresponding to the Fourier coefficients C_τ themselves (except for C_0, which does not relate to any virtual resonator and can probably be taken over unchanged into the quantum formulae). Evidently it is only the quadratic expressions $|C_\tau|^2 = C_\tau C_{-\tau}$, which have a quantum-theoretical meaning. Let us denote the quantities which correspond to them by $\Gamma(n, n')$. Since the $|C_\tau|^2$ are unchanged for a sign reversal of τ, we can assume the Γ to be symmetrical:

$$\Gamma(n, n') = \Gamma(n', n). \tag{31}$$

The problem of the determination of the Γ is closely related to the investigation into the intensities of spectral lines, and is of the greatest importance for the further development of quantum theory. When dealing with the effect of external fields, we can obviously determine the Γ, if the motion in the external field is of conditionally periodic character and can be calculated by separation of variables; for in that case the value of $W_2^{(\mathrm{qu})}$ can also be calculated from the well-established theory of quasi-periodic systems. For the case of the mutual interaction of two systems, the classical C_τ-values of the total system can be derived from those of the individual systems in external fields; it may be assumed that some corresponding argument is valid for the quantum-theoretical Γ-values, too. We shall now replace in (16a) the quantity

$$\frac{|C_\tau|^2}{(\nu\tau)} \quad \text{by} \quad \frac{\Gamma(n+\tau, n)}{\nu(n+\tau, n)}, \tag{32}$$

and the 'linear' differentiation by the difference quotient

$$\begin{aligned} W_2^{(\mathrm{qu})} &= -\sum_{\tau_k>0} \int_0^1 \sum_k \tau_k \frac{\partial}{\partial J_k}\left(\frac{\Gamma}{\nu}\right) \mathrm{d}\mu \\ &= -\frac{1}{h} \sum_{\tau_k>0} \left(\frac{\Gamma(n+\tau, n)}{\nu(n+\tau, n)} - \frac{\Gamma(n, n-\tau)}{\nu(n, n-\tau)}\right). \end{aligned} \tag{33}$$

The summation over τ_k includes all quantum jumps which are possible from state n_k, i.e. in the first term, absorption jumps, and in the second, emission jumps. We have thus obtained a separation of the actions of emission and absorption resonators, which occur with opposite signs.

The most important properties of formula (33) are as follows: First, it reduces to the corresponding classical formula for large n_k (large compared with τ_k), and therefore satisfies the correspondence principle. Secondly, the denominator contains the quantum-theoretical frequencies $\nu(n, n')$ in place of the classical $(\nu\tau)$. As we know, the latter have the property that, as one proceeds to higher harmonics, one always obtains certain $(\tau_1, \tau_2, \ldots, \tau_f)$ for which

$$(\nu\tau) = \nu_0\tau_0 + \ldots + \nu_f\tau_f$$

becomes very small; those terms in the series which correspond to such 'small denominators' then produce particularly large contributions (and are responsible for the divergence of the classical perturbation series at points which are everywhere densely distributed in $J_1, \ldots, J_f$-space). Intuitively this can be interpreted to mean that the system resonates at the extremely densely accumulated frequencies of all the upper harmonics. In the quantum-theoretical formula, the upper harmonics are replaced by quantum jumps, whose frequencies satisfy a completely different distribution law; among the allowed transitions from a stationary state, all with finite n_k, there is one with a smallest (quantum-theoretical) frequency. For this reason the quantum-theoretical series (33) does not pose quite so difficult a convergence problem as the corresponding classical series.

We want to add another remark on the perturbation energy of first order (14),

$$\overline{H}_1 = W_1 = C_0.$$

This occurs only for perturbations by external fields, and not for internal interactions. It can probably be assumed that no further quantisation procedure is needed for C_0; at least, this method has so far been successful in all cases.

4. The dispersion theory of Kramers

The dispersion formula which Kramers has recently published in *Nature* (*loc. cit.*), is now obtained from formula (24) of § 2 by means of the same quantisation procedure that was expounded above for the perturbation energy. We obtain for the electric moment created in the resonators in stationary state n_k by the light wave,

$$\mathfrak{p}_x^{(1)} = E \cos (2\pi\nu_0 t) \frac{1}{h} \Sigma \left\{ \frac{2\Gamma_{\mathrm{a}}\nu_{\mathrm{a}}}{\nu_{\mathrm{a}}^2 - \nu_0^2} - \frac{2\Gamma_{\mathrm{e}}\nu_{\mathrm{e}}}{\nu_{\mathrm{e}}^2 - \nu_0^2} \right\}, \tag{34}$$

where the abbreviations

$$\begin{aligned} \nu_{\mathrm{a}} &= \nu(n+\tau, n), & \nu_{\mathrm{e}} &= \nu(n, n-\tau), \\ \Gamma_{\mathrm{a}} &= \Gamma(n+\tau, n), & \Gamma_{\mathrm{e}} &= \Gamma(n, n-\tau) \end{aligned} \tag{35}$$

have been used, and the summation is to be taken over all absorption resonators (a) and all emission resonators (e).

Here the absorption and emission frequencies appear as resonance positions. But the two kinds of resonator have a different behaviour: only absorption resonators contribute positive amounts, as would classical resonators; the emission resonators on the other hand give negative contributions. Kramer's dispersion formula is distinguished from an older formulation of Ladenburg's [1]) by the addition of these terms; but it is just through these difference expressions that we are able to reduce the formula to the corresponding classical one in the limit of large quantum numbers (n_k large compared with τ_k), as required by the correspondence principle.

Kramers follows Ladenburg in using the relation between the

[1]) R. Ladenburg, Zs. f. Phys. **4** (1921) 451. See also R. Ladenburg and F. Reiche, Die Naturwissenschaften **11** (1923) 584.

quantities Γ and Einstein's probabilities $a_n^{n'}$ for spontaneous transitions from one stationary state, n'_k, to another, n_k, of less energy. But we shall not pursue the matter here.

5. Degeneracy and secular perturbations

By means of our formula (33), the calculation of the perturbations up to second order in λ for all non-degenerate systems reduces to a determination of the quantum variables $\Gamma(n, n')$ which correspond to the classical $|C_\tau|^2$. The practical application of the formula is not only hindered by our lack of exact knowledge of the $\Gamma(n, n')$, but above all by the fact that we are dealing with degeneracies in all cases of practical importance. The interactions then produce secular motions of the degenerate variables and the problem arises how one shall deal with them in quantum theory.

We shall only consider the very simplest cases here, and try and see whether the formalism of perturbation theory gives us any hints on how to perform a rational quantisation.

Let us denote the non-degenerate variables by the indices $\alpha, \beta, \ldots$, and the degenerate ones by $\varrho, \sigma, \ldots$.

The simplest case is that of

$$\overline{H}_1 = C_0 \neq 0, \tag{36}$$

where the averaging process is to be carried out over the non-degenerate variables. C_0 is then a function of w_ϱ, J_ϱ and the secular variables of w_ϱ, J_ϱ have to be determined from the equation

$$\overline{H}_1(w_\varrho, J_\varrho) = W_1. \tag{36a}$$

We assume that this equation can be solved by the separation method; new variables w_k, J_k can then be introduced, such that $\overline{H}_1$ becomes independent of the w_k [1]). We assume that this produces no new degeneracy, i.e. that all the secular frequencies are

$$\nu_\varrho = \frac{\partial \overline{H}_1}{\partial J_\varrho} \neq 0. \tag{37}$$

Up to this point, the classical procedure can certainly be carried over into quantum mechanics.

If the purely periodic part of a function Φ is denoted by

$$\tilde{\Phi} = \Phi - \overline{\Phi}, \tag{38}$$

[1]) See M. Born and W. Pauli Jr., Zs. f. Phys. **10** (1922) 137; see formulae (24), (25), (26), p. 151.

then S_1 is determined by the equation

$$\sum_\alpha \nu_\alpha \frac{\partial S_1}{\partial w_\alpha^0} + \tilde{H}_1 = 0 \tag{39}$$

only up to an arbitrary function S_1^* of the w_ϱ^0:

$$S_1 = S_1^0 + S_1^*. \tag{40}$$

The next-higher approximation then gives

$$\sum_\alpha \nu_\alpha \frac{\partial S_2}{\partial w_\alpha^0} + H_2 = W_2, \tag{41}$$

where the abbreviations

$$H_2 = \sum_\varrho \frac{\partial H_1}{\partial J_\varrho} \frac{\partial S_1^*}{\partial w_\varrho^0} + H_2^*, \tag{42}$$

$$H_2^* = \tfrac{1}{2} \sum_{\alpha\beta} \frac{\partial^2 H_0}{\partial J_\alpha \partial J_\beta} \frac{\partial S_1^0}{\partial w_\alpha^0} \frac{\partial S_1^0}{\partial w_\beta^0} + \sum_\alpha \frac{\partial H_1}{\partial J_\alpha} \frac{\partial S_1^0}{\partial w_\alpha^0} \tag{43}$$

have been used.

We shall denote averages over the w_α by a single bar, averages over the w_α and w_ϱ by two bars. Then

$$\bar{H}_2 = \sum_\varrho \nu_\varrho \frac{\partial S_1^*}{\partial w_\varrho^0} + \bar{H}_2^* = W_2. \tag{44}$$

Setting therefore

$$W_2 = \bar{\bar{H}}_2^*, \tag{45}$$

we obtain

$$\sum_\varrho \nu_\varrho \frac{\partial S_1^*}{\partial w_\varrho^0} + \tilde{\bar{H}}_2^* = 0, \tag{46}$$

$$\sum_\alpha \nu_\alpha \frac{\partial S_2}{\partial w_\alpha^0} + \tilde{H}_2^* = 0. \tag{47}$$

From (46) one can determine S_1^*, from (47) S_2, up to an arbitrary function S_2^* of the w_ϱ.

If we restrict ourselves to perturbations of second order, the process can be broken off at that point; comparison of formulae (45) and (43) with (13b) of § 1 shows that W_2 can now be represented by

$$W_2 = -\sum_{(\nu\tau)>0} \sum_\alpha \tau_\alpha \frac{\partial}{\partial J_\alpha} \left(\overline{\frac{|C_\tau|^2}{(\nu\tau)}} \right), \tag{48}$$

where the C_τ are the coefficients of H_1 in its Fourier expansion in the non-degenerate variables w^0_α (i.e. they are functions of the w^0_ϱ), and where the bar denotes averaging over the w^0_ϱ. It is clear from this that the quantum-theoretical coupling energy $W_2^{(\mathrm{qu})}$ must be expressed in the same form as in § 3, (33), except that the $\Gamma(n, n')$ now have to correspond to the averages $\overline{|C_\tau|^2}$.

More difficult is the case which is particularly frequent and important in practice, where

$$\bar{H}_1 = C_0 = 0 \tag{49}$$

identically for the w^0_ϱ, so that the ordinary method of determining the secular perturbations fails.

As Heisenberg and I have shown in the course of a general investigation on molecular models [1]), one will then first have to remove H_1 completely from the perturbation function (which may still contain terms $\lambda^2 H_2 + \ldots$) by means of a canonical transformation. If one restricts oneself again to perturbations of the second order, this leads to practically the same formulae that we have used above. For one has to determine a function S_1 from the following equation (which is identical with (39), since $H_1 = 0$),

$$\sum_\alpha \nu_\alpha \frac{\partial S_1}{\partial w^0_\alpha} + H_1 = 0. \tag{50}$$

The solution is of the form $S_1 = S_1^0(w^0_\alpha, w^0_\varrho) + S_1^*(w^0_\varrho)$, where S_1^0 is determined but S_1^* is arbitrary. To make S_1 unambiguous, one could e.g. postulate that $\bar{S}_1 = 0$ (averaged over w^0_α). Then

$$H = H_0 + \lambda^2 H_2 + \ldots, \tag{51}$$

where

$$H_2 = \sum_\varrho \frac{\partial H_1}{\partial J_\varrho} \frac{\partial S_1}{\partial w^0_\varrho} + H_2^*, \tag{52}$$

$$H_2^* = \tfrac{1}{2} \sum_{\alpha\beta} \frac{\partial^2 H_0}{\partial J_\alpha \, \partial J_\beta} \frac{\partial S_1}{\partial w^0_\alpha} \frac{\partial S_1}{\partial w^0_\beta} + \sum_\alpha \frac{\partial H_1}{\partial J_\alpha} \frac{\partial S_1}{\partial w^0_\alpha}. \tag{53}$$

The secular motions of the w^0_ϱ, J_ϱ are obtained from classical mechanics by means of the equation

$$\bar{H}_2(w^0_\varrho, J_\varrho; J_\alpha) = W_2. \tag{54}$$

[1]) M. Born and W. Heisenberg, Ann. d. Phys. Leipzig **74** (1924) 1.

This equation is evidently of the form

$$\bar{H}_2 = \sum_{\varrho} \overline{\frac{\partial H_1}{\partial J_\varrho} \frac{\partial S_1}{\partial w_\varrho^0}} - \sum_{(\nu\tau)>0} \sum_{\alpha} \tau_\alpha \frac{\partial}{\partial J_\alpha} \left(\frac{|C_\tau|^2}{(\nu\tau)} \right) = W_2, \quad (55)$$

which would suggest that one can perform the passage to quantum theory by taking averages, since the second term for $J_\alpha = h(n_\alpha + \mu\tau_\alpha)$ is the derivative with respect to μ. The question remains, however, what one can do with the first term. For the time being, it does not appear possible to reach any conclusions based on formal considerations.

One can gain insight into the situation by citing the formulae with which Heisenberg [1]) formally describes the multiplets and the anomalous Zeeman effect. Admittedly Heisenberg works with a highly simplified model: he considers an atom consisting of a main portion and one electron (or several electrons), and these constituent parts are dynamically specified only by the magnitudes and orientations of their angular momentum vectors. Heisenberg assumes two separate ways of deviating from the classical laws: (i) he introduces an interaction energy $H^{(\mathrm{cl})}$ between the main portion, the electrons and the magnetic field, which is calculated 'classically', except for the following changes. The Larmor precession of the main portion is assumed to be twice its classical value, and quantum numbers J, R, K_1, K_2,... are assigned to the angular momenta j of the atom as a whole, r of the main portion, k_1, k_2,... of the electrons, respectively; they, as well as their components in the direction of the field, M, P_r, P_{k_1}, P_{k_2},..., may assume half-integral values. (ii) From this quasi-classical quantity $H^{(\mathrm{cl})}$, he forms a quantum-theoretical $H^{(\mathrm{qu})}$ by using a certain averaging process; for one electron, the averaging is of the form

$$H^{(\mathrm{qu})} = \int_{-\frac{1}{2}}^{+\frac{1}{2}} H^{(\mathrm{cl})}\, \mathrm{d}J = \int_0^1 H^{(\mathrm{cl})}\, \mathrm{d}P_k,$$

depending on the choice of coordinate system. For more than one electron, one has to put

$$P_{k_1} = P_{k_1}^0 + \tau_1\mu, \quad P_{k_2} = P_{k_2}^0 + \tau_2\mu, \quad \ldots, \quad (\tau_1, \tau_2, \ldots = \pm 1),$$

and form the expression

$$H^{(\mathrm{qu})} = \int_0^1 H^{(\mathrm{cl})}\, \mathrm{d}\mu.$$

[1]) Cf. a preceding paper in this volume [W. Heisenberg, Zs. f. Phys. **26** (1924) 291].

The similarity of this rule with our formulae is immediately apparent; at the same time we cannot simply consider it as a limiting case of our theory. For, in the case of the anomalous Zeeman effect we are dealing not only with proper degeneracies, but also with accidental and limiting degeneracies of the most complicated kind. Perhaps Heisenberg's rule can be understood in the following sense:

His quantity $H^{(\mathrm{cl})}$ arises from the perturbation energy $\bar{H}_2$, defined by (55), by going over to quantum mechanics in the second term, with respect to all non-degenerate variables; if we denote it by H_2^*, Heisenberg's quasi-classical equation of motion becomes

$$H^{(\mathrm{cl})} = \bar{H}_2^* = W_2.$$

From this, the secular motions of the degenerate variables will have to be determined. If H_2^* is expressed in terms of the relevant quantum numbers, it is with respect to these that one now has to pass over to quantum mechanics (linear integration).

If we consider this interpretation as correct (it would be difficult to justify, for the time being), it would seem natural to conclude by analogy that the same procedure can *always* be applied to secular perturbations of the second order.

We have confined ourselves to show that averaging procedures over quantum integrals, such as were carried out by Heisenberg, look quite natural and lack artificiality, when considered from the point of view of general quantum mechanics.

Related papers

7a M. Born und W. Pauli jr., *Ueber die Quantelung gestörter mechanischer Systeme.* Z. f. Phys. **10**, p. 137, received May 29th, 1922.

7b M. Born und W. Heisenberg, *Ueber Phasenbeziehungen bei den Bohrschen Modellen von Atomen und Molekeln.* Z. f. Phys. **14**, p. 44, received January 16th, 1923.

7c M. Born und W. Heisenberg, *Die Elektronenbahnen im angeregten Heliumatom.* Z. f. Phys. **16**, p. 229, received May 11th, 1923.

7d M. Born und W. Heisenberg, *Zur Quantentheorie der Molekeln.* Ann. d. Phys. **74**, 9, p. 1, received December 21st, 1923.

7e M. Born und P. Jordan, *Zur Quantentheorie aperiodischer Vorgänge.* Z. Phys. **33**, p. 479, received June 11th, 1925.

Dated July 22, 1924

8

THE QUANTUM THEORY OF DISPERSION

H. A. KRAMERS

Through the courtesy of the Editor of *Nature*, I have been permitted to see Mr. Breit's letter, and I welcome the opportunity thus afforded me to add some further remarks on the theory of dispersion, in order to elucidate some points which were only briefly touched upon in my former letter.

In addition to the empirical applicability of a dispersion formula of the type (4), the arguments which led to the proposal of formula (5) rested on the classical expression for the amplitude of the secondary wavelets which an incident plane wave sets up in a system of electrified particles. Consider a system, the motion of which is of multiple periodic type, and let the electrical moment M in a given direction of the undisturbed system, which is supposed to possess u independent fundamental frequencies $\omega_1, \ldots \omega_u$, be represented by

$$M = \sum C \cos (2\pi\omega t + \gamma), \tag{1*}$$

where the frequencies $\omega = \tau_1\omega_1 + \ldots \tau_u\omega_u$ and the amplitudes C depend on the quantities $I_1, \ldots I_u$, which in the theory of stationary states are equal to integer multiples of Planck's constant h, as well as on the set of integer τ-values characteristic for the considered harmonical component of the motion. Let next the incident wave be linearly polarised with its electrical vector parallel to the given direction, and let the value of this vector at the point where the system is situated be given by $E \cos 2\pi\nu t$. The electrical moment of the forced vibrations of frequency ν set up in the system will then be equal to

$$P = \frac{E}{2} \sum \frac{\partial}{\partial I}\left(\frac{C^2\omega}{\omega^2 - \nu^2}\right) \cos 2\pi\nu t, \tag{2*}$$

Editor's note. This paper was published as Nature **114** (1924) 310–311. It **was** signed 'Institute for Theoretical Physics, Copenhagen, July 22.'

where $\partial/\partial I$ stands as an abbreviation for $\tau_1(\partial/\partial I_1)+\ldots\tau_u(\partial/\partial I_u)$.

Now in the limit of high quantum numbers the frequencies of the spectral lines connected with the different possible transitions will coincide asymptotically with the frequencies of the harmonic components of the motion, and, according to the Correspondence Principle, the energy of the spontaneous radiation per unit time combined with each of these frequencies will be asymptotically represented by the expression $(2\pi\omega)^4C^2/3c^3$. We will now make the assumption that, in this limit, formula (2*) gives an asymptotical expression for the dispersion. In order to obtain a general expression holding for all quantum numbers we note that, while the frequencies ω of the harmonic components of the motion are given by the general formula

$$\omega = \frac{\partial H}{\partial I},$$

the exact expression for the frequencies of the spectral lines is given by the general quantum relation

$$\nu_q = \frac{\Delta H}{h},$$

where ΔH signifies the difference of the energy H in two stationary states for which the values of $I_1,\ldots I_u$ differ by $\tau_1 h,\ldots\tau_u h$ respectively. The assumption presents itself that, in a generalisation of formula (2*), the symbol $\partial/\partial I$ has to be replaced by a similar difference symbol divided by h. This is just what has been done in establishing formula (5). In fact, this formula is obtained from (2*) by replacing the differential coefficient multiplied by h, by the difference between the quantities

$$\frac{3c^3A^a h}{(2\pi)^4\nu^{a2}(\nu^{a2}-\nu^2)} \quad \text{and} \quad \frac{3c^3A^e h}{(2\pi)^4\nu^{e2}(\nu^{e2}-\nu^2)}$$

referring to the two transitions coupled respectively with the absorption and emission of the spectral line which corresponds with the harmonic component under consideration.

Apart from the problem of the validity of the underlying theoretical assumptions and of any eventual restriction in the physical applicability of formula (5), the dispersion formula thus obtained possesses the advantage over a formula such as is proposed by Mr. Breit in that it contains only such quantities as allow of a direct physical interpretation on the basis of the fundamental postulates of the quantum theory of spectra and atomic constitution, and exhibits no further reminiscence of the mathematical theory of multiple periodic systems.

In this connexion it may be emphasized that the notation 'virtual oscillators' used in my former letter does not mean the introduction of any additional hypothetical mechanism, but is meant only as a terminology suitable to characterise certain main features of the connexion between the description of optical phenomena and the theoretical interpretation of spectra. This point is especially illustrated by the appearance of negative as well as positive oscillators, which helps to bring out the new feature, characteristic of the quantum theory of spectra, that the emission and absorption of a spectral line is coupled with two separated types of physical processes. The fundamental importance of this general feature for the interpretation of optical phenomena is, as mentioned in my former letter, indicated by the necessity, pointed out by Einstein, of introducing the idea of negative absorption in order to account for the law of temperature radiation.

Printed October 1924

9

THE ABSORPTION OF RADIATION BY MULTIPLY PERIODIC ORBITS, AND ITS RELATION TO THE CORRESPONDENCE PRINCIPLE AND THE RAYLEIGH-JEANS LAW

J. H. VAN VLECK

PART 1. SOME EXTENSIONS OF THE CORRESPONDENCE PRINCIPLE

Abstract

This part deals with the quantum theory aspects of the problem. In the absence of external radiation fields the distortion in the shape of the orbit is essentially the same in both the classical and quantum theories provided in the former we retain only one particular term τ_1, τ_2, τ_3 in the multiple Fourier expansion of the force $2e^2\dddot{v}/3c^3$ on the electron due to its own radiation. The term to be retained is, of course, the combination overtone asymptotically connected to the particular quantum transition under consideration. Then the changes $\Delta J_1, \Delta J_2, \Delta J_3$ in the momenta J_k which fix the orbits and which in the stationary states satisfy the relations $J_k = n_k h$, are in the ratios of the integers τ_1, τ_2, τ_3 in both the classical and quantum theories, making the character of the distortion the same in both even though the speed of the alterations may differ. One particular term in the classical radiation force is thus competent to bring an orbit from one stationary state to another.

The correspondence principle is then extended so as to include absorption as well as the spontaneous emission ordinarily considered. Commencing always with a given orbit it is possible to pair together the upward and downward transitions in such a way that in each pair the upward and downward optical frequencies (determined by the $h\nu$ relation) are nearly equal for large quantum numbers (usually long wave-lengths). That is, if s denotes the initial orbit there exist levels r and t such that the ratio $(W_r - W_s)/(W_s - W_t)$ or ν_{rs}/ν_{st} approaches unity when the quantum numbers become large. We shall define as the differential absorption the excess of positive absorption due to the upward transition $s \rightarrow r$, over the negative absorption (induced emission) for the corresponding downward transition $s \rightarrow t$. It is proved that for large quantum numbers the classical theory value for the ratio of absorption to emission approaches asymptotically the quantum theory expression for the ratio of the differential ab-

Editor's note. The original publication, Phys. Rev. **24** (1924) 330–365, consists of two parts, of which only the first is reproduced here; Part I, pp. 330–346 and Part II, pp. 347–365. References to sections 7 through 17 refer to Part II, which is entitled *Calculation of absorption by multiply periodic orbits.*

sorption to the spontaneous emission. Consequently a correspondence principle which makes the numerical values of the emission in the two theories agree asymptotically, of necessity achieves a similar connection for the absorption.

The correspondence principle basis for a dispersion formula proposed by Kramers, which assumes the dispersion to be due not to the actual orbits but to Slater's 'virtual' or 'ghost' oscillators having the spectroscopic rather than orbital frequencies, is then presented. Kramers' formula has both positive and negative terms and the differential dispersion may be defined in a manner analogous to the differential absorption. It is shown that the quantum differential dispersion approaches asymptotically the dispersion which on the classical theory would come from the actual multiply periodic orbit found in the stationary states. This asymptotic connection for the general non-degenerate multiply periodic orbit must be regarded as an important argument for Kramers' formula.

1. Introduction

According to the first postulate of the Bohr theory of atomic structure the electrons can move only in certain particular quantized non-radiating orbits or stationary states. In order that the quantum conditions may be applicable it is necessary for the motion to be of the so-called 'multiply periodic' type, which can be represented by multiple Fourier series of the form

$$x = \sum_{\tau_1\tau_2\tau_3} X(\tau_1, \tau_2, \tau_3) \cos\left[2\pi(\tau_1\omega_1 + \tau_2\omega_2 + \tau_3\omega_3)t + \gamma^{(x)}_{\tau_1\tau_2\tau_3}\right] \qquad (1)$$

with similar expansions by y and z. The constants ω_1, ω_2, ω_3 are the intrinsic orbital frequencies, and the summation is to be extended over all possible positive and negative integral values of the integers τ_1, τ_2, τ_3 subject to the restriction that $\tau_1\omega_1+\tau_2\omega_2+\tau_3\omega_3$ be positive. For simplicity in notation we have assumed that there are only three ω's, but in general the number of such frequencies can be equal to or less than the number of degrees of freedom of the atomic system to which the electron under consideration belongs. (The slight modifications necessary to extend the results of the present paper to systems with more than three frequencies will be discussed in section 17). We shall also suppose that the system is 'non-degenerate' so that the number of degrees of freedom is equal to rather than greater than the number of frequencies. This means that in the case of three ω's, the problem may be thought of as one of a single electron moving in an asymmetrical three-dimensional static force field. It will also be assumed that the orbits conform to the classical mechanics and that all solutions of the differential equations of motion can be represented by series of the form (1), thus making the complete dynamical system multiply periodic, rather than merely certain particular families of orbits.

The quantum conditions for determining the size of the stationary states consist in equating to integral multiples of h a set of orbital constants J_1, J_2, J_3 defined by the relations

$$\omega_k = \partial W/\partial J_k \quad (k = 1, 2, 3); \qquad 2\bar{T} = J_1\omega_1 + J_2\omega_2 + J_3\omega_3; \quad (2)$$

where W is the total energy and $\bar{T}$ is the average kinetic energy. We can therefore set $J_k = n_k h$ $(k = 1, 2, 3)$, where n_1, n_2, n_3 are the integers or quantum numbers characteristic of a stationary state.[1]

According to the second postulate of the Bohr theory an electron may pass from one allowed orbit to another. The frequency ν_{rs} of the quantum of light thus radiated or absorbed is determined by the familiar relation

$$h\nu_{rs} = W_r - W_s, \qquad (3)$$

where W_τ denotes the energy of the stationary state τ (with $\tau = r$, s, etc.). In the case of emission, r is the initial state and s the final state, while for absorption the significance of these symbols is just reversed, as r represents the higher energy level.

Considerable information concerning the probabilities of the various transitions between different orbits is furnished by Einstein's derivation [2] of the Planck radiation formula, $\varrho(\nu) = 8\pi\nu^3 hc^{-3}/(e^{h\nu/kT} - 1)$, for the specific energy density of black body radiation. Einstein assumed:

(I) The number of atoms N_τ in the state τ is given by the statistical mechanics formula

$$N_\tau = N\, e^{-W(\tau)/kT} / \sum_\tau e^{-W(\tau)/kT} \qquad (4)$$

where N is the total number of atoms, and the sum is to be extended over all possible states.[3]

[1] The quantum conditions have been stated above in what may be termed the correspondence principle form. Some readers may be more familiar with the formulation used by Sommerfeld and others, which consists in equating certain phase or 'quantum' integrals to integral multiples of h, but the two methods can readily be shown to yield the same results in non-degenerate systems. See, for instance, appendix 7 of Sommerfeld's 'Atombau'.

[2] Einstein, Phys. Zeit. **18** (1917) 121.

[3] The 'a priori probabilities' p_τ which ordinarily appear in front of the exponentials in eq. (4) have been omitted, as we are concerned with non-degenerate

(II) The amount of energy emitted in a time Δt by transitions from the state r to the state s is represented by a formula of the type

$$\Delta E_{r\to s} = h\nu_{rs}N_r[A_{r\to s} + B_{r\to s}\varrho(\nu_{rs})]\Delta t. \quad (5)$$

(III) The amount of energy absorbed by transitions from the state s to the state r is given by

$$\Delta E_{s\to r} = h\nu_{rs}N_s B_{s\to r}\varrho(\nu_{rs})\Delta t. \quad (6)$$

(IV) There is to be statistical equilibrium (i.e. $\Delta E_{r\to s}=\Delta E_{s\to r}$) when the energy density has the characteristic black body distribution given by the Planck formula. Using eqs. (3) and (4), one can verify that this equilibrium condition will be fulfilled provided the probability coefficients $A_{r\to s}$, $B_{r\to s}$, and $B_{s\to r}$, which are independent of the temperature and energy density, satisfy the relations

$$B_{s\to r} = B_{r\to s} = (c^3/8\pi h\nu_{rs}^3)A_{r\to s}. \quad (7)$$

Einstein's criterion of statistical equilibrium under black body radiation thus tells us how emission and absorption vary with the density, eqs. (5) and (6), and how they are related to each other, eq. (7), but still leaves undetermined the magnitude of the coefficient $A_{r\to s}$ in (7). To evaluate approximately the latter it is customary to resort to the correspondence principle, as explained below.

It is well known that according to classical electrodynamics an accelerated electron radiates energy continuously and that if in particular the orbit is of the form (1) then the light thus emitted should be resolved by spectroscopes into frequencies which are combination overtones $\tau_1\omega_1+\tau_2\omega_2+\tau_3\omega_3$ of the orbital frequencies ω_1, ω_2, ω_3. According to the quantum relation (3) there is, however, no such immediate connection between the spectroscopic frequency and actual frequencies of motion, but it can easily be proved that if the quantum numbers n_1, n_2, n_3 of the states r and s differ from each other by τ_1, τ_2, τ_3 units respectively, so that $\Delta J_k=\tau_k h$, then in the region of high quantum numbers (usually also long wave-lengths) the optical frequency ν approaches asymptotically the combination overtone $\tau_1\omega_1+\tau_2\omega_2+\tau_3\omega_3$. Since the classical and quantum mechanisms thus then give nearly the same numerical values for optical frequencies even though irreconcilably different in character, it is natural to

systems where all states have the same a priori probabilities, making the p_r's in numerator and denominator cancel.

assume that they give in the case of high quantum numbers the same numerical results for the relative intensities of different spectral lines, at least when there is no radiation field (i.e. $\varrho(\nu)=0$). It is, however, to be clearly understood that the asymptotic connection of *frequencies* is a necessary mathematical consequence of the quantum conditions, and is hence, following Ehrenfest, best termed the correspondence *theorem* for frequencies.[4] On the other hand the existence of an analogous relation for the *intensities* of lines radiated when $\varrho(\nu)=0$ must be regarded as an additional hypothesis, which we shall call the correspondence principle for emission. The latter is generally accepted not only because of its inherent reasonableness but also because of its excellent experimental verification in the 'selection principle' and in the researches of Kramers and others on the intensities of Stark effect components.

To formulate analytically the correspondence principle for emission we need simply note that from the classical expression $(2e^2/3c^3)\ddot{v}^2$ for the rate of radiation from an electron having a vector acceleration $\ddot{v}$, it follows that the amount of energy radiated as light of frequency $\tau_1\omega_1+\tau_2\omega_2+\tau_3\omega_3$ in the time interval Δt is [5]

$$\Delta E = (16\pi^4e^2/3c^3)(\tau_1\omega_1 + \tau_2\omega_2 + \tau_3\omega_3)^4[D(\tau_1, \tau_2, \tau_3)]^2\Delta t \quad (8)$$

where $D^2=X^2+Y^2+Z^2$. If we multiply by N_r to take into account the radiation from N_r electrons, then on comparing this classical expression with the quantum eq. (5) for $\varrho(\nu)=0$, and using the approximation $\nu=\tau_1\omega_1+\tau_2\omega_2+\tau_3\omega_3$, we see that the correspondence principle for emission requires that for high quantum numbers the probability coefficient $A_{r\to s}$ must have a value close to

$$A_{r\to s} = (16\pi^4e^2\nu^3_{rs}/3hc^3)[D^r(\tau_1, \tau_2, \tau_3)]^2. \quad (9)$$

where D^r denotes D evaluated for an orbit of the same size and shape as that found in the stationary state r.

4 To prove the correspondence theorem for frequencies we need simply note that $h\nu_{rs}=W_r-W_s=\Delta W$. Nearly consecutive orbits of large quantum numbers differ but little from each other in relative size and we can then without great error replace the increment ΔW of the energy by the differential

$$dW = \Sigma^3{}_1(\delta W/\delta J_k)\Delta J_k.$$

Using eq. (2) and the relation $\Delta J_k=\tau_k h$, we thus get the result that in the limit $\nu/(\tau_1\omega_1+\tau_2\omega_2+\omega_3)=1$.

5 For simplicity in printing, the arguments τ_1, τ_2, τ_3 of the amplitudes X, Y, Z, D will often be omitted; also the subscripts are often dropped from ν_{rs}.

An equation of the precise form (9) can be expected to hold only asymptotically. At ordinary wave-lengths it is more probable that $A_{r\to s}$ depends not on the orbital frequencies and amplitudes evaluated for one particular stationary state but instead involves these quantities averaged in some manner or other over the continuous succession of 'dis-allowed' orbits intermediate between the initial and final states. We might still have (9) valid in the limit if we assume, for instance, that a more exact expression is

$$A_{r\to s} = (16\pi^4 e^2 \nu^3/3hc^3)$$

$$\left(\nu^{-m}\int_0^1 (\tau_1\omega_1 + \tau_2\omega_2 + \tau_3\omega_3)^m [D(\tau_1, \tau_2, \tau_3)]^{\pm 2} d\lambda\right)^{\pm 1} \qquad (10)$$

Here m is some integer, while λ is an auxiliary parameter such that $J_k = (n_k + \tau_k\lambda)h$, where n_k and $n_k + \tau_k$ $(k = 1, 2, 3)$ denote respectively the quantum numbers of the states s and r. Either the + or − sign must be consistently used throughout. Ordinarily the + sign is taken, and then the special cases of $m = 0,4$ have been studied in some detail by F. C. Hoyt.[6]

2. A correspondence principle for orbital distortions

In the review of the correspondence principle for emission given in section 1 it must be remembered that actually a classical electron radiates simultaneously all the combination overtones in the multiple Fourier expansion rather than just one harmonic vibration component $\tau_1\omega_1 + \tau_2\omega_2 + \tau_3\omega_3$. One can, however, formally avoid this difficulty by introducing what we shall term an 'abridged' radiation force obtained by retaining only the one term τ_1, τ_2, τ_3, in the multiple Fourier expansion of the complete radiation force $2e^2\ddot{v}/3c^3$. It can then be verified that the total radiation is on the average given by eq. (8).[7]

[6] F. C. Hoyt, Phil. Mag. **46** (1923) 135; **47** (1924) 826. Eq. (10), of course, gives by no means all the possible expressions for $A_{(r\to s)}$. In fact from certain viewpoints the best arguments appear to be for a certain type of logarithmic average first introduced by Kramers [The Intensities of Spectral Lines, Dan. Acad. Memoires, 1919, p. 330] and also studied by Hoyt. Only two out of six formulas studied by Hoyt are included in eq. (10).

[7] The average value of the work $-Fv\Delta t$ done in the time Δt against the abridged radiation force F has the value (8) even though the complete expansion of v (but not of $\ddot{v}$ in F) is retained, for 'cross-terms' involving products of different combination overtones cancel out on the time average. For greater detail see section 16 of part II.

We can then imagine the emitted light in the classical theory to be monochromatic, and its frequency will approach asymptotically the optical frequency ν in the quantum theory, while an asymptotic connection of the average rates of spontaneous radiation can be secured by using some such equation as (10).

A question which naturally arises at this point is whether the abridged radiation force with only one term produces just the same distortion in the size and shape of the orbit in the classical theory as the electron experiences in the quantum theory when it passes from one stationary state to another. In the classical theory an orbit having originally the same size and shape as the initial quantized stationary state will obviously after a properly chosen lapse of time be sufficiently damped by the abridged radiation force to make the energy the same as that of the final stationary state in the quantum theory. However, we cannot immediately infer that after this interval of time has expired the classical orbit will have the same shape as the final quantized orbit unless the system has only one degree of freedom, for then only does the energy determine uniquely the shape of the orbit. In the case, for instance, of an elliptical trajectory modified by a relativity precession to make the system non-degenerate (in two dimensions) the classical and quantum orbits might have initially the same semi-major axis and eccentricity; but when the classical orbit has been sufficiently damped by the abridged radiation force to make its energy (and hence approximately the semi-major axis) the same as that of another smaller stationary state, we cannot predict off-hand that then its eccentricity will become identical with that of the latter.

The writer is not aware of any specific statement in the literature as to whether these orbital distortions are the same in the two theories, except for a brief allusion to this question in an interesting article by H. A. Senftleben, which has just appeared.[8] However, an examination

[8] H. A. Senftleben, Z. f. Phys. **22** (1924) 127. Following eq. (50) of his article Senftleben states the necessity of having the average time rate of change of the J's approach each other asymptotically in the two theories, and this implies an asymptotic connection of the orbital distortions. No detailed proof of this, however, is contained in part I of Senftleben's article (the only part available at time of writing) but an indication is given of how it might be obtained from an asymptotic connection of the rates of radiation of angular momentum and energy in the two theories provided the system has only one electron. Both the

of this point is not difficult. In a system with 3 degrees of freedom the size and shape of the orbit are completely determined if we know the values of the expressions J_1, J_2, J_3 defined in (2), for the other arbitrary constants enter only as epoch angles. Now in the quantum theory the J's always change by integral multiples of h, and their alterations in going from one state to another are in the ratio

$$\Delta J_1 : \Delta J_2 : \Delta J_3 = \tau_1 : \tau_2 : \tau_3. \tag{11}$$

On the other hand it is proved in section 16 of part II that if in the classical radiation force we keep only the term involving the combination overtone $\tau_1\omega_1+\tau_2\omega_2+\tau_3\omega_3$, then the J's are also, in the classical theory, altered in the ratio given above for all time intervals. Therefore by including only the abridged force in the classical theory, the distortion is the same in the two theories. Another way of saying this is that one particular term in the classical radiation force damps the orbit in such a way that if at the start it coincides with a given quantized stationary state it tends to pass through a succession of stationary states of smaller energy, whose quantum numbers differ from those of the initial state by $m\tau_1$, $m\tau_2$, $m\tau_3$ units respectively, where m is an integer.

The results quoted in eq. (11) hold even at ordinary wave-lengths (small quantum numbers) for it is shown in section 16 that according to the classical theory the J's always tend to change instantaneously in the ratio (11). Consequently the ΔJ's are in the ratio of integers even though the abridged radiation force be acting for such a long time that the alterations in the J's are of the order of magnitude of their initial values, making the radiated energy comparable with the total energy. The relations are also valid for systems with more than three J's (see section 17).

If we form a three-dimensional space for plotting values of J_1, J_2, J_3, the deformations of the orbit produced by one particular term in the classical radiation force may be represented by a straight line (neglecting small periodic fluctuations which cancel out on the average). The equations of these lines may be written $J_k=J_k^0+\lambda\tau_k h$, where λ

method of proof and interpretation of the result (reached independently) in the present paper differ in many respects from the above, especially the ability to generalize the results to atoms with more than one electron (section 17), and the emphasis on the validity of the relation $\Delta J_1 : \Delta J_2 : \Delta J_3 = \tau_1 : \tau_2 : \tau_3$ at ordinary wavelengths.

is a parameter. Consequently the present considerations give a semi-theoretical basis for determining the probability coefficient $A_{r\to s}$ by averaging some function of the amplitudes and orbital combination frequencies along a straight line in the three-dimensional space connecting the points corresponding to the initial and final states. This procedure, illustrated in eq. (10), was adopted by Kramers and by Hoyt, but it is not apparent from the articles of these investigators whether they realized they were actually averaging along a path which would be traversed by the orbit under the influence of the abridged radiation forces. The various methods which have been suggested for determining $A_{r\to s}$ by averaging over the straight line path mentioned above are such as to make the mean free time in the initial state different from the actual time of transit between the two states under the influence of the abridged radiation force, except, of course, for an asymptotic connection of the two times for high quantum numbers. Neglecting higher powers of $1/c^2$, the quantum mean free time and the classical time of transit can be shown to be identical at ordinary wave-lengths or small quantum numbers provided in eq. (10) we take the $-$ sign in both exponents and take the integer $m=-3$.[9] This result seems worth noting, but is it questionable how much significance should be attached to this method of procedure for there is no obvious reason why the two times should be exactly equal rather than asymptotically connected.[10]

[9] To get the result quoted we need simply integrate eq. (8) between $\lambda=0$ and $\lambda=1$, noting that $dE=(\tau_1\omega_1+\tau_2\omega_2+\tau_3\omega_3)h\,d\lambda$ by eq. (2). This gives the classical transit time. The quantum mean free time is on the other hand simply the reciprocal of $A_{r\to s}$.

[10] It is interesting to note that the classical time of transit from a 3_3 to a 2_2 orbit in hydrogen works out as 1.04×10^{-8} sec. Wien's canal ray experiments indicate a mean free time for H_α of 1.85×10^{-8} sec [Ann. der Phys. **73** (1924) 485]. The latter, Wien shows, is almost exactly the reciprocal (1.87×10^{-8} sec) of the logarithmic decrement for a linear oscillator of frequency ν (not an actual orbit), but this very close agreement is probably only a coincidence. On the other hand the value 1.04×10^{-8} sec must be corrected to allow for the transitions $3_2\to2_1$, $3_1\to2_2$, which also contribute to H_α and have longer mean free times than $3_3\to2_2$ (the 3_1 state is, in fact, almost metastable) thus making the effective mean free time much larger. The agreement is as good as can be expected, for the correct formula for $A_{r\to s}$ probably does not make the quantum mean free time exactly equal to the classical time of transit, but instead involves a different kind of average of amplitudes and frequencies than that obtained by putting $m=-3$ in eq. (10).

3. Insufficiency of a correspondence principle relating only to emission

The correspondence principle for emission reviewed in section 1, cannot in a certain sense be regarded as entirely adequate because it establishes (or rather postulates) an asymptotic numerical connection of the classical and quantum theory for intensity only in the special case that there is no radiation field (i.e. $\varrho(\nu)=0$). The term present in eq. (5) when $\varrho(\nu)=0$ we shall call the spontaneous emission (Einstein's Ausstrahlung'), while the second or remaining term proportional to the energy density we shall call the induced emission, although it is sometimes called the 'negative absorption' in distinction from the true or positive absorption given by eq. (6). The correspondence principle for emission correlates quantum theory spontaneous emission due to the radiation force $2e^2\ddot{v}/3c^3$. On the other hand an electron can according to the classical theory absorb energy from a radiation field, and there must be some kind of an asymptotic connection between this absorption in the classical electrodynamics and the induced emission and absorption in the quantum theory. This will be discussed in sections 4 and 5. There can be no question of a correlation of the classical and positive quantum absorptions alone, as this would leave the induced emission unexplained. The existence of the induced emission term in the quantum theory may at first sight appear strange, but it is well known that this is qualitatively explained in that with the proper phase relations a classical electric wave may receive energy from an atomic system although on the average (i.e. integrating over all possible phase relations) it contributes more than it receives in exchange. It is therefore the excess of positive absorption over the induced emission which one must expect to find asymptotically connected to the net absorption in the classical theory.

4. Correspondence principle for absorption for a linear oscillator

Before seeking to develop a correspondence principle for absorption for the general case of an arbitrary multiply periodic orbit, we shall first for simplicity and clarity confine our attention to a one-dimensional linear oscillator. Here the multiple Fourier expansion reduces to

$$x = D\cos(2\pi\omega t + \gamma), \qquad (y = z = 0).$$

As there is only one degree of freedom, there is just one quantum number n, and this can by the correspondence principle only change by one unit, as there are no harmonics of the fundamental frequency ω in the Fourier expansion. Now the energy W of a linear oscillator of amplitude D, mass m, and frequency ω is in general $2\pi^2\omega^2 mD^2$. Furthermore it is well known that for a linear oscillator the quantum conditions require that the energy W_n of a state of quantum number n have the value $nh\omega$. Therefore the amplitude D_n of an orbit of quantum number n is given by

$$D_n^2 = \frac{nh}{2\pi^2\nu m}. \tag{12}$$

In writing (12) we have utilized the familiar fact that for a linear oscillator $\nu=\omega$, as here the spectroscopic and orbital frequencies are identical.[11] We can now apply the correspondence principle for emission, as embodied in eq. (9), to determine the approximate value of the probability coefficient $A_{r\to s}$. Using the fact that in the present notation $D^n(\tau_1, \tau_2, \tau_3)$ is nothing but D_n, while ν_{rs} is simply ν, and taking r as an n quantum state, s as an $(n-1)$ quantum state, we thus get $A_{r\to s}=8\pi^2e^2\nu^2n/3c^3m$. From (7) it then follows that

$$B_{n\to(n-1)} = B_{(n-1)\to n} = \frac{n\pi e^2}{3hm\nu}. \tag{12a}$$

Hence by eq. (6) we see that in a time Δt the energy which is removed from a radiation field of density $\varrho(\nu)$ by the positive absorption of quanta by N oscillators all in $(n-1)$ quantum states is

$$\Delta E_{(n-1)\to n} = \tfrac{1}{3}n\pi e^2m^{-1}N\varrho(\nu)\Delta t. \tag{13}$$

On the other hand it is well known that according to the classical theory the average rate at which a linear oscillator absorbs energy in a field of radiation is independent of the amplitude and is given [12] by $\frac{1}{3}\pi e^2m^{-1}\varrho(\nu)$. Consequently in the interval Δt, N oscillators should absorb the energy

$$\Delta E_{\text{abs}} = \tfrac{1}{3}\pi e^2m^{-1}N\varrho(\nu)\Delta t. \tag{14}$$

Because of the presence of the factor n, the quantum theory expression

[11] The result $\nu=\omega$ is obvious from an inspection of eq. (3), as in the present case $W_n=nh\omega$, $\Delta n=\pm 1$.

[12] Cf. Planck, Wärmestrahlung, 4th ed., eqs. (260) and (159).

(13) for the positive absorption differs increasingly from (14) as the quantum number becomes larger. This discrepancy is not surprising, for we have not taken into account the fact that oscillators in the state $(n-1)$ may in the presence of a radiation field be induced to emit energy and pass to the state $(n-2)$ at a more rapid rate than when $\varrho(\nu)=0$ and there are only spontaneous transitions.[13] Each of these excess or induced transitions may be thought of as returning to the ether or light wave the energy $h\nu_{st}$, a sort of regenerative effect. Hence we can take as the *net* absorption of energy in the time Δt by N oscillators in the state $(n-1)$ the expression

$$\Delta F = [B_{(n-1)\to n} - B_{(n-1)\to(n-2)}]h\nu N\varrho(\nu)\Delta t.$$

This way of measuring absorption we shall term the differential rate of absorption in contrast to the true positive rate given by eq. (13). Now by (12a) we have, on lowering the integer n by one unit, the result $B_{(n-1)\to(n-2)}=(n-1)\pi e^2/3mh\nu$, and hence $\Delta F=\Delta E_{\text{abs}}$. We thus see that in the limiting case of large quantum numbers, where eq. (12a) is valid, the classical value (14) for the rate of absorption of energy is nothing but the differential rate of absorption in the quantum theory. This connection of the classical and quantum differential absorption we shall term the correspondence principle for absorption; it is a purely mathematical consequence of the correspondence principle for emission, which was used in deriving (12a). Another way of stating the results is that because of eqs. (9) and (12a) the ratio $(B_{(n-1)\to n}-B_{(n-1)\to(n-2)})\varrho(\nu)/A_{(n-1)\to(n-2)}$ of the differential absorption to the spontaneous emission for any given orbit has for large n very nearly the classical value $c^3\varrho(\nu)/16\pi^3 mD_n^2\nu^4$ for the ratio of absorption to emission.[14]

[13] If the electrons are in the orbit of lowest quantum number there can be only positive absorption because there are no still lower energy levels to which induced emission may take place. This case has been considered by Ladenburg [Z. f. Phys. **4** (1921) 451]; also by Ladenburg and Reiche [Naturw. **27** (1923) 584]. These writers note the necessity of having the factor n in comparing the classical and positive quantum absorption.

[14] Reference should be made while on the present subject to the discussion in section 158 of Planck's 'Wärmestrahlung' (4th ed.), which establishes an asymptotic connection between the classical and quantum formulas for statistical equilibrium. A correspondence principle for absorption could doubtless be derived from such considerations, but no explicit mention is made of the asymptotic connection of the classical absorption and the differential absorption

5. Generalization of the correspondence principle for absorption to an arbitrary non-degenerate multiply periodic system

We must now seek to examine whether an analogous correspondence principle for absorption holds for an arbitrary non-degenerate multiply periodic system, whose orbits are given by eq. (1) and which represent essentially the most general type of motion amenable to present quantum theory methods. Here we can no longer enjoy such simplifications as the existence of a single quantum number and the results $\Delta n = \pm 1$, $\nu = \omega$ characteristic of a linear oscillator.

Let us consider a state s of quantum numbers n_1, n_2, n_3.[15] Then when a quantum of energy is absorbed by an orbit of this type the electron will pass to some other state of quantum numbers $n_1+\tau_1$, $n_2+\tau_2$, $n_3+\tau_3$.[16] Starting with the state s (not r) there is no emissive transition to a state t such that the frequency ν_{st} of the light thus emitted is just equal to the frequency ν_{rs} of that absorbed on passage to the state r. (This amounts virtually to saying that in general energy levels are not evenly spaced, except, of course, for such an ideal case as the linear oscillator). However, if the state t has quantum numbers $n_1-\tau_1$, $n_2-\tau_2$, $n_3-\tau_3$, then for large values of n_1, n_2, n_3 the ratio ν_{rs}/ν_{st} approaches unity, for both these spectroscopic frequencies approach asymptotically the combination overtone $\tau_1\omega_1+\tau_2\omega_2+\tau_3\omega_3$ of the orbital frequencies. Hence it is natural to define as the differential rate of absorption the excess of light absorbed by transitions from the state s to the state r over the energy returned to the ether by induced emissive transitions from the state s to the state t, where, as mentioned above, the differences between the quantum numbers for the states s and t are the same as the corresponding differences for the states r and s.

With this definition it follows that in the time Δt the differential absorption of light energy of frequency approximately $\tau_1\omega_1+\tau_2\omega_2+\tau_3\omega_3$ by N atoms in the state s amounts to

$$\Delta F = [h\nu_{rs}\,\varrho(\nu_{rs})B_{s\to r} - h\nu_{st}\,\varrho(\nu_{st})B_{s\to t}]N\Delta t. \tag{15}$$

for a single orbit (where thermodynamic equilibrium need not be assumed) which is the primary concern of the present paper.

15 The generalization to cases of more than three quantum numbers is given in section 17.

16 Not all the integers τ_1, τ_2, τ_3 need be positive but the state r must have a greater energy than s.

If we admit the validity of the correspondence principle for emission then eqs. (7) and (9) show that for large quantum numbers this expression becomes approximately

$$\Delta F = \tfrac{2}{3}\pi^3 e^2 h^{-1}[\nu_{rs}\,\varrho(\nu_{rs})D_r^2 - \nu_{st}\,\varrho(\nu_{st})D_s^2]N\Delta t$$

where for brevity we have written D_r for $D^r(\tau_1, \tau_2, \tau_3)$, etc. Also for large quantum numbers the discrete succession of quantum orbits becomes so nearly consecutive that differences may be replaced by differentials, just as is done in the derivation of the correspondence theorem for frequencies.[4] Consequently

$$\Delta F = \tfrac{2}{3}\pi^3 e^2 h^{-1}\{\varrho(\nu)D_s^2(\nu_{rs}-\nu_{st})+\nu_{st}(\partial\varrho/\partial\nu)D_s^2(\nu_{rs}-\nu_{st})+\nu_{st}\varrho(\nu_{st})\delta D^2\}N\Delta t$$

where δ denotes the difference between an expression evaluated for the states s and t. It is, of course, to be understood that $\varrho(\nu)$ and $\partial\varrho/\partial\nu$ are to be evaluated with ν approximately equal to ν_{st} (or ν_{rs}). Now since we are dealing with the large quantum number limit, the correspondence theorem for frequencies tells us that ν_{st} may without sensible error be replaced by the combination overtone $\tau_1\omega_1+\tau_2\omega_2+\tau_3\omega_3$ of the orbital frequencies, and similarly $\nu_{rs}-\nu_{st}$ is to the desired approximation equal to $\delta(\tau_1\omega_1+\tau_2\omega_2+\tau_3\omega_3)$. Now since the system is non-degenerate, the amplitude D as well as the ω's and the energy W, is a function of the constants J_1, J_2, J_3 which according to the quantum conditions are equated to integral multiples of h, and consequently

$$\begin{aligned}\delta\omega_1 &= \frac{\partial\omega_1}{\partial J_1}\,\delta J_1 + \frac{\partial\omega_1}{\partial J_2}\,\delta J_2 + \frac{\partial\omega_1}{\partial J_3}\,\delta J_3 \\ &= h\left(\tau_1\frac{\partial}{\partial J_1} + \tau_2\frac{\partial}{\partial J_2} + \tau_3\frac{\partial}{\partial J_3}\right)\omega_1, \quad \text{as} \quad \delta J_k = \tau_k h.\end{aligned}$$

Similar relations hold for $\delta\omega_2$, $\delta\omega_3$, and δD^2. The expression for the differential absorption therefore becomes

$$\begin{aligned}\Delta F = \tfrac{2}{3}\pi^3 e^2 \Big[\varrho(\omega_\tau)\left(\tau_1\frac{\partial}{\partial J_1} + \tau_2\frac{\partial}{\partial J_2} + \tau_3\frac{\partial}{\partial J_3}\right)G_r \\ + \frac{\partial\varrho}{\partial\nu}\,G_r\left(\tau_1\frac{\partial}{\partial J_1} + \tau_2\frac{\partial}{\partial J_2} + \tau_3\frac{\partial}{\partial J_3}\right)\omega_\tau\Big]N\Delta t \qquad (16)\end{aligned}$$

where we have written ω_r for $\tau_1\omega_1+\tau_2\omega_2+\tau_3\omega_3$ and G for $\omega_r[D(\tau_1, \tau_2, \tau_3)]^2$.

It is shown in part II of the present paper that eq. (16) also gives the amount of energy which according to purely classical mechanics is absorbed in a time Δt by N systems, each with a multiply periodic orbit similar to the state s, when exposed to a radiation field of vanishing intensity except for frequencies in the vicinity of $\tau_1\omega_1+\tau_2\omega_2+\tau_3\omega_3$. That is to say eq. (16) also gives the part of the classical absorption due to resonance of the impressed waves with the combination overtone τ_1, τ_2, τ_3. The total classical absorption is, of course, the sum of resonance effects for all possible overtones.

We may therefore conclude that the correspondence principle for absorption holds even for an arbitrary multiply periodic orbit; i.e. assuming the validity of the correspondence principle for emission, the differential quantum absorption by a particular orbit approaches at high quantum numbers (usually long wave-lenghts) the classical absorption of light of asymptotically corresponding frequency.[17]

[17] It is clearly to be understood that although in deriving (16) we have apparently utilized eq. (9), the proof of (16) can be readily shown to be equally valid if (9) holds asymptotically, so that at ordinary wave-lengths $A_{r\to s}$ might, for instance, be determined by averaging the frequencies and amplitudes in the manner given in eq. (10). In this connection it is interesting to note that it is not difficult to prove that for the special case of a linear oscillator the differential quantum absorption and the classical absorption are exactly equal to each other even for low quantum numbers provided $A_{r\to s}$ is determined by taking the + signs in (10), with the integer m arbitrary. On the other hand if was found in section 2 that to make the mean free time of spontaneous emission in the quantum theory just equal the classical time of transit from one state to another in the absence of external radiation (i.e. $\varrho(\nu)=0$), it is necessary to take the − signs in (10) and $m=-3$.

Another way of formulating the correspondence principle for absorption is as follows. The quantity $h\nu A_{r\to s}$ plays in the quantum theory the role of a coefficient of emission, and may be denoted by α. Using arguments similar to those employed in the derivation of (16), it is easy to show that the differential absorption has for large quantum numbers very approximately the value

$$\Delta F = \frac{c^3}{8\pi}\Big[\varrho(\omega_\tau)\left(\tau_1\frac{\partial}{\partial J_1}+\tau_2\frac{\partial}{\partial J_2}+\tau_3\frac{\partial}{\partial J_3}\right)(\alpha\omega_\tau^{-3})$$
$$+\alpha\omega_\tau^{-3}\frac{\partial\nu}{\partial\varrho}\left(\tau_1\frac{\partial}{\partial J_1}+\tau_2\frac{\partial}{\partial J_2}+\tau_3\frac{\partial}{\partial J_3}\right)\omega_\tau\Big]N\Delta t.$$

This is, however, nothing but the classical theory formula for the coefficient of absorption in terms of the coefficient of emission as can be seen from (8) and (16). This way of stating the correspondence principle for absorption does not utilize (9) even asymptotically.

In the present paper we have proved the correspondence principle for absorption after assuming the validity of the ordinary correspondence principle for emission. The procedure could equally well have been reversed. That is to say, assuming an asymptotic connection of the differential absorption in the quantum theory with the corresponding absorption in the classical mechanism we could show that the spontaneous emission in the quantum theory must approach equality with the classical formula (8) for the spontaneous radiation of energy.

The classical formula (8) for spontaneous emission involves rather more electrodynamics than the analogous classical formula for absorption, given in eq. (16), for the derivation of the former is based on retarded potentials and the validity of the field equations, while the latter requires simply that there be a certain incoherence in the radiation field and that the change in the energy be equal to the scalar product of the impressed force and the velocity of the particle, a very broad mechanical principle. As pointed out in the preceding paragraph, the asymptotic values of the probability coefficients $A_{r\to s}$, $B_{r\to s}$, etc., can be derived from the absorption instead of the emission viewpoint. Because of this alternative the correspondence principle for evaluating these coefficients is in a certain sense less electrodynamical and more purely mechanical in nature than it has sometimes been considered to be. From this standpoint the connection between the quantum theory and classical electrodynamics comes through the assumption of Wien's law in obtaining eq. (7), for in Einstein's derivation of the Planck radiation formula considerations of statistical equilibrium without resorting to Wien's law do not suffice to determine $B_{r\to s}/A_{r\to s}$. If we assumed both the correspondence principle for emission and that for absorption it would be possible to dispense with the need of assuming (rather than proving) Wien's law. Virtually this procedure was adopted by Planck, who studied the special case of the linear oscillator and rotating dipole.[18] As pointed out to the writer by Prof. Ehrenfest, the content of the present paper may be regarded as placing such a procedure on a more general basis.[19]

[18] Wärmestrahlung, 4th ed., section 158. Cf. footnote [14].

[19] The writer has just learned (Sept. 1924) that in unpublished computations made at Copenhagen, J. C. Slater has independently derived an absorption formula similar to eq. (16) and has also noted the asymptotic connection of the classical and quantum absorption discussed in sections 4 and 5.

6. The general correspondence principle basis for Kramers' dispersion formula

In a recent note[20] Kramers has proposed a formula for dispersion which is a modification of an equation previously developed by Ladenburg and Reiche,[21] and which must be regarded as a distinct advance in the problem of reconciling dispersion with quantum phenomena. The physical principle underlying Kramers' formula is that of 'virtual' or 'ghost' resonators, first suggested by Slater[22] and elaborated by Bohr, Kramers, and Slater.[23] According to this viewpoint the dispersion is not to be calculated by considering the actual orbit (stationary state) as reacting classically to the impressed waves. Instead, the stationary states appear to be unaffected except for occasional quantum leaps, but the dispersion is to be computed as due to a set of hypothetical linear oscillators whose frequencies are the spectroscopic ones rather than those of the orbits. The introduction of these virtual resonators is, to be sure, in some ways very artificial, but is nevertheless apparently the most satisfactory way of combining the elements of truth in both the classical and quantum theories. In particular this avoids the otherwise almost insuperable difficulty that it is the spectroscopic rather than the orbital frequencies, i.e. ν rather than $\tau_1\omega_1+\tau_2\omega_2+\tau_3\omega_3$, which figure in dispersion.[24]

Kramers' formula for the polarization due to N_s electrons in the state s is[25]

$$P_s = \frac{c^3}{32\pi^4}\left[\sum_r \frac{A_{r\to s}}{\nu_{rs}^2(\nu_{rs}^2-\nu^2)} - \sum_t \frac{A_{s\to t}}{\nu_{st}^2(\nu_{st}^2-\nu^2)}\right] N_s \mathscr{E} \tag{17}$$

20 H. A. Kramers, Nature, May 10, 1924, p. 673; Aug. 30, 1924, p. 310.

21 Ladenburg and Reiche, Naturwissenschaften **27**, 584 (July 6, 1923). Ladenburg and Reiche's formula differs from that of Kramers, given in (17), principally in that the negative terms are not included.

22 Slater, Nature, March 1 (1924) 307.

23 Bohr, Kramers and Slater, Phil. Mag. **47** (1924) 785.

24 Reference should, however, be made to the alternative explanation proposed by Darwin, Proc. Nat. Acad. **9** (1923) 25.

25 This equation differs by a factor $\frac{1}{3}$ from the formula in Kramer's note to Nature, as (17) is written for the case where all orientations of the atom are equally probable, while Kramers' original equation is for the case that the atoms are always so oriented that their free vibrations are parallel to the impressed electric field. The assumption that all orientations are equally probable may appear at first sight contrary to the assumption that the system is non-

where $\mathscr{E}$ is the electric intensity and ν is the frequency of the impressed wave. The summation with respect to r is to be taken over all states of energy content higher than s, while that for t over all orbits of lower energy content than s. Each term in this equation represents the dispersion due to a 'virtual oscillator' and corresponds to a particular quantum transition starting from the state s. Each oscillator in the t summation gives a negative dispersion, corresponding to Einstein's negative absorption, mentioned in section 3.

Kramers states that (17) merges asymptotically for large quantum numbers into the classical formula for the polarization. It is our purpose to show that this is true not just when the quantized system is a linear oscillator, but also when it is the most general type of non-degenerate multiply periodic orbit.[26] Just as in section 5, the upward transitions

degenerate, as all three intrinsic frequencies (the criterion for non-degeneracy) can appear only when the atomic force field does not have a special symmetry, and with such a dissymmetry the quantum conditions allow only certain particular orientations. However, we may imagine the asymmetrical internal atomic force fields to have all possible orientations in the different atoms. This artifice, introduced for simplicity, enables one to have a random spacial distribution of orbits without having degenerate systems.

In this connection it is interesting to note that if the axes of the asymmetrical atomic force fields had the same orientations in all atoms, as is the case when the dissymetry is due to a constant external magnetic or electric field, then we should expect the virtual resonators to have only certain particular orientations, as only a few quantized spacial positions can be assumed by the atoms. By analogy with the classical theory results mentioned after eq. (41) in section 15 of part II, there might then be possible a polarization in the y- and z-directions even though the electric intensity coincides with the x-axis. The polarization and electric vectors would then have different directions, as in a crystal. Therefore the application of a weak constant external field would cause an abrupt dissymmetry in the index of refraction just as it creates an abrupt outstanding polarization of the spontaneously radiated light (cf. N. Bohr, the Fundamental Postulates of the Quantum Theory, Supplement of Proc. Cambr. Phil. Soc., p. 27). These sudden discontinuities are not found in the classical theory because the spacial distributions are continuous rather than quantized.

[26] In his first note Kramers did not give his proof of the desired asymptotic connection. Inasmuch as the formula for the dispersion by an arbitrary multiply periodic orbit, to be developed in section 15 of part II, did not appear to have previously been given and was not mentioned in Kramers' note, the writer supposed that Kramers' prior demonstration was somewhat less general than that of the present paper. Since writing this paper, the author has learned, however, that Kramers' unpublished computations are of the same scope and generality as those in § 15. Cf. Kramers' second note [20].

($s \to r$) and the downward ones ($s \to t$) may be mated together in such a way that in each pair ν_{rs}/ν_{st} converges to unity for large quantum numbers. This is accomplished, just as before, by making the differences in quantum numbers between the states r and s the same as those between the states s and t, and we shall denote these differences by τ_1, τ_2, τ_3. Each pair contributes a positive and a negative term in (17) and their net effect is a 'differential dispersion' very analogous to the differential absorption mentioned in sections 4 and 5. In precisely the same way that (16) is derived from (15) we find that for very large quantum numbers the contribution of one particular pair to the polarization becomes

$$P_s = \tfrac{1}{6}e^2 \left[\frac{1}{\omega_\tau^2 - \nu^2} \left(\tau_1 \frac{\partial}{\partial J_1} + \tau_2 \frac{\partial}{\partial J_2} + \tau_3 \frac{\partial}{\partial J_3} \right) G_\tau \right.$$
$$\left. - \frac{2\omega_\tau G_\tau}{(\omega_\tau^2 - \nu^2)^2} \left(\tau_1 \frac{\partial}{\partial J_1} + \tau_2 \frac{\partial}{\partial J_1} + \tau_3 \frac{\partial}{\partial J_2} \right) \omega_\tau \right] N_s \mathscr{E} \tag{18}$$

where G_τ and ω_τ are defined as in eq. (16).

It is proved in section 15 that (18) also gives the polarization which according to the classical theory would be produced by resonance between the impressed frequency and the combination overtone ω_τ in a multiply periodic orbit of the same size and shape as the state s. The differential dispersion in Kramers' theory therefore approaches asymptotically the classical dispersion for the actual orbit. The asymptotic connection is thus very similar to that previously encountered in the correspondence principle for absorption.

It is particularly interesting to note that although both the positive and negative terms in the differential dispersion taken separately represent a type of dispersion characteristic of a linear oscillator (except that oscillators with negative dispersion correspond to no ordinary physical reality, as they would have to possess a negative mass) the difference of the two terms approaches asymptotically a more complicated type of dispersion appropriate to the general multiply periodic orbit. For eq. (18) contains a term in $(\omega_\tau^2 - \nu^2)^{-2}$ as well as the familiar term in $(\omega_\tau^2 - \nu^2)^{-1}$ characteristic of the ideal resonator. The fact that the dispersion for the simple virtual oscillators thus merges asymptotically into the more complex classical dispersion for the actual orbits must be regarded as an important argument for the virtual resonator viewpoint, as such a connection is not an obvious outcome of the theory. Especially does this asymptotic connection increase our faith in the particular form of virtual oscillator theory embodied in Kramers' dispersion formula, and makes it easier to accept the rather artificial negative dispersion terms thus involved.[27]

Related papers

9a J. H. van Vleck, *A Correspondence Principle for Absorption.* J. Opt. Soc. Amer. **9**, p. 27, dated April 7, 1924.

9b K. F. Niessen, *Ableitung des Planckschen Strahlungsgesetzes für Atome mit zwei Freiheitsgraden.* Annalen Physik **75**, p. 743, received Oct. 1, 1924.

[27] It is to be noted that if $A_{r\to s}$ is determined in accordance with eq. (10), the + signs being taken, then the differential dispersion in the Kramers' formula is for the special case of a quantized linear oscillator easily proved identical even at low quantum numbers with the classical dispersion. It was pointed out in footnote [17] that this kind of formula also makes the differential and classical absorption identical for the ideal oscillator without passing to the high quantum number limit. These facts taken together furnish a limited amount of evidence for the use of the average of the form (10) (with + signs) rather than the alternative logarithmic average which Hoyt has shown explains observed intensities equally well in the instances studied by him [6].

Received January 5, 1925

10

ON THE DISPERSION OF RADIATION BY ATOMS

H. A. KRAMERS AND W. HEISENBERG

When an atom is exposed to external monochromatic radiation of frequency ν, it not only emits secondary monochromatic spherical waves of frequency ν which are coherent with the incident radiation; but the correspondence principle also demands, in general, that spherical waves of other frequencies are emitted as well. These frequencies are all of the form $|\nu \pm \nu^*|$, where $h\nu^*$ denotes the energy difference of the atom in the state under consideration and some other state. The incoherent scattered radiation corresponds in part to certain processes which have recently been discussed by Smekal, in connection with investigations which are linked to the concept of light quanta. In this paper it will be shown how a wave-theoretical analysis of the scattering effect of an atom can be carried out in a natural and apparently unambiguous manner by means of the correspondence principle. The treatment bases itself on an extension of the point of view, recently put forward in a new paper by Bohr, Kramers and Slater, that there exists a connection between the emission of waves by an atom, and its stationary states. Our conclusions, should they be justified, can be expected to constitute an interesting confirmation of this concept.

1. Introduction

It is known that the optical phenomena of dispersion and absorption, which arise from the passage of monochromatic light through a gas, can be interpreted from an atomistic point of view by saying that each irradiated atom emits secondary spherical waves, whose frequency is the same as that of the incident light, and which are coherent with it. The conclusion that, according to this view, weak scattered radiation is emitted in all directions, has been brilliantly verified in Rayleigh's theory of the blue of the sky and also in direct laboratory experiments.

Editor's note. This paper was published as Zs. f. Phys. **31** (1925) 681–708. The reception year 1924 printed in the original publication is a printing error, as is clear from the subscript 'Copenhagen, Institut for teoretisk Fysik, December 1924' at the end of the paper.

Exceptionally strong support is lent to this theoretical picture above all by the fact that the relevant experimental observations allow one to determine Avogadro's number.

The theory of electrons, based on classical electrodynamics, has succeeded in giving a more detailed description of the scattering effect of the atoms. Thus one imagines that electrons which are quasi-elastically bound in the atom can execute harmonic oscillations about an equilibrium position, and that these electrons are made to resonate by the electric forces in the radiation field. This concept has led to a theory of dispersion which reproduces all the essential features of the observed dispersion, not only in the region of normal dispersion, but also in the regions of anomalous dispersion near the absorption lines, whose frequencies are set equal to the natural frequencies of the electrons. In agreement with experiment, the theory calls for the existence of pronounced maxima in the intensity of the scattered radiation at just those frequencies (resonance radiation).

It is nevertheless known that any attempt based on classical theory to give an exact interpretation of the dispersion phenomena runs into difficulties which are closely connected with the difficulties that stand in the way of an interpretation of atomic spectra in that theory. The quantum theory of line spectra indicates how the latter difficulties are overcome. We are thus faced with the problem of describing the scattering and dispersion effects of the atom in terms of the quantum-theoretical picture of atomic structure. According to this picture, the appearance of a spectral line is not linked with the presence of elastically oscillating electrons, but with transitions from one stationary state to another. Bohr's correspondence principle however gives us a significant hint on how to describe the reaction of the atom to the radiation field, using classical concepts. In a recently published paper by Bohr, Kramers and Slater,[1]) the authors sketched in rough outline how such a description can be accomplished in a relatively simple manner. What is above all typical for this theory, is the assumption that the reaction of the atom to the radiation field should primarily be understood as a reaction of an atom existing in a given stationary state; it is also assumed that transitions between two stationary states are of very short duration, and that the detailed nature of these transitions will not play any rôle in the description of the optical phenomena. According to this view, the following would represent

[1]) Zs. f. Phys. **24** (1924) 69; Phil. Mag. **47** (1924) 785.

a first step in describing what happens when an atom in some arbitrary stationary state is irradiated by monochromatic light. Let us consider plane monochromatic polarised waves incident on the atom; let the electric vector $\mathfrak{E}(t)$ of these waves at the location of the atom be represented by the real part of the vector

$$\mathfrak{E}(t) = \mathrm{R}(\mathfrak{E}\, \mathrm{e}^{2\pi i\nu t}), \tag{1}$$

where the components of the time-independent vector $\mathfrak{E}$ are in general complex quantities, and where ν will be assumed positive throughout this paper. Under the influence of such waves, the atom now emits spherical waves into the surrounding space. Let the oscillating dipole, by means of which these spherical waves can be represented, have a moment $\mathfrak{P}(t)$, again given by the real part of an expression,

$$\mathfrak{P}(t) = \mathrm{R}(\mathfrak{P}\, \mathrm{e}^{2\pi i\nu t}), \tag{2}$$

where $\mathfrak{P}$ is in general a complex vector which depends on ν and $\mathfrak{E}$, for a given stationary state; its direction depends on the direction of $\mathfrak{E}$, whereas its absolute magnitude is proportional to the absolute magnitude of $\mathfrak{E}$, at least in the limit of weak radiation, i.e. $\mathfrak{P}$ is a linear vector function of $\mathfrak{E}$. We assume expression (2) to be valid as long as the atom remains in the stationary state under consideration.

The assumption which we have just made permits us in some wide measure to account for the phenomena of dispersion, absorption and scattering of light. This is clear for the reasons mentioned at the beginning of this article. As for the total intensity of scattered light, one might generally feel inclined to put the energy S, scattered per unit time, simply equal to

$$S = \frac{(2\pi\nu)^4}{3c^3} (\mathfrak{P}\overline{\mathfrak{P}}), \tag{3}$$

where $\overline{\mathfrak{P}}$ represents the complex conjugate vector to $\mathfrak{P}$. In those cases, however, where the atom can make one or more spontaneous jumps to stationary states of lower energy, one would have to be prepared to find that this expression may no longer be correct. From the point of view of the above-mentioned paper by Bohr, Kramers and Slater, the atom in such a state acts as a source of spherical waves, whose frequencies ν_q are associated with each jump by means of the Bohr frequency condition (spontaneous radiation), even in the absence of external radiation. The simplest starting point for a description of this radiation would be the assumption that the atom acts as a classical dipole, whose moment is represented by the real part of the expression

$$\sum_q \mathfrak{A}_q \, e^{2\pi i \nu_q t}, \tag{4}$$

where the amplitude vectors $\mathfrak{A}$ are linked to the Einstein probability coefficients a_q by means of the relation

$$a_q h\nu_q = \frac{(2\pi\nu_q)^4}{3c^3} (\mathfrak{A}_q \bar{\mathfrak{A}}_q). \tag{5}$$

In the case of irradiation by monochromatic light, the radiation (4) will in general give rise to interferences with the radiation (2). Because of the finite life time of the atom, these interferences cause further terms to appear in the expression for the energy of the scattered radiation, which may possibly not be negligible compared with the term (3) (cf. § 4, formula 50). Without considering the question of the limits within which the initial assumption (2) was valid, one of us started, some time ago, on the problem of the form which the dependence of $\mathfrak{P}$ on the frequency ν might be expected to take.[1]) Two clues for the investigation were provided: one, by the empirical applicability of the formulae for $\mathfrak{P}$ obtained from classical dispersion theory, if the atom is imagined to contain classical oscillators whose natural frequencies coincide with the frequencies of the absorption lines; the other, by the correspondence principle. According to this principle, there exists a close connection between the actual behaviour of an atomic system and that to be expected from the system in classical electron theory, because of its structure. In particular, the correspondence principle requires that in the region of high quantum numbers the actual properties of the atom can be described asymptotically with the help of the classical electrodynamical laws. Starting from this requirement, it has been possible to derive a dispersion formula adapted to quantum theory, by comparing the classical dispersion formulae with the classical behaviour of a multiply periodic system under the influence of incident radiation. For the case where the atom is in its ground state, this formula becomes identical with a

[1]) H. A. Kramers, Nature **113** (1924) 673; **114** (1924) 310.

formula which was proposed earlier by Ladenburg [1]), and which was based on different arguments.

It is the purpose of this paper to show how the correspondence idea, when pursued more closely, leads to the surprising result that the assumption (2) for the reaction of an atom to incident radiation is too restricted, and will in general have to be extended by adding a series of terms as follows:

$$\mathfrak{P}(t) = \mathrm{R}\{\mathfrak{P}\, \mathrm{e}^{2\pi i\nu t} + \sum_k \mathfrak{P}_k\, \mathrm{e}^{2\pi i(\nu+\nu_k)t} + \sum_l \mathfrak{P}_l\, \mathrm{e}^{2\pi i(\nu-\nu_l)t}\}, \tag{6}$$

where $h\nu_k$ or $h\nu_l$ denote the energy difference between two stationary states of the atom, one of which is always identical with the momentary state of the atom; the vectors $\mathfrak{P}_k$ and $\mathfrak{P}_l$ again depend on $\mathfrak{E}$ (in the form of a linear vector function) and on ν. Expressed in words, the result can be stated as follows: *Under the influence of irradiation with monochromatic light, an atom not only emits coherent spherical waves of the same frequency as that of the incident light: it also emits systems of incoherent spherical waves, whose frequencies can be represented as combinations of the incident frequency with other frequencies that correspond to possible transitions to other stationary states.* Such additional systems of spherical waves must evidently occur in the form of scattered light; but they cannot make any contribution to the dispersion and absorption of the incident light.

Some time ago, Smekal [2]) used an argument which was linked to the concept of light quanta and he arrived at the same result, i.e. the appearance of scattered radiation of frequency $\nu+\nu_k$ or $\nu-\nu_l$ at the atom. We can reproduce Smekal's line of reasoning roughly as follows. The absorption (emission) of light by an atom can be described in terms of a process in which a light quantum of frequency ν is absorbed (emitted); in this way the atom changes into a higher (lower) stationary state and alters its energy by an amount $h\nu$ and its momentum by an amount $h\nu/c$. On the other hand, the usual scattering of light by an atom can be described as the simultaneous absorption of a light quantum of frequency ν and the emission of a light quantum of frequency ν'. For this, the atom does not change into another stationary state, but it generally undergoes a change in velocity. In an arbitrarily

[1]) R. Ladenburg, Zs. f. Phys. **4** (1921) 451. See also R. Ladenburg and F. Reiche, Naturw. **11** (1923) 584.

[2]) A. Smekal, Naturwiss. **11** (1923) 873.

chosen reference system the frequencies ν and ν' are different, in general. This holds, in particular, for a reference system in which the atom is initially at rest (Compton effect). Generalising from this, Smekal conjectures that there must also exist processes in the atom for which simultaneous absorption and emission of a light quantum takes place, such that not only the state of motion of the atom is altered, but the atom also changes into another stationary state, in contrast to the above-mentioned scattering transitions. Let us first neglect the small change in velocity of the atom for the jump, and denote the energy change of the atom for the jump by $h\nu_k$ or $h\nu_l$, according to whether the change is positive or negative. Then the frequency of the light quantum emitted in this process will evidently be given by $\nu+\nu_k$ or $\nu-\nu_l$, where ν is the frequency of the incident light quantum. This result can be interpreted as follows: For incident light of frequency ν, light of frequency $\nu+\nu_k$ or $\nu-\nu_l$ is emitted by the atom, while at the same time the atom acquires a probability for a decrease in its energy content by $h\nu_k$, or an increase by $h\nu_l$.

The calculation in terms of light quanta is above all of importance because it enables us to connect the macroscopic energy- and momentum-conservation laws with the concepts of quantum theory in a simple and instructive way. But by the nature of the case, such considerations do not allow us to draw any conclusions on the possible corpuscular structure of light; we always have the requirement that we must be able to get the results so obtained to agree with the wave-theoretical description of the optical phenomena, in a way that is free from contradictions. It can be seen straightaway that this requirement is fulfilled in our case and that the wave-theoretical *Ansatz* (6) corresponds just to Smekal's result. Nevertheless it should be mentioned, even at this stage, that our subsequent considerations will show that the processes indicated by Smekal are not the only ones which can be associated with the scattering effect of atoms. For completeness we shall also have to consider processes for which, in the terminology of light quanta, the atom is excited to an emission of two light quanta under the influence of irradiation; one of these will have the frequency ν of the incident radiation, the other a frequency ν', such that the loss in energy $h(\nu+\nu')$ corresponds to a transition of the atom into a state of lower energy.

In this connection it is of interest to stress that the above requirement for a wave-theoretical description of the optical phenomena remains satisfied, even when the momentum changes of the atom for the transition are not neglected. This is seen by introducing a reference system in which the magnitude of the momentum given to the atom in its stationary state is $h\nu/c$, but whose direction is opposite to that of the incident light (however, it has to be in the same direction for those processes not considered by Smekal). The frequency of the scattered light corresponding to a given transition is then the same in all spatial directions, according to the light quantum picture. This is in agreement with the wave-theoretical concept of a train of monochromatic waves whose source is located inside a very small spatial region. We shall not discuss in any detail the curious fact that the centre of these spherical waves moves relative to the excited atom. It should only be mentioned that it provides a strong argument in favour of saying that assumptions of the type (2), (4) and (6) cannot be strictly valid and still need to be suitably modified. Such modifications will not however have any material influence on the subsequent discussions.

The idea that one could connect, by means of the correspondence principle, Smekal's scattering effect of atoms in an external radiation field with the scattering effect of the atomic system as expected from classical theory, first occurred to Kramers in connection with his work on dispersion theory. The further elaboration of this idea, which is given in the present paper, resulted from discussions between the two authors.

2. The effect of external radiation on a periodic system in classical theory

Let us consider a non-degenerate periodic system, whose motion can be described by means of the canonical generalised variables ['Uniformisierungsvariablen'] $J_1,\ldots,J_s$, $w_1,\ldots,w_\tau$. As a function of these variables let the electrical moment of the system be represented by the following multiple Fourier series:

$$\mathfrak{M}(t) = \sum_{\tau_1\ldots\tau_s} \tfrac{1}{2}\mathfrak{C}_{\tau_1\ldots\tau_s}\, e^{2\pi i(\tau_1 w_1+\ldots\tau_s w_s)}. \tag{7}$$

The summation is to be extended over all positive and negative values of the integers $\tau_1\ldots\tau_s$. The coefficients $\mathfrak{C}$ are complex vectors whose components depend only on $J_1\ldots J_s$. If, as above, we again denote conjugate complex quantities by a bar, we have

$$\mathfrak{C}_{\tau_1\ldots\tau_s} = \overline{\mathfrak{C}}_{-\tau_1\ldots-\tau_s}. \tag{8}$$

The energy H of the system, too, only depends on the J's. Let us denote the fundamental frequencies by

$$\omega_k = \frac{\partial H}{\partial J_k}, \qquad (k = 1, \ldots s), \tag{9}$$

and introduce the following abbreviation for a frequently occurring differential operator,

$$\frac{\partial}{\partial J} = \tau_1 \frac{\partial}{\partial J_1} + \ldots + \tau_s \frac{\partial}{\partial J_s}, \tag{10}$$

together with the abbreviation

$$\omega = \tau_1\omega_1 + \ldots + \tau_s\omega_s = \frac{\partial H}{\partial J} \tag{11}$$

for the frequencies of the harmonic components appearing in the motion. It is now our task to find the electrical moment of the system as a function of time, for the case when the atom is exposed to a plane monochromatic train of light waves with wave-lengths large compared with the dimensions of the system.

Let the electric vector of the incident monochromatic light again be given by expression (1). Also, let $J_1^*,\ldots,J_s^*$, $w_1^*,\ldots,w_s^*$ be a new system of generalised variables, which is produced from the old variables by an infinitesimal contact transformation

$$J_k^* - J_k = \frac{\partial K}{\partial w_k^*}, \qquad w_k^* - w_k = -\frac{\partial K}{\partial J_k^*}, \qquad (k = 1, \ldots s). \tag{12}$$

One can then choose the function $K(J_1^*\ldots J_k^*, w_1^*\ldots w_s^*, t)$ in such a way that, to a first approximation, the J_k^* become time-independent and the w_k^* increase linearly with time, so that $dw_k^*/dt = \omega_k$. It is then found that the function K can be written as the real part of a complex expression as follows:

$$K = \mathrm{R}\left\{\sum_{\tau_1\ldots\tau_s} -\frac{1}{2}\,\frac{(\mathfrak{E}\mathfrak{C}_{\tau_1\ldots\tau_s})}{2\pi i(\omega+\nu)}\,e^{2\pi i(\tau_1 w_1^* + \ldots \tau_s w_s^* + \nu t)}\right\}, \tag{13}$$

where $\mathfrak{C}$ and ω represent here the same functions of the J^*'s that they were previously of the J's. If we insert in (7) the new generalised variables defined by (12) and (13), and further replace the w_k^* by the expressions $\omega_k t$, we finally obtain the following expression for the electrical moment of the atom as a function of time,

$$\mathfrak{M}(t) = \mathfrak{M}_0(t) + \mathfrak{M}_1(t), \quad \text{where} \quad \mathfrak{M}_0(t) = \sum_{\tau_1 \ldots \tau_s} \tfrac{1}{2}\mathfrak{C}_{\tau_1 \ldots \tau_s} \, e^{2\pi i \omega t}. \tag{14}$$

Both here and in what follows, we drop the asterisks for the new generalised variables; $\mathfrak{M}_0$ corresponds to the motion of the unperturbed atom. $\mathfrak{M}_1$ can be written as the real part of a sum over $2s$ terms:

$$\mathfrak{M}_1(t) = \mathrm{R}\left\{\sum_{\tau_1 \ldots \tau_s} \sum_{\tau_1' \ldots \tau_s'} \frac{1}{4}\left[\frac{\partial \mathfrak{C}}{\partial J'} \, e^{2\pi i \omega t} \, \frac{(\mathfrak{E}\mathfrak{C}')}{\omega' + \nu} \, e^{2\pi i(\omega'+\nu)t} - \mathfrak{C} \, e^{2\pi i \omega t} \, \frac{\partial}{\partial J}\left(\frac{(\mathfrak{E}\mathfrak{C}')}{\omega' + \nu}\right) e^{2\pi i(\omega'+\nu)t}\right]\right\}. \tag{15}$$

The summation is extended over all pairs of combinations of the integral values of $\tau_1, \ldots, \tau_s, \tau_1', \ldots, \tau_s'$; $\mathfrak{C}$ and $\mathfrak{C}'$ are abbreviations for $\mathfrak{C}_{\tau_1 \ldots \tau_s}$ and $\mathfrak{C}_{\tau_1' \ldots \tau_s'}$; $\partial/\partial J'$ the abbreviation for $\tau_1' \partial/\partial J_1 + \ldots + \tau_s' \partial/\partial J_s$, in analogy with the notation (10); and ω' the abbreviation for $\tau_1'\omega_1 + \ldots + \tau_s'\omega_s$. We now want to re-write expression (15) by combining, first of all, all those terms for which the sums

$$\tau_1 + \tau_1' = \tau_1^0, \qquad \ldots, \qquad \tau_s + \tau_s' = \tau_s^0 \tag{16}$$

have the same value. Using the abbreviation

$$\tau_1^0 \omega_1 + \ldots + \tau_s^0 \omega_s = \omega_s^0, \tag{17}$$

we thus obtain

$$\mathfrak{M}_1(t) = \mathrm{R}\left\{\sum_{\tau_1^0 \ldots \tau_s^0} \sum_{\tau_1 \ldots \tau_s} \frac{1}{4}\left[\frac{\partial \mathfrak{C}}{\partial J'} \, \frac{(\mathfrak{E}\mathfrak{C}')}{\omega' + \nu} - \mathfrak{C} \, \frac{\partial}{\partial J}\left(\frac{(\mathfrak{E}\mathfrak{C}')}{\omega' + \nu}\right)\right] e^{2\pi i(\omega^0+\nu)t}\right\}. \tag{18}$$

In the summation over $\tau_1 \ldots \tau_s$, we have to replace $\tau_1' \ldots \tau_s'$ everywhere by the set of values given by (16). We also wish to draw attention to the fact that ω' and ω^0 can assume positive as well as negative values, since the summations have to be extended over all positive and negative values of τ_k and τ_k^0. It is evidently a condition for the validity of this formula that the frequency ν of the incident light should not coincide with any of the frequencies ω of the unperturbed motion.

Formula (18) tells us that, under the influence of the incident light, the system will emit scattered radiation of an intensity which is proportional to the intensity of the incident light; when separated into its harmonic components, it contains not only the frequency ν of the incident light but also frequencies which can be represented as the sum or difference of ν and a frequency ω of the form given by (17). The frequency ω^0 itself need not appear in the motion of the unperturbed system. Rather, it can be seen from (15) that ω^0 is always of the form $\pm|\omega|\pm|\omega'|$, where $|\omega|$ and $|\omega'|$ are two frequencies that actually occur in the unperturbed motion.

3. Quantum theory and coherent scattered radiation

Basing ourselves now on the quantum theory of periodic systems, we have to deal with a discrete manifold of stationary states given by the quantum conditions

$$J_k = n_k h. \tag{19}$$

The radiation emitted by the unperturbed system in a given stationary state corresponds to possible transitions to stationary states of lower energy. According to the correspondence principle it is nevertheless to be looked upon as a meaningful analogue of the radiation which one might expect from classical theory. Whereas the latter can be derived from expression (14) for the oscillating electrical moment $\mathfrak{M}_0$ of the undisturbed atom, the quantum-theoretical radiation can be considered as originating from an oscillating moment of a form such as given by expression (4). Each frequency ν_q in this expression then corresponds to a classical frequency $\tau_1\omega_1+\ldots+\tau_s\omega_s$, such that

$$\tau_k = n_k^{(1)} - n_k^{(2)}, \tag{20}$$

where $n_k^{(1)}$ and $n_k^{(2)}$ represent the values of the quantum numbers in the initial and final states, respectively. The magnitude of the classical frequency

$$\omega = \left(\tau_1 \frac{\partial}{\partial J_1} + \ldots + \tau_s \frac{\partial}{\partial J_s}\right) H = \frac{\partial H}{\partial J} \tag{21}$$

is not identical with the magnitude of the corresponding quantum-theoretical frequency ν_q, because we know that the latter is given by

$$\nu_q = \frac{1}{h}\,(H^{(1)} - H^{(2)}). \tag{22}$$

In the limit of high quantum numbers, however, this expression can be approximately written in the form

$$\nu_q = \frac{\Delta H}{h} = \left(\frac{\Delta J_1}{h}\,\frac{\partial}{\partial J_1} + \ldots + \frac{\Delta J_s}{h}\,\frac{\partial}{\partial J_s}\right) H, \tag{23}$$

from which it follows immediately, because of (19) and (20), that in this limit the quantum-theoretical frequencies are asymptotically equal to the classical frequencies. In the region of low quantum numbers, ν_q stands for a simple average of the corresponding ω's.

Furthermore, in the limit of large quantum numbers, the amplitudes $\mathfrak{A}_q$ of the harmonic components of the radiation must be asymptotically equal to the amplitudes $\mathfrak{C}$ of the classical oscillations; in the region of low quantum numbers, $\mathfrak{A}$ can be regarded symbolically as a kind of average of $\mathfrak{C}$. The complex vector $\mathfrak{A}_q$ could be described as the *characteristic amplitude* for the transition under discussion. Its value is determined to within a complex factor of absolute value 1, in view of the arbitrariness of the phase, and thus contains five constants. It is evidently related, through eq. (5), to the Einstein coefficient, a_q, for the probability of a spontaneous transition:

$$a_q h \nu_q = \frac{(2\pi\nu_q)^4}{3c^3}\,(\mathfrak{A}_q \overline{\mathfrak{A}}_q). \tag{24}$$

It has recently been stressed by Bohr [1]) that for a degenerate system, the state of polarisation of the emitted radiation is not uniquely determined by the initial state of the relevant transition; one can therefore not define a uniquely determined characteristic amplitude in such a case. However, we have confined ourselves in this paper exclusively to a discussion of non-degenerate systems, where presumably the character of the spontaneous radiation is always uniquely determined by the state of the atom.

We now have the task to set up a quantum-theoretical expression, which is to be the analogue of the classical formula (18), for the scattering effect of the system under external irradiation. In particular,

[1]) N. Bohr, Naturwiss. **12** (1924) 1115.

this will have to be done in such a way that the scattering coincides asymptotically with the classical scattering in the limit of high quantum numbers. It will be shown that this can be achieved by interpreting the differential coefficients which occur in (18) as differences between two quantities, in analogy to Bohr's procedure in the case of frequencies. In particular, we shall obtain, quite naturally, formulae which contain only the frequencies and amplitudes which are characteristic for the transitions, while all those symbols which refer to the mathematical theory of periodic systems will have disappeared.

Let us begin with that part of the scattered light which is of the same frequency as the incident radiation. In the classical case, this part corresponds to those terms in (18), for which $\tau_1^0=\ldots=\tau_s^0=0$, $\tau_k'=-\tau_k$, and which therefore belong to a scattering moment

$$\mathfrak{M}_{\mathrm{cl}}(t) = \mathrm{R}\sum_{\tau_k} \frac{1}{4}\left\{\frac{\partial\mathfrak{C}}{\partial J}\,\frac{(\mathfrak{E}\overline{\mathfrak{C}})}{\omega-\nu} + \mathfrak{C}\,\frac{\partial}{\partial J}\left(\frac{(\mathfrak{E}\overline{\mathfrak{C}})}{\omega-\nu}\right)\right\}\mathrm{e}^{2\pi i\nu t}. \tag{25}$$

The expression in the braces can obviously be written in the form of a simple differential coefficient. If, next, we always combine two terms for which the quantities τ_k are numerically equal but of different sign, (25) assumes the following form,

$$\mathfrak{M}_{\mathrm{cl}}(\nu) = \mathrm{R}\sum_{\tau_k}{}' \frac{1}{4}\,\frac{\partial}{\partial J}\left\{\frac{\mathfrak{C}(\mathfrak{E}\overline{\mathfrak{C}})}{\omega-\nu} + \frac{\overline{\mathfrak{C}}(\mathfrak{E}\mathfrak{C})}{\omega+\nu}\right\}\mathrm{e}^{2\pi i\nu t}, \tag{26}$$

where the prime on the summation sign is supposed to indicate that we are summing only over those τ-combinations for which the ω become positive. One can now obtain a quantum-theoretical expression for the scattering moment with frequency ν which acts in a given stationary state, by writing

$$\mathfrak{M}_{\mathrm{qu}}(\nu) = \mathrm{R}\Bigg[\sum_{\mathrm{a}} \frac{1}{4h}\left(\frac{\mathfrak{A}_{\mathrm{a}}(\mathfrak{E}\overline{\mathfrak{A}}_{\mathrm{a}})}{\nu_{\mathrm{a}}-\nu} + \frac{\overline{\mathfrak{A}}_{\mathrm{a}}(\mathfrak{E}\mathfrak{A}_{\mathrm{a}})}{\nu_{\mathrm{a}}+\nu}\right) - \sum_{\mathrm{e}} \frac{1}{4h}\left(\frac{\mathfrak{A}_{\mathrm{e}}(\mathfrak{E}\overline{\mathfrak{A}}_{\mathrm{e}})}{\nu_{\mathrm{e}}-\nu} + \frac{\overline{\mathfrak{A}}_{\mathrm{e}}(\mathfrak{E}\mathfrak{A}_{\mathrm{e}})}{\nu_{\mathrm{e}}+\nu}\right)\Bigg]\mathrm{e}^{2\pi i\nu t}. \tag{27}$$

where the first sum extends over all frequencies ν_{a} for which the system exhibits selective absorption, and the second over all frequencies ν_{e} which are contained in the spontaneous emission. This expression coincides asymptotically with (26) in the limit of large quantum numbers and is at the same time consistent with experiment. The quantities $\mathfrak{A}_a$ and $\mathfrak{A}_e$ are the amplitudes characteristic for the absorption and emission transitions. Formula (27) results from (26) if one tries to interpret the differential coefficients which occur in each term in (26), as the difference between two quantities that refer to two states of motion for which the values of the quantities $J_1 \dots J_s$ differ by $\tau_1 h \dots \tau_s h$. There is no point here in considering two stationary states separately. In the first place, one would not obtain in this way a quantity which could be related naturally to the reaction of the atom in a given stationary state. Secondly, one cannot ascribe a special meaning to the values of the amplitudes $\mathfrak{C}$ themselves, in the stationary states. Instead, it is a symbolic average of $\mathfrak{C}$, taken over the region between two stationary states, which can be given a meaning, namely that of a characteristic amplitude $\mathfrak{A}$ for the relevant transition, as previously remarked. We thus arrive at the interpretation of the differential coefficient in (26) as the difference, divided by h, between two quantities that refer to two transitions characterised by $\tau_1 \dots \tau_s$, where the stationary state in question represents the final state for the one, and the initial state for the other, transition.

Fig. 1 helps to illustrate the method. Let the system have two degrees of freedom, and the states of motion be represented by points in a $J_1 J_2$-plane, in the plane of the paper. The stationary states form a point lattice, P is the stationary state whose reaction is under investigation, and Q and R are two stationary states whose J-values exceed, and fall short of, that of P by the amounts $h\tau_1$, and $h\tau_2$, respectively. The differential coefficient in (26) is then interpreted as the difference between two quantities, one referring to transition a and the other to transition e.

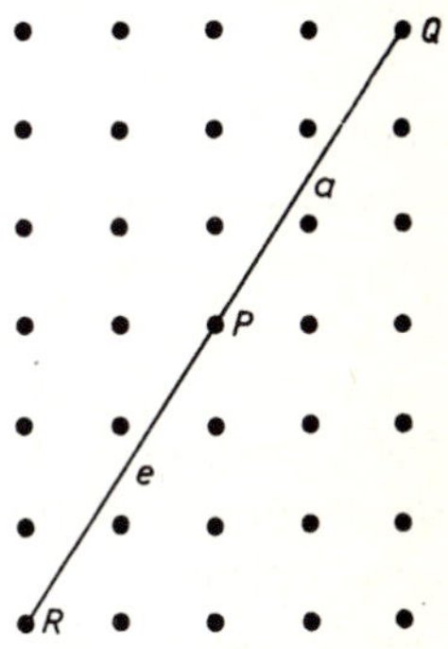

Fig. 1.

We shall write expression (27) in yet another form for the simple case where the vector $\mathfrak{E}$ and all the vectors $\mathfrak{A}_a$ and $\mathfrak{A}_e$ are real and parallel to each other, i.e. where the incident light is linearly polarised and where the electric vector is parallel to the propagation vector in all cases where the radiation corresponds to transitions a and e. We introduce the decay time τ_ν of a classically oscillating electron with frequency ν,

$$\tau_\nu = \frac{3c^3m}{8\pi^2e^2\nu^2} \tag{28}$$

and define the 'strength' of a transition by the number

$$f = a\tau_\nu$$

where a denotes Einstein's probability coefficient which occurs in (24). We can now write for (27),

$$\mathfrak{M}_{\text{qu}}(t) = \mathfrak{E}\,\frac{e^2}{4\pi^2m}\left(\sum_{\text{a}}\frac{f_{\text{a}}}{\nu_{\text{a}}^2-\nu^2} - \sum_{\text{e}}\frac{f_{\text{e}}}{\nu_{\text{e}}^2-\nu^2}\right)\cos 2\pi\nu t. \tag{29}$$

This formula, which clearly displays a similarity with the classical formulae, was given by Kramers in his first Note [1]) on the quantum theory of dispersion. In a second Note [2]), the derivation given here was briefly sketched. The terms relating to the absorption lines correspond to the formula previously obtained by Ladenburg.

In the general case where the vectors $\mathfrak{E}$ and $\mathfrak{A}$ are not real and not parallel to one another, the direction of $\mathfrak{M}$ no longer coincides with that of $\mathfrak{E}$ and it is generally not possible to write the formula in as simple a form as (29).

When the frequency of the incident radiation approaches the frequency ν_{a} of an absorption line, or ν_{e} of an emission line, the moment increases strongly; but in the immediate neighbourhood of this frequency, formula (27) must clearly lose its validity, however correct it may be elsewhere, just as the classical formula will no longer be valid when ν approaches ω. Nevertheless, we certainly obtain the result that for a close coincidence with an absorption line, light of the frequency of this line is particularly strongly scattered. Part of the resonance radiation observed by Wood and others in metal vapours must surely be ascribed to these strong, coherent, dispersion waves whose presence is made evident, after all, by the existence of the absorption itself, as well as by that of the metallic reflection at high pressures. In part, however, the resonance radiation will also originate from the excited states into which some of the atoms will have been thrown by the radiation. We shall not discuss resonance radiation in any further detail here; it was only mentioned in order

[1]) Nature **113** (1924) 673.

[2]) Nature **114** (1924) 310. See also J. H. van Vleck, Phys. Rev. **24** (1924) 344.

to stress how suitable the form of expression (27) is for representing mathematically the nature of the absorption lines as singularities for the scattering.

If, on the other hand, ν nearly coincides with an emission frequency ν_e, the expression (27) admittedly becomes very large; but, because of the presence of spontaneous radiation of frequency ν_e and because of our lack of knowledge concerning the phase of the scattered radiation for near-resonance, we cannot immediately conclude that the spherical waves of frequency ν_e are reinforced; there are certain arguments, which we will not go into here, that indicate the exact contrary.

It is immediately seen that the scattered radiation, described by (27), is coherent with the incident radiation. This follows from the fact that each term contains both the amplitude $\mathfrak{A}$ and its complex conjugate $\bar{\mathfrak{A}}$, so that the indeterminateness of the phase of $\mathfrak{A}$ itself becomes redundant. This coherence causes the dispersion, and it is easily seen that the incident ray generally divides into two polarised portions, to which correspond two different refractive indices. The terms in the second sum in (27) or (29) correspond to a negative dispersion, analogous to Einstein's 'negative absorption' at the points $\nu=\nu_e$, just as the ordinary or positive dispersion corresponds to the ordinary absorption lines at the points $\nu=\nu_a$. [1])

[1]) Wentzel has recently tried to give a quantum-theoretical treatment of dispersion in an interesting paper [Zs. f. Phys. **29** (1924) 306], which has little in common with the treatment given here. Wentzel supports a point of view according to which the quantum-theoretical dispersion formula cannot be represented in the simple Helmholtz–Ketteler form, but must be regarded as a kind of distorted classical dispersion formula. Such a possibility deserves close attention, since in our discussion (which is based on the correspondence principle) there is no question of a rigorous derivation of the formula for the induced scattering moment. Nevertheless it is our opinion that there are so far no experimental reasons why the validity of a simple formula of the type (27) or (29) should be called in question. Admittedly, Wentzel introduces as an example the dispersion for helium, since there the 'effective' absorption frequency lies on the short wave-length side of the absorption series limit. (It is almost 5% greater than the limit frequency.) But this fact in no way contradicts the classical formulae, for it is not only the absorption lines of helium which contribute to the dispersion, but also the continuous absorption which extends beyond the series limit in the direction of short wave lengths. A simple calculation can be made which is based on theoretical considerations about the magnitude of this continuous absorption [cf. H. A. Kramers, Phil. Mag. **46** (1923) 836] or is based on an extrapolation to the case of helium of the empirical formulae for

4. Incoherent scattered radiation

We now proceed to give a quantum-theoretical interpretation, consistent with the correspondence principle, for the other terms in the classical formula (18), for which the ω^0 are different from zero. For this we look at fig. 2, which again refers to a system with two degrees

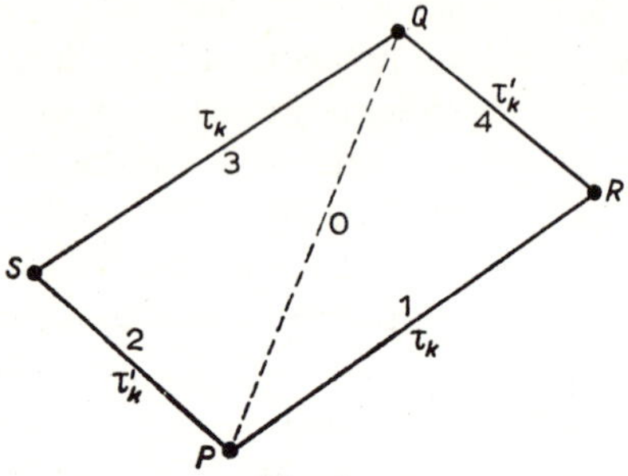

Fig. 2.

of freedom. Let the points P, Q, R and S in the J_1J_2-plane represent four stationary states such that when the quantum number $n_k = (n_k)_P$ are associated with the state P, the states Q, R, S are specified by the following quantum numbers:

$$
\begin{gathered}
(n_k)_Q = (n_k)_P + \tau_k + \tau'_k = (n_k)_P + \tau_k^0, \\
(n_k)_R = (n_k)_P + \tau_k, \qquad (n_k)_S = (n_k)_P + \tau'_k,
\end{gathered}
\tag{30}
$$

where the systems of integers τ_k, τ'_k and τ_k^0 have the same meaning as in formulae (16) and (17). The different transitions are denoted by the numbers 0, 1, 2, 3 and 4. Transitions 1 and 3, as well as 2 and 4, are assumed to represent spontaneous radiative transitions, i.e. their characteristic amplitudes are to be different from zero. On the other hand, PQ need not represent a possible transition, i.e. $\mathfrak{A}_0$ may possibly be equal to zero. We shall denote by ν_0, ν_1, ν_2, ν_3 and ν_4 the frequencies which correspond quantum-theoretically to the above transitions.

the absorption of X-rays. This shows that, by applying the classical formula, the influence of this continuous absorption in the optical region is the same as that of an absorption line whose frequency is about 1.2 times that of the series limit. Its 'intensity' can amount to several 'units', i.e. it could correspond to the presence of several dispersion electrons. For the case of helium, therefore, the experimental results do not allow us, as yet, to conclude that the classical dispersion formula is not valid. [Cf. also K. F. Herzfeld and K. L. Wolf, Ann. d. Phys. **76** (1925) 71].

Let us first of all interpret the exponential function in (18) in the following way:

$$e^{2\pi i(\omega^0+\nu)t} \approx e^{2\pi i(\nu^0+\nu)t}. \tag{31}$$

We can further interpret the expression which occurs inside the braces in (18) in a quantum-theoretical way, where the differential coefficients are suitably replaced by differences divided by h, while the frequencies ω' of the motion are replaced by quantum-theoretical frequencies, and the amplitudes $\mathfrak{C}$ by the characteristic amplitudes $\mathfrak{A}$ of the quantum-theoretical transitions. It then follows in an apparently unique way that

$$\begin{aligned}\frac{\partial\mathfrak{C}}{\partial J'}\frac{(\mathfrak{E}\mathfrak{C}')}{\omega'+\nu} &\approx \frac{\mathfrak{A}_3-\mathfrak{A}_1}{h}\,\frac{1}{2}\left(\frac{(\mathfrak{E}\mathfrak{A}_4)}{\nu_4+\nu}+\frac{(\mathfrak{E}\mathfrak{A}_2)}{\nu_2+\nu}\right),\\ \mathfrak{C}\frac{\partial}{\partial J}\left(\frac{(\mathfrak{E}\mathfrak{C}')}{\omega'+\nu}\right) &\approx \frac{\mathfrak{A}_3+\mathfrak{A}_1}{2}\,\frac{1}{h}\left(\frac{(\mathfrak{E}\mathfrak{A}_4)}{\nu_4+\nu}-\frac{(\mathfrak{E}\mathfrak{A}_2)}{\nu_2+\nu}\right).\end{aligned} \tag{32}$$

In the subtraction a number of terms cancel, and one obtains as a representation for the expression in braces in (18),

$$\{\tau_k,\tau'_k\} \approx \frac{1}{h}\left\{-\frac{\mathfrak{A}_1(\mathfrak{E}\mathfrak{A}_4)}{\nu_4+\nu}+\frac{\mathfrak{A}_3(\mathfrak{E}\mathfrak{A}_2)}{\nu_2+\nu}\right\}. \tag{33}$$

In this new formulation it was tacitly assumed that the frequencies ω, ω' and ω^0 were all positive, and that, in quantum-theoretical analogy to this, the frequencies $\nu_0, \nu_1, \nu_2, \nu_3, \nu_4$, given by the formulae

$$h\nu_0 = H(Q)-H(P),\quad \begin{matrix} h\nu_1 = H(R)-H(P), & h\nu_3 = H(Q)-H(S),\\ h\nu_2 = H(S)-H(P), & h\nu_4 = H(Q)-H(R), \end{matrix} \tag{34}$$

all turn out to be positive. Each time when one of these frequencies becomes negative, one must insert in formula (33), instead of the associated characteristic amplitude $\mathfrak{A}$, the complex conjugate vector $\overline{\mathfrak{A}}$.

Since we now have to perform the summation as in (18), it is important to note that the particular term in (18) which is obtained by interchanging the values of τ_k and τ'_k, can be associated again with exactly the same set of four stationary states P, Q, R, S. Interchanging, we obtain for the quantity inside the braces an expression which results from (33) if one interchanges 1 with 2, and 3 with 4. We can thus write

$$\{\tau'_k,\tau_k\} \approx \frac{1}{h}\left\{-\frac{\mathfrak{A}_2(\mathfrak{E}\mathfrak{A}_3)}{\nu_3+\nu}+\frac{\mathfrak{A}_4(\mathfrak{E}\mathfrak{A}_1)}{\nu_1+\nu}\right\}. \tag{35}$$

We therefore obtain for the sum of the two terms in (18) which can be associated with the quadrilateral P, Q, R, S according to their quantum-theoretical interpretation,

$$\{\tau_k, \tau_k'\} + \{\tau_k', \tau_k\} \approx \frac{1}{h}\left\{-\frac{\mathfrak{A}_1(\mathfrak{E}\mathfrak{A}_4)}{\nu_4+\nu} + \frac{\mathfrak{A}_4(\mathfrak{E}\mathfrak{A}_1)}{\nu_1+\nu} - \frac{\mathfrak{A}_2(\mathfrak{E}\mathfrak{A}_3)}{\nu_3+\nu} + \frac{\mathfrak{A}_3(\mathfrak{E}\mathfrak{A}_2)}{\nu_2+\nu}\right\}. \tag{36}$$

We now define a complex vector $\mathfrak{M}(P, Q{:}R)$, which can be associated with each set of three stationary states P, Q, R, as follows,

$$\mathfrak{M}(P, Q : R) = \frac{1}{4h}\left\{\frac{\mathfrak{A}_q(\mathfrak{E}\mathfrak{A}_p)}{\nu_p+\nu} - \frac{\mathfrak{A}_p(\mathfrak{E}\mathfrak{A}_q)}{\nu_q+\nu}\right\} e^{2\pi i(\nu_o+\nu)t}. \tag{37}$$

The small letters p, q, o refer to the transitions RP, $QR{:}QP$ (cf. fig. 3). The frequencies ν_p, ν_q and ν_o in (37) are defined by the relations

$$h\nu_p = H(R)-H(P), \quad h\nu_q = H(Q)-H(R), \quad h\nu_o = H(Q)-H(P), \tag{38}$$

and can therefore be negative in certain cases. Whenever ν_p (ν_q) becomes negative, the characteristic amplitude $\mathfrak{A}_p$ ($\mathfrak{A}_q$) of the corresponding transition will have to be replaced by the complex conjugate vector $\overline{\mathfrak{A}}_p$ ($\overline{\mathfrak{A}}_q$).

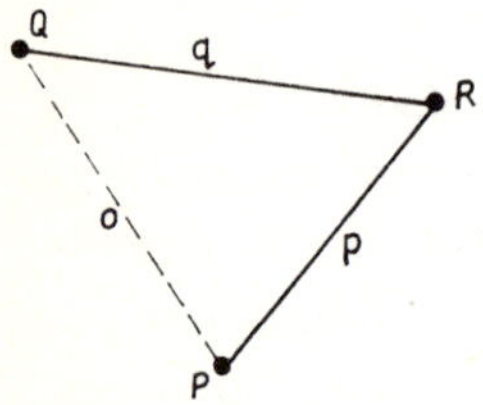

Fig. 3.

The expression on the right-hand side of (36), when multiplied by $\frac{1}{4}e^{2\pi i\nu t}$, can evidently be written as the sum of $\mathfrak{M}(P,Q{:}R)$ and $\mathfrak{M}(P,Q{:}S)$. It can therefore be expected from (18) that the scattering moment of the atoms can be represented by a sum of terms, each of which contains the expression $\mathrm{R}\mathfrak{M}(P, Q{:}R)$. One difficulty now arises, in that it is not known whether this expression refers to the reaction of the atom in state P or in state Q (R is not affected). This question can only be resolved if other considerations are introduced which do not directly touch on the content of the correspondence principle. A similar situation is met with when one looks for the spontaneous emission of radiation in the case of an unperturbed periodic system. Such an emission consists of harmonic components, each of which is linked to a combination of two stationary states, and whose frequencies (22) and amplitudes can be interpreted in accordance with the correspondence principle; the correspondence principle, however, does not immediately allow us to decide, from which of these two states it is that the atom emits this spontaneous radiation; this point can only be settled by an argument which is linked to the energy principle and the nature of the emission.[1]) As we know, this argument states that emission always takes place from a state whose energy content is the greater, in keeping with Bohr's radiation postulate. A similar decision will now have to be made also in our case. We shall have to assume, indeed, that the scattering moment (37) refers to the state Q when $\nu_0+\nu$ is positive, and to the state P when $\nu_0+\nu$ is negative. We can justify this statement as follows. From the fundamental laws governing the energy exchange between radiation field and atoms, the scattered radiation of frequency $\nu_0+\nu$ has the meaning ascribed to it that there exists a probability for the atom in this state to lose an amount of energy $h(|\nu_0+\nu|)$ for a change of state. On the other hand, an irradiation of frequency ν always gives rise to scattered radiation of frequency ν as well. The joint effect of these two causes a reaction to the radiation field such that the atom has an inherent probability for either a loss, or a gain, of an amount of energy $h\nu$ during a change of state. But an actual change in the state of the atom will always consist of a transition to another stationary state; in order that scattered radiation of frequency $|\nu+\nu_0|$ should produce such a transition, it must therefore be accompanied by an effect with frequency ν. If $(\nu+\nu_0)$ is positive, this is only possible if the atom simultaneously gains an energy $h\nu$ and loses an energy $h(\nu+\nu_0)$. During the transition, therefore, the atom has a net loss of energy of amount $h\nu_0$, i.e. it must be in state Q before, and in state P after, the transition. (This holds both for positive and negative values of ν_0.) If, on the other hand, $\nu+\nu_0$ is negative (in which case ν_0 is always negative) the transition can only be produced through the loss by the atom of energy $h\nu$ as well as $h(-[\nu+\nu_0])$, i.e. of a total energy $-h\nu_0$. Before the transition the atom must therefore be in state P, afterwards in state Q. In Smekal's paper, quoted above, transitions of the first kind were considered. Transitions of the latter kind, which hardly present themselves in such a natural way from the point of view of light quanta, were not discussed.

We are now able to give a quantum-theoretical interpretation of formula (18) in its full generality, using formulae (36) and (37) and with the help of the decision just reached on the state of atom to

[1]) See N. Bohr, Zs. f. Phys. **18** (1923) 164.

which (37) refers. If the atom is in a stationary state, we obtain for the scattering moment induced by the external radiation,

$$\mathfrak{M}(t) = \mathrm{R}\{\sum_Q \sum_R \mathfrak{M}(Q, P : R) + \sum_Q \sum_R \mathfrak{M}(P, Q : R)\}, \tag{39}$$

where in both sums the summation is taken over all those stationary states R of the atom which are different from P and Q. In the first summation one will further have to sum over all stationary states Q of the atom for which $H(Q) < H(P) + h\nu$; in the second sum, over all those states Q for which $H(Q) < H(P) - h\nu$. Just as before, $H(P)$ and $H(Q)$ stand for the values of the energy of the atom in states P and Q, respectively.

For clarity, we wish to write the general formulae (37) and (39) in a somewhat more special form. If ν^* is to be always a positive quantity, we want to write down that particular contribution $\mathfrak{M}(|\nu \pm \nu^*|)$ to the total scattering moment of the atom in state P, which corresponds to a definite frequency $|\nu \pm \nu^*|$ in the scattered light, due to the presence of a state Q. We shall have to distinguish between different cases.

Case I. $H(Q) > H(P), \qquad H(Q) - H(P) = h\nu^*.$

State Q contributes to the scattered radiation when $\nu > \nu^*$, and the frequency of this scattered radiation is equal to $\nu - \nu^*$. States R, which are different from P and Q, divide into three groups R_b, R_a and R_c according as $H(R)$ is greater than $H(Q)$; smaller than $H(Q)$ but greater than $H(P)$; or smaller than $H(P)$. Let us denote the absolute magnitudes of the frequencies which correspond to the transitions between P or Q on the one hand, and R on the other, by ν_1, ν_2, etc., corresponding to the numbers in fig. 4. Only terms which occur in the first sum in (39) will appear and we obtain for the scattering moment $\mathfrak{M}(\nu - \nu^*)$ the expression

$$\mathfrak{M}(\nu - \nu^*) = \mathrm{R}\frac{1}{4h}\Bigg\{\sum_{R_a}\left(\frac{\mathfrak{A}_2(\mathfrak{E}\overline{\mathfrak{A}}_1)}{\nu_1 - \nu} + \frac{\overline{\mathfrak{A}}_1(\mathfrak{E}\mathfrak{A}_2)}{\nu_2 + \nu}\right) +$$

$$\sum_{R_b}\left(\frac{\overline{\mathfrak{A}}_4(\mathfrak{E}\overline{\mathfrak{A}}_3)}{\nu_3 - \nu} - \frac{\overline{\mathfrak{A}}_3(\mathfrak{E}\overline{\mathfrak{A}}_4)}{\nu_4 - \nu}\right) + \sum_{R_c}\left(-\frac{\overline{\mathfrak{A}}_6(\mathfrak{E}\mathfrak{A}_5)}{\nu_5 + \nu} - \frac{\mathfrak{A}_5(\mathfrak{E}\overline{\mathfrak{A}}_6)}{\nu_6 - \nu}\right)\Bigg\} e^{2\pi i(\nu - \nu^*)t}. \tag{40}$$

In the ground state of the atom, only scattered radiation of this type occurs.

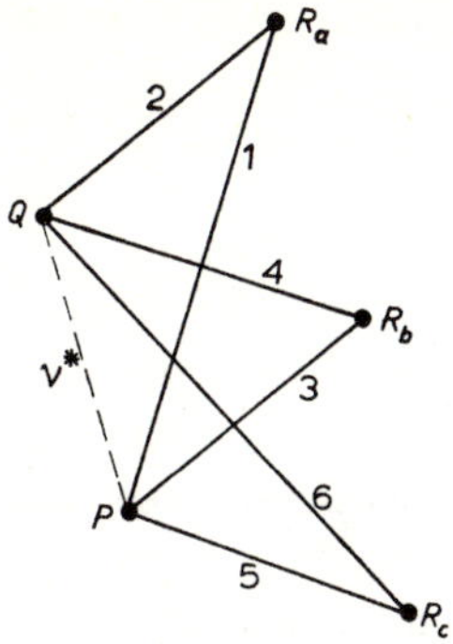

Fig. 4.

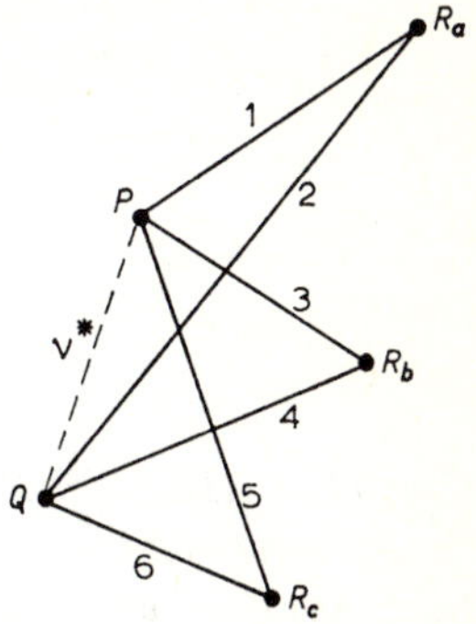

Fig. 5.

Case II. $H(Q) < H(P) \qquad H(P) - H(Q) = h\nu^*$.

(a) Scattered radiation of frequency $\nu + \nu^*$ appears for all values of the frequency ν. The relevant terms in formula (39) all belong to the first sum. The stationary states different from P and Q split into states R_a, for which $H(R_a) > H(P)$; states R_b, for which $H(P) > H(R_b) > H(Q)$; and states R_c, for which $H(Q) > H(R_c)$. The corresponding frequencies will be labelled as indicated in Fig. 5 and will always be taken to be positive. For the scattering moment $\mathfrak{M}(\nu + \nu^*)$ we obtain the expression

$$\mathfrak{M}(\nu + \nu^*) = \mathrm{R}\frac{1}{4h}\Big\{\sum_{R_a}\left(\frac{\mathfrak{A}_2(\mathfrak{E}\bar{\mathfrak{A}}_1)}{\nu_1 - \nu} + \frac{\bar{\mathfrak{A}}_1(\mathfrak{E}\mathfrak{A}_2)}{\nu_2 + \nu}\right)$$

$$+ \sum_{R_b}\left(-\frac{\mathfrak{A}_4(\mathfrak{E}\mathfrak{A}_3)}{\nu_3 + \nu} + \frac{\mathfrak{A}_3(\mathfrak{E}\mathfrak{A}_4)}{\nu_4 + \nu}\right)$$

$$+ \sum_{R_c}\left(-\frac{\bar{\mathfrak{A}}_6(\mathfrak{E}\mathfrak{A}_5)}{\nu_5 + \nu} - \frac{\mathfrak{A}_5(\mathfrak{E}\bar{\mathfrak{A}}_6)}{\nu_6 - \nu}\right)\Big\} e^{2\pi i(\nu+\nu^*)t}. \tag{41}$$

(b) Scattered radiation of frequency $\nu^* - \nu$ occurs for all those values of the incident frequency ν which are smaller than ν^*. The relevant terms in (39) all belong to the second sum. Using the same notations as in Case II(a), we obtain for the dispersive moment $\mathfrak{M}(\nu - \nu^*)$ the expression

$$\mathfrak{M}(\nu-\nu^*) = \mathrm{R}\frac{1}{4h}\Big\{\sum_{R_a}\Big(\frac{\overline{\mathfrak{A}}_2(\mathfrak{E}\mathfrak{A}_1)}{\nu_1+\nu}+\frac{\mathfrak{A}_1(\mathfrak{E}\overline{\mathfrak{A}}_2)}{\nu_2-\nu}\Big)$$

$$+\sum_{R_b}\Big(-\frac{\overline{\mathfrak{A}}_4(\mathfrak{E}\overline{\mathfrak{A}}_3)}{\nu_3-\nu}+\frac{\overline{\mathfrak{A}}_3(\mathfrak{E}\overline{\mathfrak{A}}_4)}{\nu_4-\nu}\Big)$$

$$+\sum_{R_c}\Big(-\frac{\mathfrak{A}_6(\mathfrak{E}\overline{\mathfrak{A}}_5)}{\nu_5-\nu}-\frac{\overline{\mathfrak{A}}_5(\mathfrak{E}\mathfrak{A}_6)}{\nu_6+\nu}\Big)\Big\}\,\mathrm{e}^{2\pi\mathrm{i}(\nu-\nu^*)t}. \qquad (42)$$

Case III. $H(Q)=H(P)$. The scattered radiation has the frequency of the incident light.

(a) State Q is not identical with state P (cf. fig. 6). This case can be obtained both from I and II(a), when ν^* is set equal to zero. The summation over states R_b has of course disappeared. The frequencies ν_1 and ν_2, as well as the frequencies ν_5 and ν_6 coincide: we shall denote them by ν_{12} and ν_{56}, respectively.

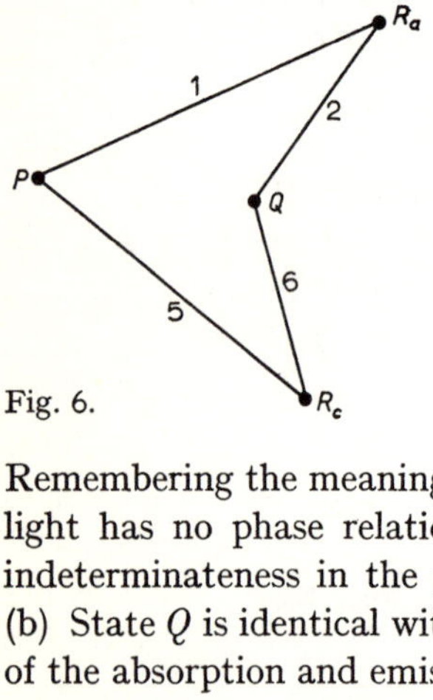

Fig. 6.

$$\mathfrak{M}(\nu) = \mathrm{R}\frac{1}{4h}\Big\{\sum_{R_a}\Big(\frac{\mathfrak{A}_2(\mathfrak{E}\overline{\mathfrak{A}}_1)}{\nu_{12}-\nu}+\frac{\overline{\mathfrak{A}}_1(\mathfrak{E}\mathfrak{A}_2)}{\nu_{12}+\nu}\Big)$$

$$+\sum_{R_b}\Big(-\frac{\mathfrak{A}_5(\mathfrak{E}\overline{\mathfrak{A}}_6)}{\nu_{56}-\nu}-\frac{\overline{\mathfrak{A}}_6(\mathfrak{E}\mathfrak{A}_5)}{\nu_{56}+\nu}\Big)\Big\}\mathrm{e}^{2\pi\mathrm{i}\nu t}. \quad (43)$$

Remembering the meaning of the vectors $\mathfrak{A}$, we see that the scattered light has no phase relation with the incident light, because of the indeterminateness in the phases of the amplitudes $\mathfrak{A}$.

(b) State Q is identical with P (cf. fig. 7). If we denote the frequencies of the absorption and emission lines in state P by ν_a and ν_e, we obtain

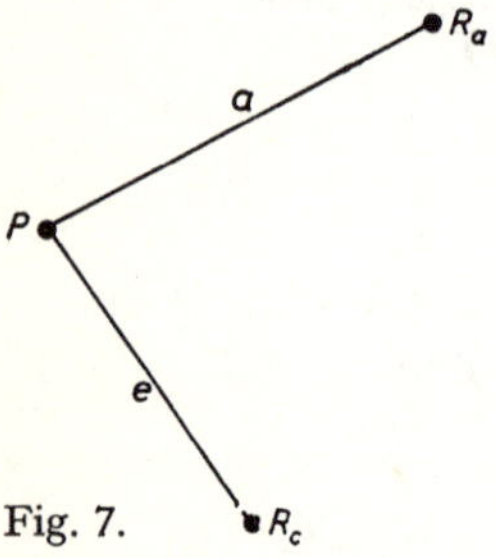

$$\mathfrak{M}(\nu) = \mathrm{R}\frac{1}{4h}\Big\{\sum_{R_a}\Big(\frac{\mathfrak{A}_a(\mathfrak{E}\overline{\mathfrak{A}}_a)}{\nu_a-\nu}+\frac{\overline{\mathfrak{A}}_a(\mathfrak{E}\mathfrak{A}_a)}{\nu_a+\nu}\Big)$$

$$-\sum_{R_e}\Big(\frac{\mathfrak{A}_e(\mathfrak{E}\overline{\mathfrak{A}}_e)}{\nu_e-\nu}+\frac{\overline{\mathfrak{A}}_e(\mathfrak{E}\mathfrak{A}_e)}{\nu_e+\nu}\Big)\Big\}\mathrm{e}^{2\pi\mathrm{i}\nu t}. \qquad (44)$$

Fig. 7.

The undetermined character of the phases for the quantities $\mathfrak{A}$ is obviously no longer of account in this expression, and the scattered radiation is therefore coherent with the incident radiation. Formula (44) expresses the same thing as formula (27).

It is of particular interest to study the singular behaviour of the expressions (40), (41) and (42) in the neighbourhood of certain critical values of the frequency. Thus we see in Case I from formula (40), that the dispersive moment tends to very large values when the frequency approaches the value of the absorption frequency ν_1 and when the scattered frequency approaches the value ν_2. Naturally the scattered moment cannot become infinitely large, as the formula would demand; in analogy to the classical formula (18), which only holds for values ν that differ from the natural frequencies ω of the system, we can expect formula (40) only to be valid as long as $\nu - \nu_1$ is large compared with the line width of the absorption line. Nevertheless we may assume that formula (40) is still correct, in that it is just the frequency ν_1 which represents a critical position for the dispersion. In fact, the scattered radiation will display a pronounced maximum at frequency ν_2 when irradiated by light containing frequency ν_1. Quite parallel to this is now the fact that for an irradiation of frequency ν_1, the atom acquires a probability for a transition into the state R_a. Once arrived in this state, it may possibly drop spontaneously to Q; to the possibility of such transitions there corresponds the emission of spherical waves of the frequency ν_2 in state R_a. If therefore a scattered frequency ν_2 is observed for incident radiation of frequency ν_1, this could arise from two distinct types of processes. In part, the scattered radiation is emitted by all atoms in state P, which could, e.g., be thought of as the ground state; in part, the radiation stems from those few atoms which were transformed into state R_a by irradiation and which now spontaneously emit the frequency ν_2 as well, among other frequencies. The case is obviously analogous to the previously mentioned fact that a scattered radiation, whose frequency coincides with that of the incident light, partly originates from atoms in the ground state P, partly from excited atoms Q. Our two contributions to the observable scattered radiation of frequency ν_2 can presumably be of the same order of magnitude. We shall not, at this point, go into the question how large their ratio can be under given circumstances, a question which is once again closely linked with the structure of the absorption lines.

Attention should however be drawn to two important points. The first concerns the ratio of the intensities with which the frequencies ν_1 and ν_2 in state P are scattered when the incident light contains frequency ν_1. The ratio can of course not be calculated by means of formulae (40) and (44), precisely because these formulae fail for very small values of $\nu_1 - \nu_2$ that are of the same order of magnitude as the line width. But we want to consider the case where the frequency ν of the incident light differs from ν_1 in such a way that $\nu - \nu_1$ is still large compared with the line width, but where at the same time $\nu - \nu_1 = -\delta$ is so small that the relevant term in expressions (40) and (44) is large compared with the sum of all the other terms. The scattering moment of the frequency $\nu_1 - \delta$ will then be approximately equal to

$$\mathrm{R}\,\frac{\mathfrak{A}_1(\mathfrak{E}\overline{\mathfrak{A}}_1)}{4h\delta}\,\mathrm{e}^{2\pi\mathrm{i}(\nu_1-\delta)t},$$

and the scattering moment of the frequency $\nu_2 - \delta$ equal to

$$\mathrm{R}\,\frac{\mathfrak{A}_2(\mathfrak{E}\overline{\mathfrak{A}}_1)}{4h\delta}\,\mathrm{e}^{2\pi\mathrm{i}(\nu_2-\delta)t}.$$

The intensity ratio of these two radiations will therefore become independent of δ, and in fact be equal to $\nu_1^4(\mathfrak{A}_1\overline{\mathfrak{A}}_1)/\nu_2^4(\mathfrak{A}_2\overline{\mathfrak{A}}_2)$. But this is exactly the ratio of the intensities with which the atom in state R_a emits frequencies ν_1 and ν_2. We may therefore expect quite generally that the intensity ratio of the scattered radiations that occur when an atom is irradiated by light containing an absorption frequency, will always be exactly the same as the intensity ratio of the spontaneous radiations of the corresponding frequencies which occur in the excited state. This conjecture, or assumption, is an essential point in the discussion of the problem of the polarisation of resonance radiation in metallic vapours.[1]) It will be realised that the above arguments furnish, by the correspondence principle, strong support for these assumptions, which are independent of the elaboration of more detailed concepts concerning the origins of the scattered radiation, such as has in part been attempted in this paper.

The second point refers to the possibility of certain transitions

[1]) Cf. N. Bohr, Naturwiss. **12** (1924) 1115, and particularly W. Heisenberg, Zs. f. Phys. **31** (1925) 617, where an attempt has been made to set up a quantitative theory of observations, on the basis of the correspondence principle.

taking place between stationary states, with which the existence of scattered radiation is inextricably linked. From the discussion on p. 241 it follows immediately that in all the special cases considered on pp. 242–245 the production of scattered radiation is connected with the possibility of a jump of the atom from state P to state Q, in such a way that the probability for a jump per unit time is exactly equal to the energy, divided by $h|\nu \pm \nu^*|$, which is emitted by the scattered radiation per unit time with frequency $|\nu \pm \nu^*|$. If, in Case I, the incident radiation contains the absorption frequency ν_1, the probability for a direct jump to state Q becomes very great. From the foregoing it may be expected that the ratio of this probability to the probability of a jump to some other state Q', also accessible from R_a, will be the same as that of the probabilities for atomic jumps from R_a partly to Q and partly to Q'.

So far we have only discussed the critical rôle played by the frequency ν_1 of an absorption line in relation to the scattered radiation for Case I. But we can deduce formula (40) that the expression for the scattering moment also increases strongly when ν approaches the frequency ν_6. This situation strikes one at first as very peculiar, since it might be thought that for an irradiation with light containing frequency ν_6, the scattered radiation should grow very strong and that therefore the atom should exhibit an absorption line of frequency ν_6, for energy balance reasons. Such a conclusion would however be in plain contradiction to (44). This formula expresses the fact that for the atom in its reaction to monochromatic irradiation, only those frequencies are critical which correspond to a transition to another stationary state. But it can be shown very simply that formula (40) can reproduce the situation very well; there is no need for the coincidence $\nu = \nu_6$ to make itself felt in any particular way during the absorption of dispersion of the incident light, as demanded by formula (44). When ν approaches ν_6, the frequency of the scattered radiation $\nu - \nu^*$ approaches more and more the frequency ν_5. But this frequency was contained in the spontaneous emission of the atom from the start. The part of the electrical moment of the atom which corresponds to this frequency is given by the expression $\mathrm{R}\mathfrak{A}_5\, e^{2\pi i \nu_5 t}$. For small values of the difference $\nu - \nu_6 = -\delta$, the induced moment

$$\mathrm{R}\left\{-\frac{\mathfrak{A}_5(\mathfrak{E}\overline{\mathfrak{A}}_6)}{4h\delta}\, e^{2\pi i(\nu_5 - \delta)t}\right\},$$

according to (40), is more and more able to interfere with the existing spontaneous radiation: extra terms will make their appearance in the expression for the scattered intensity, which contain the electric vector $\mathfrak{E}$ of the incident light linearly. The natural reason why these terms do not become vanishingly small in spite of the difference δ between the frequencies, lies in the fact that the atom has only a finite life time, by the very fact of its spontaneous emission. The real intensity of the scattered radiation induced by the radiation incident on the atom depends on the phase relations between the induced scattering moment and the moment of the spontaneous emission which is already present. The calculation shows that for suitable phase relations the intensity of the induced scattered radiation will not display any peculiar features for an incident frequency in the neighbourhood $\nu=\nu_q$. At the same time it shows that equation (40) is justified. A short calculation may clarify this point.

Let us assume for simplicity that the characteristic amplitude of the spontaneous scattered radiation of frequency ν_0 in a stationary state P is real, and that its absolute value is p. The mean life time of P shall be given by $1/a$. Let the frequency of the induced scattering moment be $\nu_0-\delta$, where δ is small compared with ν_0, but large compared with a. Let the absolute value of its amplitude be q. Let the instant at which the atom arrives in state P be denoted by $t=0$, and the phase difference between the two scattering moments at this time be φ. Then the energy emitted during the life time T is proportional to the expression

$$s(T, \varphi) = \int_0^T \{p \cos 2\pi\nu_0 t + q \cos [2\pi(\nu_0 - \delta)t + \varphi]\}^2 \, \mathrm{d}t, \tag{45}$$

and the mean energy emitted during the life time of state P is proportional to

$$S(\varphi) = a \int_0^\infty s(T, \varphi) \, \mathrm{e}^{-aT} \, \mathrm{d}T. \tag{46}$$

Neglecting small quantities of order δ/ν_0, the calculation gives

$$S(\varphi) = \frac{p^2}{2a} + \frac{pq}{a^2 + (2\pi\delta)^2} (a \cos \varphi + 2\pi\delta \sin \varphi) + \frac{q^2}{2a}, \tag{47}$$

or

$$S(\varphi) = \frac{p^2}{2a} + \frac{pq}{\sqrt{a^2 + (2\pi\delta)^2}} \sin (\varphi + \alpha) + \frac{q^2}{2a}, \tag{48}$$

where

$$\tan\alpha = \frac{a}{2\pi\delta}.$$

Let the moment q now be given by

$$q = -\frac{E'pp}{4h\delta}, \tag{49}$$

in agreement with (40), where E denotes the amplitude of the electric vector of the incident waves and p' is of the same order of magnitude as p. Neglecting terms of order δ^2/a^2, we obtain the following form for (48):

$$S(\varphi) = \frac{p^2}{2a} - \frac{p^2p'E}{8\pi h}\,\frac{\sin(\varphi+\alpha)}{\delta^2} + \frac{p^2p'^2E^2}{32h^2a}\,\frac{1}{\delta^2}. \tag{50}$$

It follows from this expression that for sufficiently weak irradiation (E very small), the second term can always be of the same order of magnitude as the third, because both terms vary in the same way with δ if the angle $\varphi+\alpha$ is held constant. The third term would be exactly compensated by the second if, in the mean, $\sin(\varphi+\alpha)$ were equal to $\pi p'E/4ha$. For our argument to be valid, this quantity must be small compared with 1. By comparison with (49) and by the argument that for very small values of $2\pi\delta$ this quantity in (49) can be replaced by an expression roughly of the form $\sqrt{a^2+(2\pi\delta)^2}$, we see that our discussion can only be valid when the induced scattering moments are smaller than the moments of the spontaneous scattered radiation in the atom, even in cases where the induced moment attains a maximum for certain critical frequencies. Such a restriction seems to be a natural one to make from the start.

Our correspondence formulae (40) and (44) have therefore led us to assume the existence of a relation between the phases, such as we considered here. The fact that the phase of the induced scattering moment depends not only on $\mathfrak{A}_5$ but also on $\overline{\mathfrak{A}}_6$, and that this vector in state P is intrinsically undeterminable as far as its phase is concerned, shows that this can hardly be considered an artificial assumption.

So far we have only dealt with the critical values of the frequency ν of the incident radiation for Case I, where the scattered radiation has frequency $\nu-\nu^*$ and is connected with a transition to a state of higher energy. But the discussion assumes an essentially similar form in

Cases II(a) and II(b). In Case II(a) there exist two kinds of critical frequencies. The one arises again from the absorption frequencies ν_1. Irradiation with light which contains frequency ν_1 gives rise to a scattered radiation which reaches a maximum intensity for the frequency ν_2 (cf. fig. 5). In analogy with this we have the fact that the atoms can be raised to the state R_a by irradiation, and from this state ν_2 is emitted spontaneously. Furthermore, ν_6 is a critical frequency that must not, however, give rise to any maximum in the intensity of the scattered radiation of frequency ν_5, which is already present as spontaneous radiation from the beginning. In Case II(b), only the critical frequencies ν_3 and ν_4 have to be considered (ν_2 and ν_5 are always greater than ν^*, whereas ν is always supposed to be smaller than ν^*). Irradiation by light containing frequency ν_3 gives rise to a scattered radiation which possesses a maximum at ν_4. This corresponds to the fact that the atom acquires an increased probability for the transition to state R, and in this state it is just ν_4 which is emitted spontaneously. On the other hand, a strong scattering moment with frequency ν_3 will appear in the atom when it is irradiated with frequency ν_4. But since we must assume that the scattering moment does not show a maximum at ν_4 as a function of the frequency of the incident light, the intensity of the scattered radiation can also have no intensity maximum at the frequency ν_3, which had been present in the atom from the start.

Summarising, we can say that for irradiation with light containing an absorption (emission) frequency of the atom, the atom simulates the radiative properties of that state which, together with the given state, is characteristic for the relevant absorption (emission) frequency.

5. Final remarks

It must be a requirement that the different types of scattered radiation, discussed in this paper, and the jumps between stationary states to which they give rise, should leave the energy distribution in the black-body radiation and the statistical equilibrium distribution of the atoms unchanged. The postulates by means of which this can be achieved will evidently have to reveal a great similarity to the postulates which enabled Pauli to describe the thermal equilibrium between free electrons and black-body radiation. [1]) In Case I, e.g., the probability for a direct transition from P to Q would have to be represented in the form of a series which contains, among others, a term propor-

[1]) W. Pauli Jr., Zs. f. Phys. **18** (1923) 272.

tional to the radiation density $\varrho(\nu_1)$ for frequency ν_1, and a term proportional to the product $\varrho(\nu_1)\,\varrho(\nu_2)$. But it follows from the above that no term proportional to $\varrho(\nu_6)$ may appear.

The discussion in this paper shows that there can hardly be any way other than through a formula of the type (39), where $\mathfrak{M}$ is of the characteristic form (37), of satisfying the 'correspondence' requirement that the scattered radiation induced in an atomic system by external radiation should coincide in the limit of large quantum numbers with the scattered radiation demanded by classical theory. Even if the description of the true scattering moment should require a less simple expression for $\mathfrak{M}$, it would be difficult to escape the conclusion that the moment of the scattered radiation will assume a particularly large value whenever a condition of the form $\nu+\nu_p=0$ or $\nu+\nu_q=0$ is satisfied. Furthermore it seems impossible to set up the formulae in accordance with the correspondence principle, and yet avoid having the critical values of the frequency ν correspond to transitions for which, in general, neither the initial nor the final state coincide with the atomic state under discussion. The difficulty which would thus seem to arise in the problem of the energy exchange between atoms and radiation field, can be resolved in a natural way, as has been shown, by assuming that there exists a spontaneous emission by the atom which is effective during the whole of the life time of the state under consideration. It can be deduced from this that an interpretation which regards the scattering effect of an atom under external radiation as if it were in a given stationary state, need not necessarily lead to the view that the spontaneous emission from an atom, too, must be described as radiation from a given stationary state, rather than as an effect which the atom develops only during a transition between stationary states. But this is just the hypothesis put forward by Slater, and which gave rise to the views that Bohr, Kramers and Slater developed in more detail. As was mentioned in the introduction, the basic idea underlying these arguments can be formulated in just such a way that the rôle played by the atom in optical phenomena can always be reduced to interactions between the radiation field and the atom in some arbitrary stationary state; the special processes which we describe as 'transitions from one stationary state to another', may be regarded as of very short duration. No information about their inherent properties is given by the optical phenomena, in so far as they are known today and have been analysed.

Related papers

10a W. Heisenberg, *Ueber eine Anwendung des Korrespondenzprinzips auf die Frage nach der Polarisation des Fluoreszenzlichtes*. Z. f. Phys. **31**, p. 617, received Nov. 30, 1924.

Received May 14, 1925

11

ON THE TOTAL INTENSITY OF ABSORPTION LINES EMANATING FROM A GIVEN STATE

W. KUHN

Assuming far-reaching validity for the hypotheses of classical dispersion theory, a quantum-theoretical sum rule is formulated. In this, the total intensity of the lines comprised in an absorption region is related to the number of electrons connected with this absorption region through the totality of all the transition processes that depend on them intrinsically.

All atoms in every stationary state seem to have the property in common that light of an appropriate frequency can be absorbed. This property can be modified, but never completely destroyed, by the influence exerted by external fields. One may be led to assume that this state of affairs is based on some exact rule. An attempt will be made in this note to specify such a rule more closely. The discussion has as its starting point the hypotheses of dispersion theory and is based on the assumption that, for short wave lengths, these hypotheses yield a result which is in agreement with the classical dispersion theory for X-rays.

We consider, first of all, a hydrogen atom in its ground state (0). Light of frequencies $\nu_1, \nu_2, \ldots \nu_r$ can induce transitions of the atom into a number of states of higher energy $(\varepsilon_1, \varepsilon_2, \ldots \varepsilon_r)$, where the quantum numbers which are characteristic for the motions of the electrons run through a range of values. For irradiation with frequency ν, not close to one of the frequencies $\nu_1, \nu_2, \ldots \nu_r$, we assume that the oscillating dipole moment of the atom can be represented by

$$\frac{\mathfrak{P}}{\mathfrak{E}} = \frac{1}{4\pi^2} \frac{e^2}{m} \sum_{i=1}^{r} \frac{p_i}{\nu_i^2 - \nu^2}. \tag{1}$$

in analogy with classical dispersion theory. Here, $\mathfrak{P}$ and $\mathfrak{E}$ are the

Editor's note. This paper was published as Zs. f. Phys. **33** (1925) 408–412. It was signed 'Copenhagen, Universitetets Institut for teoretisk Fysik, May 1925.'

amplitudes of the dipole moment and electric vector of the incident light, respectively, and p_i can be regarded as the number of dispersion electrons which is appropriate for the transition $(0 \rightarrow i)$ and which is assumed identical with the number of absorption electrons.[1]) For frequencies $\nu \gg \nu_0$, this expression reduces to

$$\frac{\mathfrak{P}}{\mathfrak{E}} = -\frac{e^2}{4\pi^2 m} \frac{1}{\nu^2} \sum_{i=1}^{r} p_i.$$

Now it is known that the dispersion associated with the dispersion vector $\mathfrak{P}$ is equal to

$$-\frac{\mathrm{d}E}{\mathrm{d}t} = \frac{(2\pi\nu)^4}{3c^3} \mathfrak{P}^2.$$

One therefore obtains for the energy dispersed per unit time per atom,

$$-\frac{\mathrm{d}E}{\mathrm{d}t} = \frac{e^4}{3c^3 m^2} \left(\sum_{i=1}^{r} p_i\right)^2 \mathfrak{E}^2. \tag{2}$$

Substituting $\sum_i p_i = 1$ in (2), we obtain the theoretical expression derived by J. J. Thomson, which has found such significant application in X-ray dispersion experiments. Its general validity from the point of view of quantum theory should probably be sought in the fact that in the X-ray region, the energy quantum $h\nu$ is large compared with the binding energy of the electrons in the atoms.

The assumption that the dispersion formulae (1) are generally valid, at least for the shorter wave length values of ν_i and up to fairly high frequencies ν, therefore leads one to expect that $\sum p_i = 1$. It should be noted here that the summation will have to be replaced by an integral for the continuous absorption band beyond the series limit. As for a comparison with experiment, the sum rule cannot be tested in the case of hydrogen, since the relevant absorption measurements are not available.

[1]) Cf. R. Ladenburg, Zs. f. Phys. **4** (1921) 451 and R. Ladenburg and F. Reiche, Naturwiss. **11** (1923) 584. According to these authors' postulates, the relation

$$B_0^i = p_i \frac{\pi e^2}{m} \frac{1}{h\nu}$$

links the p_i which are defined by (1) with Einstein's transition probabilities and thus with the amount of absorption within the regions of selective absorption.

In some cases, e.g. neutral alkali atoms, it seems possible to associate a part of the absorption spectrum with a certain group of transition processes. Such a group would, on the one hand, only be comprised of processes for which the state of motion of only a single electron suffers an appreciable alteration; but they would, at the same time, also contain essentially all those transition processes which determine the state of motion of the relevant electron; in such cases it would seem natural to put $\sum p_i = 1$ for this part of the spectrum, in analogy to the case of hydrogen.[1])

So far, we have considered atoms in their ground state. But by its very nature, the proof of the sum rule ought to be valid for all states of an atom. Only, one would have to remember in performing the summation that one has to insert negative p-values for those spectral lines which correspond to quantum jumps into lower states.[2]) One thus obtains, for the intensity of the series lines and of the continuous absorption in the different states, as many relations as there are stationary states.

In the region of high quantum numbers, the above theorem on the p-summation can be directly interpreted as a condition imposed by the correspondence principle. In the region of low quantum numbers, our rule represents a strengthening of the requirements of the correspondence principle, provided the initial assumptions which were made about dispersion are proved correct. In this, it displays a certain affinity with Ornstein and Burger's rules about the intensity of

[1]) The p-value for the two D-lines in non-luminous sodium vapour was measured by R. Minkowski using magnetic rotation dispersion. [Cf. R. Minkowski, Ann. d. Phys. **66** (1921) 206 and R. Ladenburg and R. Minkowski, Zs. f. Phys. **6** (1921) 153.] He obtained the value $p=1$ for the common intensity of both D-lines. If the sum rule formulated above is correct, this value would seem a little too high, since the subsequent series terms and continuous absorption should be included in the sum. As for the continuous absorption beyond the series limit, it can be deduced from measurements of the absorption coefficient for the case of Na, by G. R. Harrison [Phys. Rev. **24** (1921) 206], that $\int dp_{\text{cont.}}$ is probably equal to, or a little smaller than, 0.1; a contribution of the same order of magnitude is to be expected from the higher series terms.

[2]) Cf. H. A. Kramers, Nature **113** (1924) 673. In this paper a general theory is formulated of the connection between the absorption (emission) and dispersion properties of atoms, which takes Einstein's negative absorption into account. See also H. A. Kramers and W. Heisenberg, Zs. f. Phys. **31** (1925) 681.

multiplet lines, which gave rise to a sharpening of the correspondence principle, formulated by Heisenberg.[1])

Whereas the latter rules refer to the relative intensities of connected spectral lines, we are dealing here in our sum rule with a theorem which concerns the absolute values of the intensities and would permit one to calculate them, once the relative values are known.

For substances other than hydrogen one will generally have to take different groups of electrons into account.

In that case, we know that the classical dispersion depends on the ratio of the wave length of the incident light to the distance between the oscillating electrons in the atom. A formula such as (1) can then still be considered valid, provided this ratio is large compared with unity. In such a case, where practically the whole of the absorption in the atom is associated with wave lengths that satisfy this condition, it would seem natural to extend the above discussion by setting $\sum p_i$ equal to the number, n, of the electrons in the atom. In particular, it may be possible to associate part of the absorption spectrum with a group of τ electrons in the atom, similarly to the case where it appeared possible to associate part of the absorption spectrum of, e.g., alkali atoms, with the transitions of a single electron. One would then like to assume for this part of the absorption spectrum that $\sum p_i = \tau$. One could reasonably expect that this relation is valid to a very good approximation, provided the dimensions of the group of electrons in question are small compared with the wave lengths of the corresponding absorption range, even though these wave lengths may not be large compared with the dimensions of the atom as a whole.

Provisional verification of this relation seems to be possible for the case of helium, where we should have $\sum p_i = 2$. Dispersion measurements [2]) are available for He between $\lambda = 6438$ A.U. and $\lambda = 2379$ A.U. The resonance line of He is situated at 584.4 A.U., the series limit at 503 A.U. If we assume the wave-length dependence of the continuous absorption coefficient to be of the form [3]) $C\lambda^3$ ($\lambda < \lambda$ of series limit), we obtain $p_{584} = 0.462$; $\int_{\lambda=503}^{0} dp_{\text{cont.}} = 1.03$, after fitting the p-values

1) W. Heisenberg, Zs. f. Phys. **31** (1925) 617.

2) A detailed review with a critical assessment of the various measurements has recently been given by Herzfeld and Wolf, Ann. d. Phys. **76** (1925) 71; **76** (1925) 576.

3) On the theoretical confirmation of this behaviour, experimentally found in the X-ray region, see H. A. Kramers, Phil. Mag. **46** (1923) 836.

at the first and last points of the dispersion curve. Since the initial assumption is not on a firm basis for this case of very noticeable continuous absorption, and since further regions of selective absorption may exist for wave lengths shorter than 500 A.U., this result is not in contradiction with the expected value 2 of the p-sum for He.

As for the application of the rule to spectra of complicated systems, such as optical molecular spectra, one could try, conversely, to measure p-sums or-integrals in order to obtain an indication of the number of electrons producing an absorption region.

More recently, refraction measurements have been carried out in the X-ray region and the attempt has been made to represent the refraction by means of relations similar to (1) and to determine the occupation numbers of the inner electron groups, cf. e.g. B. R. v. Nardorff, Phys. Rev. 24 (1924) 113, and M. Siegbahn, Naturwiss. 12 (1924) 1212. The possibility of such a description and use of dispersion measurements lies well within the meaning of the sum rules formulated here, and this could be of importance for their subsequent experimental verification.

I should like to thank Prof. Bohr and Dr. Kramers at this point for their kind interest in this work, and should also like to take this opportunity to thank the International Education Board for having rendered my stay in Copenhagen possible.

Related papers

11a W. Thomas, *Über die Zahl der Dispersionselektronen, die einem stationären Zustande zugeordnet sind* (*Vorläufige Mitteilung*). Naturwiss. **13** (1925) 627.

PART II

THE BIRTH OF QUANTUM MECHANICS

Received July 29, 1925

12

QUANTUM-THEORETICAL RE-INTERPRETATION OF KINEMATIC AND MECHANICAL RELATIONS

W. HEISENBERG

The present paper seeks to establish a basis for theoretical quantum mechanics founded exclusively upon relationships between quantities which in principle are observable.

It is well known that the formal rules which are used in quantum theory for calculating observable quantities such as the energy of the hydrogen atom may be seriously criticized on the grounds that they contain, as basic element, relationships between quantities that are apparently unobservable in principle, e.g., position and period of revolution of the electron. Thus these rules lack an evident physical foundation, unless one still wants to retain the hope that the hitherto unobservable quantities may later come within the realm of experimental determination. This hope might be regarded as justified if the above-mentioned rules were internally consistent and applicable to a clearly defined range of quantum mechanical problems. Experience however shows that only the hydrogen atom and its Stark effect are amenable to treatment by these formal rules of quantum theory. Fundamental difficulties already arise in the problem of 'crossed fields' (hydrogen atom in electric and magnetic fields of differing directions). Also, the reaction of atoms to periodically varying fields cannot be described by these rules. Finally, the extension of the quantum rules to the treatment of atoms having several electrons has proved unfeasible.

It has become the practice to characterize this failure of the quantum-theoretical rules as a deviation from classical mechanics, since the rules themselves were essentially derived from classical mechanics. This characterization has, however, little meaning when one realizes

Editor's note. This paper was published as Zs. Phys. **33** (1925) 879–893. It was signed 'Göttingen, Institut für theoretische Physik'.

that the *Einstein–Bohr* frequency condition (which is valid in all cases) already represents such a complete departure from classical mechanics, or rather (using the viewpoint of wave theory) from the kinematics underlying this mechanics, that even for the simplest quantum-theoretical problems the validity of classical mechanics simply cannot be maintained. In this situation it seems sensible to discard all hope of observing hitherto unobservable quantities, such as the position and period of the electron, and to concede that the partial agreement of the quantum rules with experience is more or less fortuitous. Instead it seems more reasonable to try to establish a theoretical quantum mechanics, analogous to classical mechanics, but in which only relations between observable quantities occur. One can regard the frequency condition and the dispersion theory of *Kramers*[1] together with its extensions in recent papers[2] as the most important first steps toward such a quantum-theoretical mechanics. In this paper, we shall seek to establish some new quantum-mechanical relations and apply these to the detailed treatment of a few special problems. We shall restrict ourselves to problems involving one degree of freedom.

1. In classical theory, the radiation emitted by a moving electron (in the wave zone, i.e., in the region where $\mathfrak{E}$ and $\mathfrak{H}$ are of the same order of magnitude as $1/r$) is not entirely determined by the expressions

$$\mathfrak{E} = \frac{e}{r^3c^2}[\mathfrak{r}[\mathfrak{r}\dot{\mathfrak{v}}]], \qquad \mathfrak{H} = \frac{e}{r^2c^2}[\dot{\mathfrak{v}}\mathfrak{r}],$$

but additional terms occur in the next order of approximation, e.g. terms of the form $e\dot{\mathfrak{v}}\mathfrak{v}/rc^3$ which can be called ‘quadrupole radiation’. In still higher order, terms such as $e\dot{\mathfrak{v}}\mathfrak{v}^2/rc^4$ appear. In this manner the approximation can be carried to arbitrarily high order. (The following symbols, have been employed: $\mathfrak{E}$, $\mathfrak{H}$ are field strengths at a given point, $\mathfrak{r}$ the vector between this point and the position of the electron, $\mathfrak{v}$ the velocity and e the charge of the electron).

One may inquire about the form these higher order terms would assume in quantum theory. The higher order approximations can easily be calculated in classical theory if the motion of the electron is

[1] H. A. Kramers, Nature **113** (1924) 673.

[2] M. Born, Zs. f. Phys. **26** (1924) 379. H. A. Kramers and W. Heisenberg, Zs. f. Phys. **31** (1925) 681. M. Born and P. Jordan, Zs. f. Phys. (in course of publication) [**33** (1925) 479; paper 7a].

given in Fourier expansion, and one would expect a similar result in quantum theory. This point has nothing to do with electrodynamics but rather – and this seems to be particularly important – is of a purely kinematic nature. We may pose the question in its simplest form thus: If instead of a classical quantity $x(t)$ we have a quantum-theoretical quantity, what quantum-theoretical quantity will appear in place of $x(t)^2$?

Before we can answer this question, it is necessary to bear in mind that in quantum theory it has not been possible to associate the electron with a point in space, considered as a function of time, by means of observable quantities. However, even in quantum theory it is possible to ascribe to an electron the emission of radiation. In order to characterize this radiation we first need the frequencies which appear as functions of two variables. In quantum theory these functions are of the form

$$\nu(n, n-\alpha) = \frac{1}{h}\{W(n) - W(n-\alpha)\},$$

and in classical theory of the form

$$\nu(n, \alpha) = \alpha\nu(n) = \alpha\frac{1}{h}\frac{\mathrm{d}W}{\mathrm{d}n}.$$

(Here one has $nh=J$, where J is one of the canonical constants).

As characteristic for the comparison between classical and quantum theory with respect to frequency, one can write down the combination relations:

Classical:

$$\nu(n, \alpha) + \nu(n, \beta) = \nu(n, \alpha+\beta).$$

Quantum-theoretical:

$$\nu(n, n-\alpha) + \nu(n-\alpha, n-\alpha-\beta) = \nu(n, n-\alpha-\beta)$$

or

$$\nu(n-\beta, n-\alpha-\beta) + \nu(n, n-\beta) = \nu(n, n-\alpha-\beta).$$

In order to complete the description of radiation it is necessary to have not only the frequencies but also the amplitudes. The amplitudes may be treated as complex vectors, each determined by six independent components, and they determine both the polarization and the phase. As the amplitudes are also functions of the two variables

n and α, the corresponding part of the radiation is given by the following expressions:

Quantum-theoretical:

$$\mathrm{Re}\{\mathfrak{A}(n, n-\alpha)\, e^{i\omega(n, n-\alpha)t}\}. \tag{1}$$

Classical:

$$\mathrm{Re}\{\mathfrak{A}_\alpha(n)\, e^{i\omega(n)\alpha t}\}. \tag{2}$$

At first sight the phase contained in $\mathfrak{A}$ would seem to be devoid of physical significance in quantum theory, since in this theory frequencies are in general not commensurable with their harmonics. However, we shall see presently that also in quantum theory the phase has a definite significance which is analogous to its significance in classical theory. If we now consider a given quantity $x(t)$ in classical theory, this can be regarded as represented by a set of quantities of the form

$$\mathfrak{A}_\alpha(n)\, e^{i\omega(n)\alpha t},$$

which, depending upon whether the motion is periodic or not, can be combined into a sum or integral which represents $x(t)$:

$$x(n, t) = \sum_{-\infty}^{+\infty}{}_\alpha\, \mathfrak{A}_\alpha(n)\, e^{i\omega(n)\alpha t}$$

or (2a)

$$x(n, t) = \int_{-\infty}^{+\infty} \mathfrak{A}_\alpha(n)\, e^{i\omega(n)\alpha t} d\alpha.$$

A similar combination of the corresponding quantum-theoretical quantities seems to be impossible in a unique manner and therefore not meaningful, in view of the equal weight of the variables n and $n-\alpha$. However, one may readily regard the ensemble of quantities $\mathfrak{A}(n, n-\alpha)e^{i\omega(n, n-\alpha)t}$ as a representation of the quantity $x(t)$ and then attempt to answer the above question: how is the quantity $x(t)^2$ to be represented?

The answer in classical theory is obviously:

$$\mathfrak{B}_\beta(n)\, e^{i\omega(n)\beta t} = \sum_{-\infty}^{+\infty}{}_\alpha\, \mathfrak{A}_\alpha \mathfrak{A}_{\beta-\alpha}\, e^{i\omega(n)(\alpha+\beta-\alpha)t} \tag{3}$$

or

$$= \int_{-\infty}^{+\infty} \mathfrak{A}_\alpha \mathfrak{A}_{\beta-\alpha}\, e^{i\omega(n)(\alpha+\beta-\alpha)t} d\alpha, \tag{4}$$

so that

$$x(t)^2 = \sum_{-\infty}^{+\infty}{}_\beta\, \mathfrak{B}_\beta(n)\; e^{i\omega(n)\beta t} \tag{5}$$

or, respectively,

$$= \int_{-\infty}^{+\infty} \mathfrak{B}_\beta(n)\; e^{i\omega(n)\beta t} d\beta. \tag{6}$$

In quantum theory, it seems that the simplest and most natural assumption would be to replace equations (3) and (4) by:

$$\mathfrak{B}(n, n-\beta)\, e^{i\omega(n,\, n-\beta)t} = \sum_{-\infty}^{+\infty}{}_\alpha\, \mathfrak{A}(n, n-\alpha)\mathfrak{A}(n-\alpha, n-\beta)\, e^{i\omega(n,\, n-\beta)t} \tag{7}$$

or

$$= \int_{-\infty}^{+\infty} \mathfrak{A}(n, n-\alpha)\mathfrak{A}(n-\alpha, n-\beta)\, e^{i\omega(n,\, n-\beta)t} d\alpha, \tag{8}$$

and in fact this type of combination is an almost necessary consequence of the frequency combination rules. On making assumptions (7) and (8), one recognizes that the phases of the quantum-theoretical $\mathfrak{A}$ have just as great a physical significance as their classical analogues. Only the origin of the time scale and hence a phase factor common to all the $\mathfrak{A}$ is arbitrary and accordingly devoid of physical significance, but the phases of the individual $\mathfrak{A}$ enter in an essential manner into the quantity $\mathfrak{B}$.[1] A geometrical interpretation of such quantum-theoretical phase relations in analogy with those of classical theory seems at present scarcely possible.

If we further ask for a representation for the quantity $x(t)^3$ we find without difficulty:

Classical:

$$\mathfrak{C}(n, \gamma) = \sum_{-\infty}^{+\infty} \sum_{-\infty}^{+\infty}{}_{\alpha,\beta}\, \mathfrak{A}_\alpha(n)\mathfrak{A}_\beta(n)\mathfrak{A}_{\gamma-\alpha-\beta}(n). \tag{9}$$

Quantum-theoretical:

$$\mathfrak{C}(n, n-\gamma) =$$

$$= \sum_{-\infty}^{+\infty} \sum_{-\infty}^{+\infty}{}_{\alpha,\beta}\, \mathfrak{A}(n, n-\alpha)\mathfrak{A}(n-\alpha, n-\alpha-\beta)\mathfrak{A}(n-\alpha-\beta, n-\gamma) \tag{10}$$

or the corresponding integral forms.

[1] Cf. also H. A. Kramers and W. Heisenberg, loc.cit. The phases enter essentially into the expressions used there for the induced scattering moment.

In a similar manner, one can find a quantum-theoretical representation for all quantities of the form $x(t)^n$, and if any function $f[x(t)]$ is given, one can always find the corresponding quantum-theoretical expression, provided the function can be expanded as a power series in x. A significant difficulty arises, however, if we consider two quantities $x(t)$, $y(t)$, and ask after their product $x(t)y(t)$. If $x(t)$ is characterized by $\mathfrak{A}$, and $y(t)$ by $\mathfrak{B}$, we obtain the following representations for $x(t)y(t)$:

Classical:

$$\mathfrak{C}_\beta(n) = \sum_{-\infty}^{+\infty}{}_\alpha \mathfrak{A}_\alpha(n)\mathfrak{B}_{\beta-\alpha}(n).$$

Quantum-theoretical:

$$\mathfrak{C}(n, n-\beta) = \sum_{-\infty}^{+\infty}{}_\alpha \mathfrak{A}(n, n-\alpha)\mathfrak{B}(n-\alpha, n-\beta).$$

Whereas in classical theory $x(t)y(t)$ is always equal to $y(t)x(t)$, this is not necessarily the case in quantum theory. In special instances, e.g., in the expression $x(t)x(t)^2$, this difficulty does not arise.

If, as in the question posed at the beginning of this section, one is interested in products of the form $v(t)\dot{v}(t)$, then in quantum theory this product $v\dot{v}$ should be replaced by $\frac{1}{2}(v\dot{v}+\dot{v}v)$, in order that $v\dot{v}$ be the differential coefficient of $\frac{1}{2}v^2$. In a similar manner it would always seem possible to find natural expressions for the quantum-theoretical mean values, though they may be even more hypothetical than the formulae (7) and (8).

Apart from the difficulty just mentioned, formulae of the type (7), (8) should quite generally also suffice to express the interaction of the electrons in an atom in terms of the characteristic amplitudes of the electrons.

2. After these considerations which were concerned with the kinematics of quantum theory, we turn our attention to the dynamical problem which aims at the determination of the $\mathfrak{A}$, ν, W from the given forces of the system. In earlier theory this problem was solved in two stages:

1. Integration of the equation of motion

$$\ddot{x} + f(x) = 0. \tag{11}$$

2. Determination of the constants for periodic motion through

$$\oint p \,\mathrm{d}q = \oint m\dot{x} \,\mathrm{d}x = J(= nh). \tag{12}$$

If one seeks to construct a quantum-mechanical formalism corresponding as closely as possible to that of classical mechanics, it is very natural to take over the equation of motion (11) directly into quantum theory. At this point, however, it is necessary – in order not to depart from the firm foundation provided by those quantities that are in principle observable – to replace the quantities $\ddot{x}$ and $f(x)$ by their quantum-theoretical representatives, as given in § 1. In classical theory it is possible to obtain the solution of (11) by first expressing x as a Fourier series or Fourier integral with undetermined coefficients (and frequencies). In general, we then obtain an infinite set of equations containing infinitely many unknowns, or integral equations, which can be reduced to simple recursive relations for the $\mathfrak{A}$ in special cases only. In quantum theory we are at present forced to adopt this method of solving equation (11) since, as has been said before, it was not possible to define a quantum-theoretical function directly analogous to the function $x(n, t)$.

Consequently the quantum-theoretical solution of (11) is only possible in the simplest cases. Before we consider such simple examples, let us give a quantum-theoretical re-interpretation of the determination, from (12), of the constant of periodic motion. We assume that (classically) the motion is periodic:

$$x = \sum_{-\infty}^{+\infty}{}_{\alpha}\, a_\alpha(n) e^{i\alpha\omega_n t}; \tag{13}$$

hence

$$m\dot{x} = m \sum_{-\infty}^{+\infty}{}_{\alpha}\, a_\alpha(n) i\alpha\omega_n e^{i\alpha\omega_n t}$$

and

$$\oint m\dot{x}\,\mathrm{d}x = \oint m\dot{x}^2 \,\mathrm{d}t = 2\pi m \sum_{-\infty}^{+\infty}{}_{\alpha}\, a_\alpha(n) a_{-\alpha}(n) \alpha^2 \omega_n.$$

Furthermore, since $a_{-\alpha}(n) = \overline{a_\alpha(n)}$, as x is to be real, it follows that

$$\oint m\dot{x}^2 \,\mathrm{d}t = 2\pi m \sum_{-\infty}^{+\infty}{}_{\alpha}\, |a_\alpha(n)|^2 \alpha^2 \omega_n. \tag{14}$$

In the earlier theory this phase integral was usually set equal to an integer multiple of h, i.e., equal to nh, but such a condition does not fit naturally into the dynamical calculation. It appears, even when regarded from the point of view adopted hitherto, arbitrary in the sense of the correspondence principle, because from this point of view the J are determined only up to an additive constant as multiples of h. Instead of (14) it would be more natural to write

$$\frac{\mathrm{d}}{\mathrm{d}n}(nh) = \frac{\mathrm{d}}{\mathrm{d}n}\oint m\dot{x}^2\,\mathrm{d}t,$$

that is,

$$h = 2\pi m \sum_{-\infty}^{+\infty} \alpha\,\frac{\mathrm{d}}{\mathrm{d}n}(\alpha\omega_n . |a_\alpha|^2). \tag{15}$$

Such a condition obviously determines the a_α only to within a constant, and in practice this indeterminacy has given rise to difficulties due to the occurrence of half-integral quantum numbers.

If we look for a quantum-theoretical relation corresponding to (14) and (15) and containing observable quantities only, the uniqueness which had been lost is automatically restored.

We have to admit that only equation (15) has a simple quantum-theoretical reformulation which is related to *Kramers'* dispersion theory:[1]

$$h = 4\pi m \sum_{0}^{\infty}{}_{\alpha} \{|a(n, n+\alpha)|^2\omega(n, n+\alpha) - |a(n, n-\alpha)|^2\omega(n, n-\alpha)\}. \tag{16}$$

Yet this relation suffices to determine the a uniquely since the undetermined constant contained in the quantities a is automatically fixed by the condition that a ground state should exist, from which no radiation is emitted. Let this ground state be denoted by n_0; then we should have $a(n_0, n_0-\alpha)=0$ (for $\alpha>0$). Hence we may expect that the question of half-integer or integer quantization does not arise in a theoretical quantum mechanics based only upon relations between observable quantities.

Equations (11) and (16), if soluble, contain a complete determination not only of frequencies and energy values, but also of quantum-theoretical transition probabilities. However, at present the actual mathematical solution can be obtained only in the simplest cases. In many systems, e.g. the hydrogen atom, a particular complication

[1] This relation has already been derived from dispersion considerations by W. Kuhn, Zs. Phys. **33** (1925) 408, and W. Thomas, Naturwiss. **13** (1925) 627.

arises because the solutions correspond to motion which is partly periodic and partly aperiodic. As a consequence of this property, the quantum-theoretical series (7), (8) and equation (16) decompose into a sum and an integral. Quantum-mechanically such a decomposition into 'periodic and aperiodic motion' cannot be carried out in general.

Nevertheless, one could regard equations (11) and (16) as a satisfactory solution, at least in principle, of the dynamical problem if it were possible to show that this solution agrees with (or at any rate does not contradict) the quantum-mechanical relationships which we know at present. It should, for instance, be established that the introduction of a small perturbation into a dynamical problem leads to additional terms in the energy, or frequency, of the type found by *Kramers* and *Born* – but not of the type given by classical theory. Furthermore, one should also investigate whether equation (11) in the present quantum-theoretical form would in general give rise to an energy integral $\frac{1}{2}m\dot{x}^2+U(x)=\text{const.}$, and whether the energy so derived satisfies the condition $\Delta W=h\nu$, in analogy with the classical condition $\nu=\partial W/\partial J$. A general answer to these questions would elucidate the intrinsic connections between previous quantum-mechanical investigations and pave the way toward a consistent quantum-mechanics based solely upon observable quantities. Apart from a general connection between Kramer's dispersion formula and equations (11) and (16), we can answer the above questions only in very special cases which may be solved by simple recursion relations.

The general connection between *Kramers'* dispersion theory and our equations (11) and (16) is as follows. From equation (11) (more precisely, from the quantum-theoretical analogue) one finds, just as in classical theory, that the oscillating electron behaves like a free electron when acted upon by light of much higher frequency than any eigenfrequency of the system. This result also follows from Kramers' dispersion theory if in addition one takes account of equation (16). In fact, *Kramers* finds for the moment induced by a wave of the form $E \cos 2\pi\nu t$:

$$M = e^2E \cos 2\pi\nu t \, \frac{2}{h} \sum_0^\infty{}_\alpha \left\{ \frac{|a(n, n+\alpha)|^2\nu(n, n+\alpha)}{\nu^2(n, n+\alpha)-\nu^2} - \right.$$

$$\left. - \frac{|a(n, n-\alpha)|^2\nu(n, n-\alpha)}{\nu^2(n, n-\alpha)-\nu^2} \right\},$$

so that for $\nu \gg \nu(n, n+\alpha)$,

$$M = -\frac{2Ee^2 \cos 2\pi\nu t}{\nu^2 h} \sum_0^\infty{}_\alpha \{|a(n, n+\alpha)|^2 \nu(n, n+\alpha) - |a(n, n-\alpha)|^2 \nu(n, n-\alpha)\},$$

which, due to equation (16), becomes

$$M = -\frac{e^2 E \cos 2\pi\nu t}{4\pi^2 m \nu^2}.$$

3. As a simple example, the anharmonic oscillator will now be treated:

$$\ddot{x} + \omega_0^2 x + \lambda x^2 = 0. \tag{17}$$

Classically, this equation is satisfied by a solution of the form

$$x = \lambda a_0 + a_1 \cos \omega t + \lambda a_2 \cos 2\omega t + \lambda^2 a_3 \cos 3\omega t + \ldots \lambda^{\tau-1} a_\tau \cos \tau\omega t,$$

where the a are power series in λ, the first terms of which are independent of λ. Quantum-theoretically we attempt to find an analogous expression, representing x by terms of the form

$$\lambda a(n, n); \qquad a(n, n-1) \cos \omega(n, n-1)t;$$
$$\lambda a(n, n-2) \cos \omega(n, n-2)t;$$
$$\ldots \lambda^{\tau-1} a(n, n-\tau) \cos \omega(n, n-\tau)t \ldots.$$

The recursion formulae which determine the a and ω (up to, but excluding, terms of order λ) according to equations (3), (4) or (7), (8) are:

Classical:

$$\begin{aligned} \omega_0^2 a_0(n) + \tfrac{1}{2} a_1^2(n) &= 0; \\ -\omega^2 + \omega_0^2 &= 0; \\ (-4\omega^2 + \omega_0^2) a_2(n) + \tfrac{1}{2} a_1^2 &= 0; \\ (-9\omega^2 + \omega_0^2) a_3(n) + a_1 a_2 &= 0; \\ \cdots\cdots\cdots\cdots\cdots \end{aligned} \tag{18}$$

Quantum-theoretical:

$$\begin{aligned} \omega_0^2 a_0(n) + \tfrac{1}{4}[a^2(n+1, n) + a^2(n, n-1)] &= 0; \\ -\omega^2(n, n-1) + \omega_0^2 &= 0; \\ [-\omega^2(n, n-2) + \omega_0^2] a(n, n-2) + \tfrac{1}{2}[a(n, n-1) a(n-1, n-2)] &= 0; \\ [-\omega^2(n, n-3) + \omega_0^2] a(n, n-3) & \\ + \tfrac{1}{2}[a(n, n-1) a(n-1, n-3)] + \tfrac{1}{2}[a(n, n-2) a(n-2, n-3)] &= 0; \\ \cdots\cdots\cdots\cdots\cdots \end{aligned} \tag{19}$$

The additional quantum condition is:
Classical ($J=nh$):

$$1 = 2\pi m \frac{d}{dJ} \sum_{-\infty}^{+\infty} \tfrac{1}{4}\tau^2 |a_\tau|^2 \omega.$$

Quantum-theoretical:

$$h = \pi m \sum_{0}^{\infty} [|a(n+\tau, n)|^2 \, \omega(n+\tau, n) - |a(n, n-\tau)|^2 \, \omega(n, n-\tau)].$$

We obtain in first order, both classically and quantum-mechanically

$$a_1^2(n) \quad \text{or} \quad a^2(n, n-1) = \frac{(n + \text{const})h}{\pi m \omega_0}. \tag{20}$$

In quantum theory, the constant in equation (20) can be determined from the condition that $a(n_0, n_0-1)$ should vanish in the ground state. If we number the n in such a way that in the ground state n is zero, i.e. $n_0=0$, then $a^2(n, n-1)=nh/\pi m\omega_0$.

It thus follows from the recursive relations (18) that in classical theory the coefficient a_τ has (to first order in λ) the form $\varkappa(\tau)n^{\frac{1}{2}\tau}$ where $\varkappa(\tau)$ represents a factor independent of n. In quantum theory, equation (19) implies

$$a(n, n-\tau) = \varkappa(\tau) \sqrt{\frac{n!}{(n-\tau)!}}, \tag{21}$$

where $\varkappa(\tau)$ is the same proportionality factor, independent of n. Naturally, for large values of n the quantum-theoretical value of a_τ tends asymptotically to the classical value.

An obvious next step would be to try inserting the classical expression for the energy $\frac{1}{2}m\dot{x}^2+\frac{1}{2}m\omega_0^2x^2+\frac{1}{3}m\lambda x^3=W$, because in the present first-approximation calculation it actually is constant, even when treated quantum-theoretically. Its value is given by (19), (20) and (21) as:

Classical:

$$W = nh\omega_0/2\pi. \tag{22}$$

Quantum-theoretical, from (7) and (8):

$$W = (n + \tfrac{1}{2})h\omega_0/2\pi \tag{23}$$

(terms of order λ^2 have been excluded).

Thus from the present viewpoint, even the energy of a harmonic oscillator is not given by 'classical mechanics', i.e., by equation (22), but has the form (23).

The more precise calculation, taking into account higher order approximations in W, a, ω will now be carried out for the simpler example of an anharmonic oscillator $\ddot{x}+\omega_0^2 x+\lambda x^3=0$.

Classically, one can in this case set

$$x = a_1 \cos \omega t + \lambda a_3 \cos 3\omega t + \lambda^2 a_5 \cos 5\omega t + \ldots;$$

quantum-theoretically we attempt to set by analogy

$$a(n, n-1) \cos \omega(n, n-1)t; \qquad \lambda a(n, n-3) \cos \omega(n, n-3)t; \qquad \ldots$$

The quantities a are once more power series in λ whose first term has the form, as in equation (21),

$$a(n, n-\tau) = \varkappa(\tau) \sqrt{\frac{n!}{(n-\tau)!}},$$

as one finds by evaluating the equations corresponding to (18) and (19).

If the evaluation of ω and a from equations (18) and (19) is carried out to order λ^2 or λ respectively, one obtains

$$\omega(n, n-1) = \omega_0 + \lambda \frac{3nh}{8\pi\omega_0^2 m} - \lambda^2 \frac{3h^2}{256\omega_0^5 m^2 \pi^2}(17n^2+7) + \ldots \tag{24}$$

$$a(n, n-1) = \sqrt{\frac{nh}{\pi\omega_0 m}} \left(1 - \lambda \frac{3nh}{16\pi\omega_0^3 m} + \ldots\right). \tag{25}$$

$$a(n, n-3) = \frac{1}{32} \sqrt{\frac{h^3}{\pi^3\omega_0^7 m^3} n(n-1)(n-2)} \cdot \left(1 - \lambda \frac{39(n-1)h}{32\pi\omega_0^3 m}\right). \tag{26}$$

The energy, defined as the constant term in the expression

$$\tfrac{1}{2}m\dot{x}^2 + \tfrac{1}{2}m\omega_0^2 x^2 + \tfrac{1}{4}m\lambda x^4,$$

(I could not prove in general that all periodic terms actually vanish, but this was the case for all terms evaluated) turns out to be

$$W = \frac{(n + \frac{1}{2})h\omega_0}{2\pi} + \lambda \frac{3(n^2 + n + \frac{1}{2})h^2}{8 \cdot 4\pi^2\omega_0^2 m}$$

$$- \lambda^2 \frac{h^3}{512\pi^3\omega_0^5 m^2} (17n^3 + \tfrac{51}{2}n^2 + \tfrac{59}{2}n + \tfrac{21}{2}). \quad (27)$$

This energy can also be determined using the *Kramers–Born* approach by treating the term $\frac{1}{4}m\lambda x^4$ as a perturbation to the harmonic oscillator. The fact that one obtains exactly the same result (27) seems to me to furnish remarkable support for the quantum-mechanical equations which have here been taken as basis. Furthermore, the energy calculated from (27) satisfies the relation (cf. eq. 24):

$$\frac{\omega(n, n-1)}{2\pi} = \frac{1}{h}[W(n) - W(n-1)],$$

which can be regarded as a necessary condition for the possibility of a determination of the transition probabilities according to equations (11) and (16).

In conclusion we consider the case of a rotator and call attention to the relationship of equations (7), (8) to the intensity formulae for the Zeeman effect[1] and for multiplets.[2]

Consider the rotator as represented by an electron which circles a nucleus with constant distance a. Both classically and quantum-theoretically, the 'equations of motion' simply state that the electron describes a plane, uniform rotation at a distance a and with angular velocity ω about the nucleus. The 'quantum condition' (16) yields, according to (12),

$$h = \frac{\mathrm{d}}{\mathrm{d}n}(2\pi m a^2 \omega),$$

and according to (16)

$$h = 2\pi m\{a^2\omega(n+1, n) - a^2\omega(n, n-1)\},$$

[1] S. Goudsmit and R. de L. Kronig, Naturwiss. **13** (1925) 90; H. Hönl, Zs. f. Phys. **31** (1925) 340.

[2] R. de L. Kronig, Zs. f. Phys. **31** (1925) 885; A. Sommerfeld and H. Hönl, Sitzungsber. d. Preuss. Akad. d. Wiss. (1925) 141; H. N. Russell, Nature **115** (1925) 835.

from which, in both cases, it follows that

$$\omega(n, n-1) = \frac{h(n + \text{const})}{2\pi m a^2}.$$

The condition that the radiation should vanish in the ground state ($n_0=0$) leads to the formula

$$\omega(n, n-1) = \frac{hn}{2\pi m a^2}. \tag{28}$$

The energy is

$$W = \tfrac{1}{2} m v^2,$$

or, from equations (7), (8),

$$W = \frac{m}{2} a^2 \frac{\omega^2(n, n-1) + \omega^2(n+1, n)}{2} = \frac{h^2}{8\pi^2 m a^2} (n^2+n+\tfrac{1}{2}), \tag{29}$$

which again satisfies the condition $\omega(n, n-1) = (2\pi/h)[W(n) - W(n-1)]$.

As support for the validity of the formulae (28) and (29), which differ from those of the usual theory, one might mention that, according to Kratzer,[1] many band spectra (including spectra for which the existence of an electron momentum is improbable) seem to require formulae of type (28), (29), which, in order to avoid rupture with the classical theory of mechanics, one had hitherto endeavoured to explain through half-integer quantization.

In order to arrive at the *Goudsmit–Kronig–Hönl* formula for the rotator we have to leave the field of problems having one degree of freedom. We assume that the rotator has a direction in space which is subject to a very slow precession $\mathfrak{o}$ about the z-axis of an external field. Let the quantum number corresponding to this precession be m. The motion is then represented by the quantities

$$\begin{aligned} z: &\quad a(n, n-1; m, m) \cos \omega(n, n-1)t; \\ x + iy: &\quad b(n, n-1; m, m-1)\, e^{i[\omega(n,n-1)+\mathfrak{o}]t}; \\ &\quad b(n, n-1; m-1, m)\, e^{i[-\omega(n,n-1)+\mathfrak{o}]t}. \end{aligned}$$

The equations of motion are simply

$$x^2 + y^2 + z^2 = a^2.$$

[1] Cf. for example, B. A. Kratzer, Sitzungsber. d. Bayr. Akad. (1922) p. 107

Because of (7) this leads to[1]

$$\tfrac{1}{2}\{\tfrac{1}{2}a^2(n, n-1; m, m)+b^2(n, n-1; m, m-1)+b^2(n, n-1; m, m+1) + \tfrac{1}{2}a^2(n+1, n; m, m) + b^2(n+1, n; m-1, m) + b^2(n+1, n; m+1, m)\} = a^2. \quad (30)$$

$$\tfrac{1}{2}a(n, n-1; m, m)a(n-1, n-2; m, m) = b(n, n-1; m, m+1)b(n-1, n-2; m+1, m) + b(n, n-1; m, m-1)b(n-1, n-2; m-1, m). \quad (31)$$

One also has the quantum condition from (16):

$$2\pi m\{b^2(n, n-1; m, m-1)\omega(n, n-1) - b^2(n, n-1; m-1, m)\omega(n, n-1)\} = (m + \text{const})h. \quad (32)$$

The classical relations corresponding to these equations are

$$\begin{aligned} \tfrac{1}{2}a_0^2 + b_1^2 + b_{-1}^2 &= a^2; \\ \tfrac{1}{4}a_0^2 &= b_1 b_{-1}; \\ 2\pi m(b_{+1}^2 - b_{-1}^2)\omega &= (m + \text{const})h. \end{aligned} \quad (33)$$

They suffice (up to the unknown constant added to m) to determine a_0, b_1, b_{-1} uniquely.

The simplest solution of the quantum-theoretical equations (30), (31), (32) which presents itself is:

$$b(n, n-1; m, m-1) = a\sqrt{\frac{(n+m+1)(n+m)}{4(n+\frac{1}{2})n}};$$

$$b(n, n-1; m-1, m) = a\sqrt{\frac{(n-m)(n-m+1)}{4(n+\frac{1}{2})n}};$$

$$a(n, n-1; m, m) = a\sqrt{\frac{(n+m+1)(n-m)}{(n+\frac{1}{2})n}}.$$

These expressions agree with the formulae of *Goudsmit*, *Kronig* and *Hönl*. It is, however, not easily seen that these expressions represent the *only* solution of equations (30), (31), (32), though this would seem likely to me from consideration of the boundary conditions (vanishing

[1] Equation (30) is essentially identical with the *Ornstein–Burger* sum rules.

of *a* and *b* at the 'boundary'; cf. the papers of *Kronig, Sommerfeld* and *Hönl, Russell* quoted above).

Considerations similar to the above, applied to the multiplet intensity formulae, lead to the result that these intensity rules are in agreement with equations (7) and (16). This finding may again be regarded as furnishing support for the validity of the kinematic equation (7).

Whether a method to determine quantum-theoretical data using relations between observable quantities, such as that proposed here, can be regarded as satisfactory in principle, or whether this method after all represents far too rough an approach to the physical problem of constructing a theoretical quantum mechanics, an obviously very involved problem at the moment, can be decided only by a more intensive mathematical investigation of the method which has been very superficially employed here.

Received September 27, 1925

13

ON QUANTUM MECHANICS

M. BORN AND P. JORDAN

The recently published theoretical approach of Heisenberg is here developed into a systematic theory of quantum mechanics (in the first place for systems having one degree of freedom) with the aid of mathematical matrix methods. After a brief survey of the latter, the mechanical equations of motion are derived from a variational principle and it is shown that using Heisenberg's quantum condition, the principle of energy conservation and Bohr's frequency condition follow from the mechanical equations. Using the anharmonic oscillator as example, the question of uniqueness of the solution and of the significance of the phases of the partial vibrations is raised. The paper concludes with an attempt to incorporate electromagnetic field laws into the new theory.

Introduction

The theoretical approach of Heisenberg[1] recently published in this Journal, which aimed at setting up a new kinematical and mechanical formalism in conformity with the basic requirements of quantum theory, appears to us of considerable potential significance. It represents an attempt to render justice to the new facts by setting up a new and really suitable conceptual system instead of adapting the customary conceptions in a more or less artificial and forced manner. The physical reasoning which led Heisenberg to this development has been so clearly described by him that any supplementary remarks appear superfluous. But, as he himself indicates, in its formal, mathematical aspects his approach is but in its initial stages. His hypotheses have been applied only to simple examples without being fully carried through to a generalized theory. Having been in an advantageous position to familiarize ourselves with his ideas throughout their formative stages, we now strive (since his investigations have been

Editor's note. This paper was published as Zs. f. Phys. **34** (1925) 858–888. Chapter 4 (pp. 883–888) of the original paper is not reproduced here.

[1] W. Heisenberg, Zs. f. Phys. **33** (1925) 879.

concluded) to clarify the mathematically formal content of his approach and present some of our results here. These indicate that it is in fact possible, starting with the basic premises given by Heisenberg, to build up a closed mathematical theory of quantum mechanics which displays strikingly close analogies with classical mechanics, but at the same time preserves the characteristic features of quantum phenomena.

In this we at first confine ourselves, like Heisenberg, to systems having *one degree of freedom* and assume these to be – from a classical standpoint – *periodic*. We shall in the continuation of this publication concern ourselves with the generalization of the mathematical theory to systems having an arbitrary number of degrees of freedom, as also to aperiodic motion. A noteworthy generalization of Heisenberg's approach lies in our confining ourselves neither to treatment of nonrelativistic mechanics nor to calculations involving Cartesian systems of coordinates. The only restriction which we impose upon the choice of coordinates is to base our considerations upon *libration coordinates*, which in classical theory are *periodic* functions of time. Admittedly, in some instances it might be more reasonable to employ other coordinates: for example, in the case of a rotating body to introduce the angle of rotation φ, which becomes a linear function of time. Heisenberg also proceeded thus in his treatment of the rotator; however, it remains undecided whether the approach applied there can be justified from the standpoint of a consistent quantum mechanics.

The mathematical basis of Heisenberg's treatment is the *law of multiplication* of quantum-theoretical quantities, which he derived from an ingenious consideration of correspondence arguments. The development of his formalism, which we give here, is based upon the fact that this rule of multiplication is none other than the well-known mathematical rule of *matrix multiplication*. The infinite square array (with discrete or continuous indices) which appears at the start of the next section, termed a *matrix*, is a representation of a physical quantity which is given in classical theory as a function of time. The mathematical method of treatment inherent in the new quantum mechanics is thereby characterized through the employment of *matrix analysis* in place of the usual number analysis.

Using this method, we have attempted to tackle some of the simplest problems in mechanics and electrodynamics. A *variational principle*, derived from correspondence considerations, yields *equations of motion* for the most general Hamilton function which are in closest analogy with the classical canonical equations. The quantum condition conjoined with one of the relations which proceed from the equations of motion permits a simple matrix notation. With the aid of this, one can prove the general validity of the *law of conservation of energy* and the *Bohr frequency relation* in the sense conjectured by Heisenberg: this proof could not be carried through in its entirety by him even for the simple examples which he considered. We shall later return in more detail to one of these examples in order to derive a basis for consideration of the part played by the phases of the partial vibrations in the new theory. We show finally that the basic laws of the electromagnetic field in a vacuum can readily be incorporated and we furnish substantiation for the assumption made by Heisenberg that the squares of the absolute values of the elements in a matrix representing the electrical moment of an atom provide a measure for the transition probabilities.

CHAPTER 1. MATRIX CALCULATION

1. Elementary operations. Functions

We consider square infinite matrices,[1] which we shall denote by heavy type to distinguish them from ordinary quantities which will throughout be in light type,

$$\mathbf{a} = (a(nm)) = \begin{pmatrix} a(00) & a(01) & a(02) \dots \\ a(10) & a(11) & a(12) \dots \\ a(20) & a(21) & a(22) \dots \\ \cdot \;\; \cdot \;\; \cdot & \cdot \;\; \cdot \;\; \cdot & \cdot \;\; \cdot \;\; \cdot \end{pmatrix}.$$

Equality of two matrices is defined as equality of corresponding components:

$$\mathbf{a} = \mathbf{b} \quad \text{means} \quad a(nm) = b(nm). \tag{1}$$

Matrix addition is defined as addition of corresponding components:

$$\mathbf{a} = \mathbf{b} + \mathbf{c} \quad \text{means} \quad a(nm) = b(nm) + c(nm). \tag{2}$$

[1] Further details of matrix algebra can be found, e.g., in M. Bôcher, Einführung in die höhere Algebra (translated from the English by Hans Beck; Teubner, Leipzig, 1910) § 22–25; also in R. Courant and D. Hilbert, Methoden der mathematischen Physik **1** (Springer, Berlin, 1924) Chapter I.

Matrix multiplication is defined by the rule 'rows times columns', familiar from the theory of determinants:

$$a = bc \quad \text{means} \quad a(nm) = \sum_{k=0}^{\infty} b(nk)\, c(km). \tag{3}$$

Powers are defined by repeated multiplication. The associative rule applies to multiplication and the distributive rule to combined addition and multiplication:

$$(ab)c = a(bc); \tag{4}$$

$$a(b + c) = ab + ac. \tag{5}$$

However, the commutative rule does *not* hold for multiplication: it is not in general correct to set $ab=ba$. If a and b do satisfy this relation, they are said to commute.

The *unit matrix* defined by

$$\mathbf{1} = (\delta_{nm}), \qquad \begin{cases} \delta_{nm} = 0 & \text{for} \quad n \neq m, \\ \delta_{nn} = 1 \end{cases} \tag{6}$$

has the property

$$a\mathbf{1} = \mathbf{1}a = a. \tag{6a}$$

The *reciprocal matrix* to a, namely a^{-1}, is defined by[1]

$$a^{-1}a = aa^{-1} = \mathbf{1}. \tag{7}$$

As *mean value* of a matrix a we denote that matrix whose diagonal elements are the same as those of a whereas all other elements vanish:

$$\bar{a} = (\delta_{nm} a(nn)). \tag{8}$$

The sum of these diagonal elements will be termed the *diagonal sum* of the matrix a and written as $D(a)$, viz.

$$D(a) = \sum_{n} a(nn). \tag{9}$$

From (3) it is easy to prove that if the diagonal sum of a product $y=x_1 x_2 \cdots x_m$ be finite, then it is unchanged by cyclic rearrangement

[1] As is known, a^{-1} is uniquely defined by (7) for *finite* square matrices when the determinant A of the matrix a is non-zero. If $A=0$ there is no matrix reciprocal to a.

of the factors:

$$D(x_1 x_2 \cdots x_m) = D(x_r x_{r+1} \cdots x_m x_1 x_2 \cdots x_{r-1}). \tag{10}$$

Clearly, it suffices to establish the validity of this rule for *two* factors.

If the elements of the matrices a and b are functions of a parameter t, then

$$\frac{\mathrm{d}}{\mathrm{d}t} \sum_k a(nk)\, b(km) = \sum_k \{\dot{a}(nk)\, b(km) + a(nk)\, \dot{b}(km)\},$$

or from the definition (3):

$$\frac{\mathrm{d}}{\mathrm{d}t}(ab) = \dot{a}b + a\dot{b}. \tag{11}$$

Repeated application of (11) gives

$$\frac{\mathrm{d}}{\mathrm{d}t}(x_1 x_2 \cdots x_n) = \dot{x}_1 x_2 \cdots x_n + x_1 \dot{x}_2 \cdots x_n + \ldots + x_1 x_2 \cdots \dot{x}_n. \tag{11'}$$

From the definitions (2) and (3) we can define *functions* of matrices. To begin with, we consider as the most general function of this type, $f(x_1, x_2, \ldots x_m)$, one which can formally be represented as a sum of a finite or infinite number of products of powers of the arguments x_k weighted by numerical coefficients.

Through the equations

$$\begin{aligned} f_1(y_1, \ldots y_n; x_1, \ldots x_n) &= 0, \\ \cdots\cdots\cdots\cdots \\ f_n(y_1, \ldots y_n; x_1, \ldots x_n) &= 0 \end{aligned} \tag{12}$$

we can then also define functions $y_l(x_1, \ldots x_n)$; namely, in order to obtain functions y_l having the above form and satisfying equation (12), the y_l need only be set in form of a series in increasing power products of the x_k and the coefficients determined through substitution in (12). It can be seen that one will always derive as many equations as there are unknowns. Naturally, the number of equations and unknowns exceeds that which would ensue from applying the method of undetermined coefficients in the normal type of analysis incorporating *commutative* multiplication. In each of the equations (12), upon substituting the series for the y_l and gathering together like terms one obtains not only a sum term $C'x_1x_2$ but also a term $C''x_2x_1$ and thereby has to bring both C' and C'' to vanish (e.g., not only $C'+C''$). This is, however, made possible by the fact that in the expansion of each of the y_l, two terms x_1x_2 and x_2x_1 appear, with two available coefficients.

2. Symbolic differentiation

At this stage we have to examine in detail the process of *differentiation* of a matrix function, which will later be employed frequently in calculation. One should at the outset note that only in a few respects does this process display similarity to that of differentiation in ordinary analysis. For example, the rules for differentiation of a product or of a function of a function here no longer apply in general. Only if all the matrices which occur *commute* with one another can one apply all the rules of normal analysis to this differentiation.

Suppose

$$y = \prod_{m=1}^{s} x_{l_m} = x_{l_1}x_{l_2}\cdots x_{l_s}. \tag{13}$$

We define

$$\frac{\partial y}{\partial x_k} = \sum_{r=1}^{s} \delta_{l_r k} \prod_{m=r+1}^{s} x_{l_m} \prod_{m=1}^{m=r-1} x_{l_m}, \qquad \begin{cases} \delta_{jk} = 0 \quad \text{for} \quad j \neq k, \\ \delta_{kk} = 1. \end{cases} \tag{14}$$

This rule may be expressed as follows: In the given product, one regards all factors as written out *individually* (e.g., not as $x_1^3x_2^2$, but as $x_1x_1x_1x_2x_2$); one then picks out any factor x_k and builds the product of all the factors which follow this and which precede (in this sequence). The sum of all such expressions is the differential coefficient of the product with respect to this x_k.

The procedure may be illustrated by some examples:

$$y = x^n, \qquad \frac{dy}{dx} = nx^{n-1}$$

$$y = x_1^n x_2^m, \qquad \frac{\partial y}{\partial x_1} = x_1^{n-1}x_2^m + x_1^{n-2}x_2^m x_1 + \ldots + x_2^m x_1^{n-1},$$

$$y = x_1^2 x_2 x_1 x_3, \qquad \frac{\partial y}{\partial x_1} = x_1x_2x_1x_3 + x_2x_1x_3x_1 + x_3x_1^2x_2.$$

If we further stipulate that

$$\frac{\partial(y_1 + y_2)}{\partial x_k} = \frac{\partial y_1}{\partial x_k} + \frac{\partial y_2}{\partial x_k}, \tag{15}$$

then the derivative $\partial y/\partial x$ is defined for the most general analytical functions y.

With the above definitions, together with that of the diagonal sum (9), there follows the relation

$$\frac{\partial D(y)}{\partial x_k(nm)} = \frac{\partial y}{\partial x_k}(mn), \tag{16}$$

on the right-hand side of which stands the *mn*-component of the matrix $\partial y/\partial x_k$. This relation can also be used to define the derivative $\partial y/\partial x_k$. In order to prove (16), it obviously suffices to consider a function y having the form (13). From (14) and (3) it follows that

$$\frac{\partial y}{\partial x_k}(mn) = \sum_{r=1}^{s} \delta_{l_r k} \sum_{\tau} \prod_{p=r+1}^{s} x_{l_p}(\tau_p \tau_p + 1) \prod_{p=1}^{r-1} x_{l_p}(\tau_p \tau_p + 1); \tag{17}$$

$$\tau_{r+1} = m, \qquad \tau_{s+1} = \tau_1, \qquad \tau_r = n.$$

On the other hand, from (3) and (9) ensues

$$\frac{\partial D(y)}{\partial x_k(mn)} = \sum_{r=1}^{s} \delta_{l_r k} \sum_{\tau} \prod_{p=1}^{r-1} x_{l_p}(\tau_p \tau_p + 1) \prod_{p=r+1}^{s} x_{l_p}(\tau_p \tau_p + 1); \tag{17'}$$

$$\tau_1 = \tau_{s+1}, \qquad \tau_r = n, \qquad \tau_{r+1} = m.$$

Comparison of (17) with (17′) yields (16).

We here pick out a fact which will later assume importance and which can be deduced from the definition (14): *the partial derivatives of a product are invariant with respect to cyclic rearrangement of the factors*. Because of (16) this can also be inferred from (10).

To conclude this introductory section, some additional description is devoted to functions $g(pq)$ of *two* variables. For

$$y = p^s q^r \tag{18}$$

it follows from (14) that

$$\frac{\partial y}{\partial p} = \sum_{l=1}^{s-1} p^{s-1-l} q^r p^l, \qquad \frac{\partial y}{\partial q} = \sum_{j=1}^{r-1} q^{r-1-j} p^s q^j. \tag{18'}$$

The most general function $g(pq)$ to be considered is to be represented in accordance with § 1 by a linear aggregate of terms

$$z = \prod_{j=1}^{k} (p^{s_j} q^{r_j}). \tag{19}$$

With the abbreviation

$$P_l = \prod_{j=l+1}^{k} (p^{s_j} q^{r_j}) \prod_{j=1}^{l-1} (p^{s_j} q^{r_j}), \tag{20}$$

one can write the derivatives as

$$\left.\begin{aligned} \frac{\partial z}{\partial p} &= \sum_{l=1}^{k} \sum_{m=0}^{s_l-1} p^{s_l-1-m} q^{r_l} P_l p^m, \\ \frac{\partial z}{\partial q} &= \sum_{l=1}^{k} \sum_{m=0}^{r_l-1} q^{r_l-1-m} P_l p^{s_l} q^m. \end{aligned}\right\} \tag{21}$$

From these equations we find an important consequence. We consider the matrices

$$d_1 = q \frac{\partial z}{\partial q} - \frac{\partial z}{\partial q} q, \qquad d_2 = p \frac{\partial z}{\partial p} - \frac{\partial z}{\partial p} p. \tag{22}$$

From (21) we have

$$d_1 = \sum_{l=1}^{k} (q^{r_l} P_l p^{s_l} - P_l p^{s_l} q^{r_l}),$$

$$d_2 = \sum_{l=1}^{k} (p^{s_l} q^{r_l} P_l - q^{r_l} P_l p^{s_l}),$$

and thus it follows that

$$d_1 + d_2 = \sum_{l=1}^{k} (p^{s_l} q^{r_l} P_l - P_l p^{s_l} q^{r_l}).$$

Herein the second member of each term cancels the first member of the following, and the first and last member of the overall sum also cancel, so that

$$d_1 + d_2 = 0. \tag{23}$$

Because of its linear character in z, this relation holds not only for expressions z having the form (19), but indeed for arbitrary analytical functions $g(pq)$.[1]

In concluding this brief survey of matrix analysis, we establish the following rule: *Every matrix equation*

$$F(x_1, x_2, \dots x_r) = 0$$

remains valid if in all the matrices x_j one and the same permutation of all rows and columns is undertaken. To this end, it suffices to show that for two matrices a, b which thereby become transposed to a', b', the following invariance conditions apply:

$$a' + b' = (a + b)', \qquad a'b' = (ab)',$$

wherein the right-hand sides denote those matrices which are formed from $a+b$ and ab respectively by such an interchange.

We set forth this proof by replacing the procedure of permutation by that of multiplication with a suitable matrix.[2]

We write a permutation as

$$\begin{pmatrix} 0 & 1 & 2 & 3 & \dots \\ k_0 & k_1 & k_2 & k_3 & \dots \end{pmatrix} = \begin{pmatrix} n \\ k_n \end{pmatrix},$$

and to this we assign a *permutation matrix,*

$$p = (p(nm)), \qquad p(nm) = \begin{cases} 1 \text{ when } m = k_n \\ 0 \text{ otherwise.} \end{cases}$$

The transposed matrix to p is

$$\tilde{p} = (\tilde{p}(nm)), \qquad \tilde{p}(nm) = \begin{cases} 1 \text{ when } n = k_m \\ 0 \text{ otherwise.} \end{cases}$$

1 More generally, for functions of r variables, one has

$$\sum_r \left(x_r \frac{\partial g}{\partial x_r} - \frac{\partial g}{\partial x_r} x_r \right) = 0.$$

2 The method of proof adopted here possesses the merit of revealing the close connection of permutations with an important class of more general transformations of matrices. The validity of the rule in question can however also be established directly on noting that in the definitions of *equality*, as also of *addition* and *multiplication* of matrices, no use was made of order relationships between the rows or the columns.

On multiplying the two together, one has

$$\mathsf{p}\tilde{\mathsf{p}} = (\sum_k p(nk)\, \tilde{p}(km)) = (\delta_{nm}) = \mathbf{1},$$

since the two factors $p(nk)$ and $\tilde{p}(km)$ differ from zero simultaneously only if $k=k_n=k_m$, i.e., when $n=m$. Hence $\tilde{\mathsf{p}}$ is reciprocal to p:

$$\tilde{\mathsf{p}} = \mathsf{p}^{-1}.$$

If now a be any given matrix, then

$$\mathsf{pa} = (\sum_k p(nk)\, a(km)) = (a(k_n, m))$$

is a matrix which arises from the permutation $\binom{n}{k_n}$ of the rows of a, and equivalently

$$\mathsf{ap}^{-1} = (\sum_k a(nk)\, \tilde{p}(km)) = (a(n, k_m))$$

is the matrix arising from permutation of the columns of a. One and the same permutation applied both to the rows and the columns of a thus yields the matrix

$$\mathsf{a}' = \mathsf{pap}^{-1}.$$

Thence follows directly

$$\begin{aligned} \mathsf{a}' + \mathsf{b}' &= \mathsf{p}(\mathsf{a} + \mathsf{b})\mathsf{p}^{-1} = (\mathsf{a} + \mathsf{b})', \\ \mathsf{a}'\mathsf{b}' &= \mathsf{pabp}^{-1} \qquad\quad = (\mathsf{ab})', \end{aligned}$$

which proves our original contention.

It is thus apparent that from matrix equations one can never determine any given sequence or order of rank of the matrix elements.

Moreover, it is evident that a much more general rule applies, namely that every matrix equation is invariant with respect to transformations of the type

$$\mathsf{a}' = \mathsf{bab}^{-1},$$

where b denotes an *arbitrary* matrix. We shall see later that this does not necessarily always apply to matrix differential equations.

CHAPTER 2. DYNAMICS

3. The basic laws

The dynamic system is to be described by the spatial coordinate q and the momentum p, these being represented by matrices

$$q = \left(q(nm)e^{2\pi i\nu(nm)t}\right), \qquad p = \left(p(nm)e^{2\pi i\nu(nm)t}\right). \tag{24}$$

Here the $\nu(nm)$ denote the quantum-theoretical frequencies associated with transitions between states described by the *quantum numbers* n and m. The matrices (24) are to be Hermitian, e.g., on transposition of the matrices, each element is to go over into its complex conjugate value, a condition which should apply for all real t. We thus have

$$q(nm)\, q(mn) = |q(nm)|^2 \tag{25}$$

and

$$\nu(nm) = -\,\nu(mn). \tag{26}$$

If q be a *Cartesian* coordinate, then the expression (25) is a measure of the *probabilities*[1] of the transitions $n \rightleftarrows m$.

Further, we shall require that

$$\nu(jk) + \nu(kl) + \nu(lj) = 0. \tag{27}$$

This can be expressed together with (26) in the following manner: there exist quantities W_n such that

$$h\nu(nm) = W_n - W_m. \tag{28}$$

From this, with equations (2), (3), it follows that a function $g(pq)$ invariably again takes on the form

$$g = \left(g(nm)e^{2\pi i\nu(nm)t}\right) \tag{29}$$

and the matrix $(g(nm))$ therein results from identically the same process applied to the matrices $(q(nm))$, $(p(nm))$ as was employed to find g from q, p. For this reason we can henceforth abandon the representation (24) in favour of the shorter notation

$$q = (q(nm)), \qquad p = (p(nm)). \tag{30}$$

For the *time derivative* of the matrix $g = (g(nm))$, recalling to mind

[1] In this connection see § 8.

(24) or (29), we obtain the matrix

$$\dot{g} = 2\pi i(\nu(nm)g(nm)). \tag{31}$$

If $\nu(nm) \neq 0$ when $n \neq m$, a condition which we wish to assume, then the formula $\dot{g}=0$ denotes that g is a diagonal matrix with $g(nm)=\delta_{nm}g(nn)$.

A matrix differential equation $\dot{g}=a$ is invariant with respect to that process in which the same permutation is carried out on rows and columns of all the matrices and also upon the numbers W_n. In order to realize this, consider the diagonal matrix

$$W = (\delta_{nm}W_n).$$

Then

$$Wg = (\sum_k \delta_{nk}W_n g(km)) = (W_n g(nm)),$$

$$gW = (\sum_k g(nk)\delta_{km}W_k) = (W_m g(nm)),$$

i.e., according to (31),

$$\dot{g} = \frac{2\pi i}{h}((W_n - W_m)g(nm)) = \frac{2\pi i}{h}(Wg - gW).$$

If now p be a permutation matrix, then the transform of W,

$$W' = pWp^{-1} = (\delta_{n_k m}W_{n_k})$$

is the diagonal matrix with the permuted W_n along the diagonal. Thence one has

$$p\dot{g}p^{-1} = \frac{2\pi i}{h}(W'g' - g'W') = \dot{g}',$$

where $g'=pgp^{-1}$ and $\dot{g}'$ denotes the time derivative of g' constructed in accordance with the rule (31) with permuted W_n.

The rows and columns of $\dot{g}$ thus experience the same permutation as those of g, and hence our contention is vindicated.

It is to be noted that a corresponding rule does *not* apply to arbitrary transformations of the form $a'=bab^{-1}$ since for these W' is no longer a diagonal matrix. Despite this difficulty, a thorough study of these general transformations would seem to be called for, since it offers promise of insight into the deeper connections intrinsic to this new theory: we shall later revert to this point.[1]

[1] Cf. the continuation of this work, to be published forthwith.

In the case of a Hamilton function having the form

$$H = \frac{1}{2m} p^2 + U(q)$$

we shall assume, as did Heisenberg, that the *equations of motion* are just of the same form as in classical theory, so that using the notation of § 2 we can write:

$$\left.\begin{aligned} \dot{q} &= \frac{\partial H}{\partial p} = \frac{1}{m} p, \\ \dot{p} &= -\frac{\partial H}{\partial q} = -\frac{\partial U}{\partial q}. \end{aligned}\right\} \qquad (32)$$

We now use correspondence considerations to try more generally to elucidate the equations of motion belonging to an arbitrary Hamilton function $H(pq)$. This is required from the standpoint of relativistic mechanics and in particular for the treatment of electron motion under the influence of magnetic fields. For in this latter case, the function H cannot in a Cartesian coordinate system any longer be represented by the sum of two functions of which one depends only on the momenta and the other on the coordinates.

Classically, equations of motion can be derived from the action principle

$$\int_{t_0}^{t_1} L \, \mathrm{d}t = \int_{t_0}^{t_1} \{p\dot{q} - H(pq)\} \, \mathrm{d}t = \text{extremum}. \qquad (33)$$

If we now envisage the Fourier expansion of L substituted in (33) and the time interval $t_1 - t_0$ taken sufficiently large, we find that only the constant term of L supplies a contribution to the integral. The form which the action principle thence acquires suggests the following translation into quantum mechanics:

The diagonal sum $D(\mathbf{L}) = \sum_k L(kk)$ *is to be made an extremum:*

$$D(\mathbf{L}) = D(\mathbf{p}\dot{\mathbf{q}} - \mathbf{H}(\mathbf{pq})) = \text{extremum}, \qquad (34)$$

namely, by suitable choice of $\mathbf{p}$ *and* $\mathbf{q}$, *with* $\nu(nm)$ *kept fixed.*

Thus, by setting the derivatives of $D(\mathbf{L})$ with respect to the elements of p and q equal to zero, one obtains the equations of motion

$$2\pi i\nu(nm)\, q(nm) = \frac{\partial D(H)}{\partial p(mn)},$$

$$2\pi i\nu(mn)\, p(mn) = \frac{\partial D(H)}{\partial q(mn)}.$$

From (26), (31) and (16) one observes that these equations of motion can always be written in *canonical* form,

$$\left.\begin{aligned} \dot{q} &= \frac{\partial H}{\partial p}, \\ \dot{p} &= -\frac{\partial H}{\partial q}. \end{aligned}\right\} \qquad (35)$$

For the quantization condition, Heisenberg employed a relation proposed by Thomas[1] and Kuhn.[2] The equation

$$J = \oint p\, dq = \int_0^{1/\nu} p\dot{q}\, dt$$

of 'classical' quantum theory can, on introducing the Fourier expansions of p and q,

$$p = \sum_{\tau=-\infty}^{\infty} p_\tau e^{2\pi i\nu\tau t}, \qquad q = \sum_{\tau=-\infty}^{\infty} q_\tau e^{2\pi i\nu\tau t},$$

be transformed into

$$1 = 2\pi i \sum_{\tau=-\infty}^{\infty} \tau \frac{\partial}{\partial J}(q_\tau p_{-\tau}). \qquad (36)$$

If therein one has $p=m\dot{q}$, one can express the p_τ in terms of q_τ and thence obtain that classical equation which on transformation into a difference equation according to the principle of correspondence yields the formula of Thomas and Kuhn. Since here the assumption that $p=m\dot{q}$ should be avoided, we are obliged to translate equation (36) directly into a difference equation.

[1] W. Thomas, Naturwiss. **13** (1925) 627.
[2] W. Kuhn, Zs. f. Phys. **33** (1925) 408.

The following expressions should correspond:

$$\sum_{\tau=-\infty}^{\infty} \tau \frac{\partial}{\partial J}(q_\tau p_{-\tau}) \quad \text{with}$$

$$\frac{1}{h}\sum_{\tau=-\infty}^{\infty} q(n+\tau, n)p(n, n+\tau) - q(n, n-\tau)p(n-\tau, n));$$

where in the right-hand expression those $q(nm)$, $p(nm)$ which take on a negative index are to be set equal to zero. In this way we obtain the quantization condition corresponding to (36) as

$$\sum_k (p(nk)q(kn) - q(nk)p(kn)) = \frac{h}{2\pi i}. \tag{37}$$

This is a system of infinitely many equations, namely one for each value of n.

In particular, for $p = m\dot{q}$ this yields

$$\sum_k \nu(kn)|q(nk)|^2 = \frac{h}{8\pi^2 m},$$

which, as may easily be verified, agrees with Heisenberg's form of the quantization condition, or with the Thomas–Kuhn equation. The formula (37) has to be regarded as the appropriate generalization of this equation.

Incidentally one sees from (37) that the diagonal sum $D(\mathbf{pq})$ necessarily becomes infinite. For otherwise one would have $D(\mathbf{pq}) - D(\mathbf{qp}) = 0$ from (10), whereas (37) leads to $D(\mathbf{pq}) - D(\mathbf{qp}) = \infty$. Thus the matrices under consideration are never finite.[1]

4. Consequences. Energy-conservation and frequency laws

The content of the preceding paragraphs furnishes the basic rules of the new quantum mechanics in their entirety. All other laws of quantum mechanics, whose general validity is to be verified, must be *derivable* from these basic tenets. As instances of such laws to be proved, the law of energy conservation and the Bohr frequency condition primarily enter into consideration. The law of conservation of energy states that if $\mathbf{H}$ be the energy, then $\dot{\mathbf{H}} = 0$, or that $\mathbf{H}$ is a

[1] Further, they do not belong to the class of 'bounded' infinite matrices hitherto almost exclusively investigated by mathematicians.

diagonal matrix. The diagonal elements $H(nn)$ of H are interpreted, according to Heisenberg, as the *energies of the various states of the system* and the Bohr frequency condition requires that

$$h\nu(nm) = H(nn) - H(mm),$$

or

$$W_n = H(nn) + \text{const.}$$

We consider the quantity

$$\mathsf{d} = \mathsf{pq} - \mathsf{qp}.$$

From (11), (35) one finds

$$\begin{aligned}\dot{\mathsf{d}} &= \dot{\mathsf{p}}\mathsf{q} + \mathsf{p}\dot{\mathsf{q}} - \dot{\mathsf{q}}\mathsf{p} - \mathsf{q}\dot{\mathsf{p}} \\ &= \mathsf{q}\frac{\partial \mathsf{H}}{\partial \mathsf{q}} - \frac{\partial \mathsf{H}}{\partial \mathsf{q}}\mathsf{q} + \mathsf{p}\frac{\partial \mathsf{H}}{\partial \mathsf{p}} - \frac{\partial \mathsf{H}}{\partial \mathsf{p}}\mathsf{p}.\end{aligned}$$

Thus from (22), (23) it follows that $\dot{\mathsf{d}}=0$ and d is a diagonal matrix. The diagonal elements of d are, however, specified just by the quantum condition (27). Summarizing, we obtain the equation

$$\mathsf{pq} - \mathsf{qp} = \frac{h}{2\pi i}\mathbf{1}, \tag{38}$$

on introducing the unit matrix $\mathbf{1}$ defined by (6). We term the equation (38) the 'stronger quantum condition' and base all further conclusions upon it.

From the form of this equation, we deduce the following: If an equation (A) be derived from (38), then (A) remains valid if p be replaced by q and simultaneously h by $-h$. For this reason one need for instance derive only one of the following two equations from (38), which can readily be performed by induction

$$\mathsf{p}^n\mathsf{q} = \mathsf{q}\mathsf{p}^n + n\frac{h}{2\pi i}\mathsf{p}^{n-1}, \tag{39}$$

$$\mathsf{q}^n\mathsf{p} = \mathsf{p}\mathsf{q}^n - n\frac{h}{2\pi i}\mathsf{q}^{n-1}. \tag{39'}$$

We shall now prove the energy-conservation and frequency laws, as expressed above, in the first instance for the case

$$\mathsf{H} = \mathsf{H}_1(\mathsf{p}) + \mathsf{H}_2(\mathsf{q}).$$

From the statements of § 1, it follows that we may formally replace $H_1(p)$ and $H_2(q)$ by power expansions

$$H_1 = \sum_s a_s p^s, \qquad H_2 = \sum_s b_s q^s.$$

Formulae (39) and (39′) indicate that

$$\left.\begin{aligned} Hq - qH &= \frac{h}{2\pi i} \frac{\partial H}{\partial p}, \\ Hp - pH &= -\frac{h}{2\pi i} \frac{\partial H}{\partial q}. \end{aligned}\right\} \tag{40}$$

Comparison with the equations of motion (35) yields

$$\left.\begin{aligned} \dot{q} &= \frac{2\pi i}{h} (Hq - qH). \\ \dot{p} &= \frac{2\pi i}{h} (Hp - pH). \end{aligned}\right\} \tag{41}$$

Denoting the matrix $Hg-gH$ by $\left|{H \atop g}\right|$ for brevity, one has

$$\left|\begin{matrix} H \\ ab \end{matrix}\right| = \left|\begin{matrix} H \\ a \end{matrix}\right| b + a \left|\begin{matrix} H \\ b \end{matrix}\right|; \tag{42}$$

from which generally for $g=g(pq)$ one may conclude that

$$\dot{g} = \frac{2\pi i}{h} \left|\begin{matrix} H \\ g \end{matrix}\right| = \frac{2\pi i}{h} (Hg - gH). \tag{43}$$

To establish this result, one need only conceive $\dot{g}$ as expressed in function of p, q and $\dot{p}$, $\dot{q}$ with the aid of (11), (11′), and $\left|{H \atop g}\right|$ as evaluated by means of (42) in function of p, q and $\left|{H \atop p}\right|$, $\left|{H \atop q}\right|$, followed by application of the relations (41). In particular, if in (43) one sets $g=H$, one obtains

$$\dot{H} = 0. \tag{44}$$

Now that we have verified the energy-conservation law and recognized the matrix H to be diagonal, equation (41) can be put into the form

$$h\nu(nm)\, q(nm) = (H(nn) - H(mm))\, q(nm),$$
$$h\nu(nm)\, p(nm) = (H(nn) - H(mm))\, p(nm),$$

from which the frequency condition follows.

If we now go over to consideration of more general Hamilton functions $H^*=H^*(pq)$, it can easily be seen that in general $\dot{H}^*$ no longer vanishes (examples such as $H^*=p^2q$, readily reveal this). It can however be observed that the Hamilton function $H=\frac{1}{2}(p^2q+qp^2)$ yields the same equations of motion as H^* and that $\dot{H}$ again vanishes. In consequence we may express the energy-conservation and frequency laws in the following way: *To each function $H^*=H^*(pq)$ there can be assigned a function $H=H(pq)$ such that as Hamiltonians H^* and H yield the same equations of motion and that for these equations of motion H assumes the role of an energy which is constant in time and which fulfils the frequency condition.*

On bearing in mind the considerations discussed above, it suffices to show that the function H to be specified satisfies not only the conditions

$$\frac{\partial H}{\partial p} = \frac{\partial H^*}{\partial p}, \qquad \frac{\partial H}{\partial q} = \frac{\partial H^*}{\partial q}, \tag{45}$$

but in addition satisfies equations (40). From § 1, the matrix H^* is formally to be represented as a sum of products of powers of p and q. Because of the linearity of equations (40), (45) in H, H^* we have simply to specify the commensurate sum term in H as counterpart to each individual sum term in H^*. Thus we need consider solely the case

$$H^* = \prod_{j=1}^{k} (p^{s_j}q^{r_j}). \tag{46}$$

It follows from the remarks of § 2 that equations (45) can be satisfied by specifying H as a linear form of those products of powers of p, q which arise from H^* through cyclic interchange of the factors; herein the sum of the coefficients must be held to unity. The question as to how these coefficients are to be chosen so that equations (40) may also be satisfied is less easy to answer. It may at this juncture suffice to dispose of the case $k=1$, namely

$$H^* = p^s q^r. \tag{47}$$

The formula (39) can be generalized[1] to

$$p^m q^n - q^n p^m = m \frac{h}{2\pi i} \sum_{l=0}^{n-1} q^{n-1-l} p^{m-1} q^l. \tag{48}$$

For $n=1$ this reverts to (39); in general (48) ensues from the fact that because of (39) one has

$$p^m q^{n+1} - q^{n+1} p^m = (p^m q^n - q^n p^m) q + m \frac{h}{2\pi i} q^n p^{m+1}.$$

The new formula

$$p^m q^n - q^n p^m = n \frac{h}{2\pi i} \sum_{j=0}^{m-1} p^{m-1-j} q^{n-1} p^j \tag{48'}$$

is obtained on interchanging p and q and reversing the sign of h. Comparison with (48) yields

$$\frac{1}{s+1} \sum_{l=0}^{s} p^{s-l} q^r p^l = \frac{1}{r+1} \sum_{j=0}^{r} q^{r-j} p^s q^j. \tag{49}$$

We now assert: The matrix H belonging to H^* as given by (47) is:

$$H = \frac{1}{s+1} \sum_{l=0}^{s} p^{s-l} q^r p^l. \tag{50}$$

We need only prove equations (40), to which end we recall the derivatives, (18′) § 2.

From (50), we now obtain the relation

$$Hp - pH = \frac{1}{s+1} (q^r p^{s+1} - p^{s+1} q^r),$$

and according to (48) this is equivalent to the lower of equations (40).

Further, using (49) we find

$$Hq - qH = \frac{1}{r+1} (p^s q^{r+1} - q^{r+1} p^s),$$

[1] A different generalization is furnished by the formulae

$$p^m q^n = \sum_{j=0}^{m,n} j! \binom{m}{j} \binom{n}{j} \left(\frac{h}{2\pi i}\right)^j q^{n-j} p^{m-j},$$

$$q^n p^m = \sum_{j=0}^{m,n} j! \binom{m}{j} \binom{n}{j} \left(\frac{-h}{2\pi i}\right)^j p^{m-j} q^{n-j},$$

where j runs to the lesser of the two integers m, n.

and by (48′) this is equivalent to the upper of equations (40). This completes the requisite proof.

Whereas in classical mechanics energy conservation ($\dot{H}=0$) is directly apparent from the canonical equations, the same law of energy conservation in quantum mechanics, $\dot{H}=0$ lies, as one can see, more deeply hidden beneath the surface.

That its demonstrability from assumed postulates is far from being trivial will be appreciated if, following more closely the classical method of proof, one sets out to prove H to be constant simply by evaluating $\dot{H}$. To this end, one first has to express $\dot{H}$ as function of p, q and $\dot{p}$, $\dot{q}$ with the aid of (11), (11′), whereupon for $\dot{p}$ and $\dot{q}$ the values $-\partial H/\partial q$, $\partial H/\partial p$ have to be introduced. This yields $\dot{H}$ in function of p and q. Equation (38) or the formulae quoted in the footnote to equation (48) which were derived from (38) permit this function to be converted into a sum of terms of the type $ap^s q^r$ and one then has to prove that the coefficient a in each of such terms vanishes. This calculation for the most general case, as considered above along different lines, becomes so exceedingly involved[1] that it seems hardly feasible. The fact that nonetheless energy-conservation and frequency laws could be proved in so general a context would seem to us to furnish strong grounds to hope that this theory embraces truly deep-seated physical laws.

In conclusion, we append a result here which can easily be derived from the formulae of this section, namely: *Equations* (35), (37) *can be replaced by* (38) *and* (44) (*with* H *representing the energy*)*; the frequencies are thereby to be derived from the frequency condition.*

In the continuation to this paper, we shall examine the important applications to which this theorem gives rise.

CHAPTER 3. INVESTIGATION OF THE ANHARMONIC OSCILLATOR

The anharmonic oscillator, having

$$H = \tfrac{1}{2}p^2 + \tfrac{1}{2}\omega_0^2 q^2 + \tfrac{1}{3}\lambda q^3 \tag{51}$$

has already been considered in detail by Heisenberg. Nevertheless, its

[1] For the case $H=(1/2m)p^2+U(q)$ it can immediately be carried out with the aid of (39′).

investigation will here be renewed with the aim of determining the *most general* solution of the fundamental equations for this case. If the basic equations of the present theory are indeed complete and do not require to be supplemented any further, then the absolute values $|q(nm)|$, $|p(nm)|$ of the elements of the matrices q and p must *uniquely* be determined by these equations, and thus it becomes important to check this for the example (51). On the other hand, it is to be expected that an uncertainty will still persist with respect to the phases φ_{nm}, ψ_{nm} in the relations

$$q(nm) = |q(nm)|e^{i\varphi_{nm}},$$
$$p(nm) = |p(nm)|e^{i\psi_{nm}}.$$

For the statistical theory, e.g., of the interaction of quantized atoms with external radiation fields, it becomes of fundamental importance to ascertain the precise degree of such uncertainty.

5. Harmonic oscillator

The starting point in our considerations is the theory of the harmonic oscillator; for small λ, one can regard the motion as expressed by equation (51) to be a perturbation of the normal harmonic oscillation having energy

$$H = \tfrac{1}{2}p^2 + \tfrac{1}{2}\omega_0^2 q^2. \tag{52}$$

Even for this simple problem it is necessary to supplement Heisenberg's analysis. This latter employs correspondence considerations to arrive at significant deductions as to the form of the solution: namely, since classically only a *single* harmonic component is present, Heisenberg selects a matrix which represents transitions between adjacent states only, and which thus has the form

$$q = \begin{pmatrix} 0 & q^{(01)} & 0 & 0 & 0 & \ldots \\ q^{(10)} & 0 & q^{(12)} & 0 & 0 & \ldots \\ 0 & q^{(21)} & 0 & q^{(23)} & 0 & \ldots \\ \cdot & \cdot & \cdot & \cdot & \cdot & \cdot \end{pmatrix}. \tag{53}$$

We here strive to build up the entire theory self-dependently, without invoking assistance from classical theory on the basis of the principle of correspondence. We shall therefore investigate whether the form of the matrix (53) cannot itself be derived from the basic formulae or, if this proves impossible, which additional postulates are required.

From what has been stated in § 3 regarding the invariance with respect to permutation of rows and columns, one can see right away that the exact form of the matrix (53) can never be deduced from the fundamental equations, since if rows and columns be subjected to the same permutation, the canonical equations and the quantum condition remain invariant and thereby one obtains a new and apparently different solution. But all such solutions naturally differ only in the notation, i.e., in the way the elements are numbered. We seek to prove that through a mere renumbering of its elements, the solution can always be brought into the form (53). The equation of motion

$$\ddot{q} + \omega_0^2 q = 0 \tag{54}$$

runs as follows for the elements:

$$(\nu^2(nm) - \nu_0^2)q(nm) = 0, \tag{55}$$

where

$$\omega_0 = 2\pi\nu_0, \qquad h\nu(nm) = W_n - W_m.$$

From the stronger quantum condition

$$pq - qp = \frac{h}{2\pi i}\mathbf{1}, \tag{56}$$

it follows that for each n there must exist a corresponding n' such that $q(nn') \neq 0$, since if there were a value of n for which all $q(nn')$ were equal to zero, then the nth diagonal element of $pq-qp$ would be zero, which contradicts the quantum condition. Hence equation (55) implies that there is always an n' for which

$$|W_n - W_{n'}| = h\nu_0.$$

But since we have assumed in our basic principles that when $n \neq m$, the energies are always unequal $(W_n \neq W_m)$, it follows that at most *two* such indices n' and n'' can exist, for the corresponding $W_{n'}$, $W_{n''}$ are solutions of the quadratic equation

$$(W_n - x)^2 = h^2\nu_0^2;$$

and if indeed *two* such indices n', n'' exist, it follows that the corresponding frequences must be related as:

$$\nu(nn') = -\nu(nn''). \tag{57}$$

Now from (56) we get

$$\sum_k \nu(kn)|q(nk)|^2 = \nu(n'n)\{|q(nn')|^2 - |q(nn'')|^2\} = h/8\pi^2, \tag{58}$$

and the energy (52) ensues as

$$H(nm) = \tfrac{1}{2} \times 4\pi^2 \sum_k \{-\nu(nk)\,\nu(km)\,q(nk)\,q(km) + \nu_0^2 q(nk)\,q(km)\}$$

$$= 2\pi^2 \sum_k q(nk)\,q(km)\{\nu_0^2 - \nu(nk)\,\nu(km)\}.$$

In particular, for $m=n$ we have

$$H(nn) = W_n = 4\pi^2\nu_0^2(|q(nn')|^2 + |q(nn'')|^2). \tag{59}$$

Moreover, we can now distinguish between three possible cases:
(a) no n'' exists and one has $W_{n'} > W_n$;
(b) no n'' exists and one has $W_{n'} < W_n$;
(c) n'' exists.

In case (b) we now consider n' in place of n; to this there belong at most two indices $(n')'$ and $(n')''$ and of these, one has to equal n. We thereby revert to one of the cases (a) or (c) and can accordingly omit further consideration of (b).

In case (a), $\nu(n'n) = +\nu_0$ and from (58) it follows that

$$\nu_0|q(nn')|^2 = h/8\pi^2, \tag{60}$$

and thus from (59) that

$$W_n = H(nn) = 4\pi^2\nu_0^2|q(nn')|^2 = \tfrac{1}{2}\nu_0 h.$$

Because of the assumption that $W_n \neq W_m$ for $n \neq m$ there is thus at most *one* index $n=n_0$ for which the case (a) applies.

If such an n_0 exists, we can specify a series of numbers n_0, n_1, n_2, n_3, ..., such that $(n_k)' = n_{k+1}$ and $W_{k+1} > W_k$. Then invariably $(n_{k+1})'' = n_k$. Hence for $k>0$, equations (58) and (59) give

$$H(n_k n_k) = 4\pi^2\nu_0^2\{|q(n_k, n_{k+1})|^2 + |q(n_k, n_{k-1})|^2\}, \tag{61}$$

$$\tfrac{1}{2}h = 4\pi^2\nu_0\{|q(n_k, n_{k+1})|^2 - |q(n_k, n_{k-1})|^2\}. \tag{62}$$

From (60) and (62) it follows that

$$|q(n_k, n_{k+1})|^2 = \frac{h}{8\pi^2\nu_0}(k+1), \tag{63}$$

and thence from (61) that

$$W_{n_k} = H(n_k, n_k) = \nu_0 h(k + \tfrac{1}{2}). \tag{64}$$

Now, we still have to check whether it be possible that there is no value of n for which case (a) applies. Beginning with an arbitrary n_0 we can then build $n_0'=n$, and $n_0''=n_{-1}$ and with each of these latter write $n_1'=n_2$, $n_1''=n_0$ and $n_{-1}'=n_0$, $n_{-1}''=n_{-2}$ etc. In this manner we obtain a series of numbers ... $n_{-2}, n_{-1}, n_0, n_1, n_2$ ..., and equations (61), (62) hold for every k between $-\infty$ and $+\infty$. But this is impossible, since by (62) the quantities $x_k=|q(n_{k+1}, n_k)|^2$ form an equispaced series of numbers, and since they are positive, there must be a least value. The relevant index can then again be designated as n_0 and we thereby revert to the previous case – thus here also, the formulae (63), (64) apply.

One can further see that every number n must be contained within the numbers n_k, since otherwise one could construct a new series (65) proceeding from n, and for this formula (60) would again hold. The starting terms of both series would then have the same value $W_n=H(nn)$, which is not possible.

This proves that the indices 0, 1, 2, 3 ... can be rearranged into a new sequence n_0, n_1, n_2, n_3 ... such that formulae (63), (64) apply: with these new indices, the solution then takes on Heisenberg's form (53). Hence this appears as the 'normal form' of the general solution. By virtue of (64), it possesses the property that

$$W_{n_{k+1}} > W_{n_k}.$$

If, inversely, one stipulate that $W_n=H(nn)$ should always increase with n, then it necessarily follows that $n_k=k$; this principle thus uniquely establishes the normal form of the solution. But thereby only the notation becomes fixed and the calculation more transparent: nothing new is conferred *physically*.

Therein lies the big difference between this and the previously adopted semiclassical methods of determining the stationary states. The classically calculated orbits merge into one another continuously; consequently the quantum orbits selected at a later stage have a particular sequence right from the outset. The new mechanics presents itself as an essentially discontinuous theory in that herein there is no question of a sequence of quantum states defined by the physical process, but rather of quantum numbers which are indeed no more than distinguishing indices which can be ordered and normalized according to any practical standpoint whatsoever (e.g., according to increasing energy W_n).

6. Anharmonic oscillator

The equations of motion

$$\ddot{q} + \omega_0^2 q + \lambda q^2 = 0, \tag{66}$$

together with the quantum condition yield the following system of equations for the elements:

$$\begin{aligned} (\omega_0^2 - \omega^2(nm))q(nm) + \lambda \sum_k q(nk)\, q(km) &= 0, \\ \sum_k \omega(nk)\, q(nk)\, q(kn) &= -\, h/4\pi. \end{aligned} \tag{67}$$

We introduce series expansions

$$\begin{aligned} \omega(nm) &= \omega^0(nm) + \lambda\omega^{(1)}(nm) + \lambda^2\omega^{(2)}(nm) + \ldots \\ q(nm) &= q_0(nm) + \lambda q^{(1)}(nm) + \lambda^2 q^{(2)}(nm) + \ldots \end{aligned} \tag{68}$$

in seeking the solution.

When $\lambda = 0$, one has the case of the harmonic oscillator considered in the previous section; we write the solution (53) in the form

$$q^0(nm) = a_n\delta_{n,m-1} + \overline{a_m}\delta_{n-1,m}, \tag{69}$$

where the bar denotes the conjugate complex value. If one builds the square or higher powers of the matrix $\mathbf{q}^0 = (q^0(nm))$, one arrives at matrices of similar form, being composed of sums of terms

$$(\xi)^{(p)}_{nm} = \xi_n\delta_{n,m-p} + \overline{\xi_m}\delta_{n-p,m}. \tag{70}$$

This prompts us to try a solution of the form

$$\begin{aligned} q^0(nm) &= (a)^{(1)}_{nm}, \\ q^{(1)}(nm) &= (x)^0_{nm} + (x')^{(2)}_{nm}, \\ q^{(2)}(nm) &= (y)^{(1)}_{nm} + (y')^{(3)}_{nm}, \\ &\ldots\ldots\ldots\ldots\ldots \end{aligned} \tag{71}$$

n which odd and even values of the index p always alternate. If one actually inserts this in the approximation equations

$$\lambda:\left\{\begin{aligned}&\left(\omega_0^2-\omega^0(nm)^2\right)q^{(1)}(nm)-2\omega^0(nm)\,\omega^{(1)}(nm)\,q^0(nm)\\&\qquad+\sum_k q^0(nk)\,q^0(km)=0,\\&\sum_k\{\omega^0(nk)\,\left(q^0(nk)\,q^{(1)}(kn)+q^{(1)}(nk)\,q^0(kn)\right)\\&\qquad+\omega^{(1)}(nk)\,q^0(nk)\,q^0(kn)\}=0,\end{aligned}\right\}\tag{72}$$

$$\lambda^2:\left\{\begin{aligned}&\left(\omega_0^2-\omega^0(nm)^2\right)q^{(2)}(nm)-2\omega^0(nm)\,\omega^{(1)}(nm)\,q^{(1)}(nm)\\&\qquad-\left(\omega^{(1)}(nm)^2+2\omega^0(nm)\,\omega^{(2)}(nm)\right)q^0(nm)\\&\qquad+\sum_k\left(q^0(nk)\,q^{(1)}(km)+q^{(1)}(nk)\,q^0(km)\right)=0,\\&\sum_k\{\omega^0(nk)\,\left(q^0(nk)\,q^{(2)}(km)+q^{(1)}(nk)\,q^{(1)}(km)\right.\\&\qquad\left.+q^{(2)}(nk)\,q^0(km)\right)+\omega^{(1)}(nk)\,\left(q^0(nk)\,q^{(1)}(km)\right.\\&\qquad\left.+q^{(1)}(nk)\,q^0(km)\right)+\omega^{(2)}(nk)\,q^0(nk)\,q^0(km)\}=0\end{aligned}\right\}\tag{73}$$

and notes the multiplication rule

$$\begin{aligned}\sum_k \Omega_{nkm}(\xi)^{(p)}_{nk}(\eta)^{(q)}_{km}=&\ \Omega_{n,\,n+p,\,n+p+q}\,\xi_n\,\eta_{n+p}\,\delta_{n,\,m-p-q}\\&+\Omega_{n,\,n+p,\,n+p-q}\,\xi_n\,\bar{\eta}_{n+p-q}\,\delta_{n,\,m-p+q}\\&+\Omega_{n,\,n-p,\,n-p+q}\,\bar{\xi}_{n-p}\,\eta_{n-p}\,\delta_{n,\,m+p-q}\\&+\Omega_{n,\,n-p,\,n-p-q}\,\bar{\xi}_{n-p}\,\bar{\eta}_{n-p-q}\,\delta_{n,\,m+p+q},\end{aligned}\tag{74}$$

one sees, in setting each of the factors of $\delta_{n,\,m-s}$ singly to zero, that through the substitution (71) all conditions can in fact be satisfied and that higher terms in (71) would identically vanish.

In detail, the calculation yields the following:

The first of the equations (72) gives, after substitution of the expressions (71),

$$\left.\begin{aligned}2\omega_0^2x_n+|a_n|^2+|a_{n-1}|^2&=0,\\-3\omega_0^2x_n'+a_na_{n+1}&=0,\\\omega^{(1)}_{n,\,n-1}&=0,\end{aligned}\right\}\tag{75}$$

and the second is identically satisfied. One thus has

$$\left.\begin{aligned}x_n&=-\frac{|a_n|^2+|a_{n-1}|^2}{2\omega_0^2},\\x_n'&=\frac{a_na_{n+1}}{3\omega_0^2}.\end{aligned}\right\}\tag{76}$$

The first of the equations (73) yields

$$\left.\begin{aligned} 2\omega_0 a_n \omega^{(2)}_{n,n+1} + 2a_n x_{n+1} + 2a_n x_n + \bar{a}_{n-1} x'_{n-1} + \bar{a}_{n+1} x'_n &= 0, \\ -\, 8\omega_0^2 y'_n + a_n x'_{n+1} + a_{n+2} x'_n &= 0, \\ \omega^{(1)}_{n,n-2} &= 0, \end{aligned}\right\} \qquad (77)$$

whereas the second equation is not identically satisfied, but furnishes a relation from which y_n can be determined:

$$a_n\bar{y}_n + \bar{a}_n y_n - a_{n-1}\bar{y}_{n-1} - \bar{a}_{n-1} y_{n-1} + 2|x'_n|^2 - 2|x'_{n-2}|^2$$
$$- \frac{\omega^{(2)}_{n,n+1}}{\omega_0}|a_n|^2 - \frac{\omega^{(2)}_{n,n-1}}{\omega_0}|a_{n-1}|^2 = 0. \qquad (78)$$

The solution is:

$$\left.\begin{aligned} \omega^{(2)}_{n,n+1} &= \frac{1}{3\omega_0^3}\left(|a_{n+1}|^2 + |a_{n-1}|^2 + 3|a_n|^2\right), \\ y'_n &= \frac{1}{12\omega_0^4} a_n a_{n+1} a_{n+2}. \end{aligned}\right\} \qquad (79)$$

Further, if for brevity one introduces

$$\eta_n = a_n\bar{y}_n + \bar{a}_n y_n, \qquad (80)$$

then the η are determined by the equation

$$\eta_n - \eta_{n-1} = \frac{1}{\omega_0^4}\left(|a_n|^4 - |a_{n-1}|^4 + \tfrac{1}{9}|a_n|^2|a_{n+1}|^2 - \tfrac{1}{9}|a_{n-1}|^2|a_{n-2}|^2\right). \qquad (81)$$

Expressions (76) and (79) show that the quantities x_n, x'_n, y'_n can be expressed through the solution of the zero-th order approximation a_n. Thus their phases are determined by those of the harmonic oscillator. For the quantities y_n, the situation seems to be different, since although η_n can uniquely be determined from (81), y_n cannot be obtained absolutely from (80). It is probable that the next higher order of approximation gives rise to an auxiliary determining equation for y_n. We have to leave this question open here but we should like to indicate its significance as a point of principle in regard to the completeness of the entire theory. All questions of statistics invariably depend finally upon whether or not our supposition that of the phases of the $q(nm)$ *one* in each row (or each column) of the matrix remains undetermined be valid.

In conclusion we present the explicit formulae which are obtained by substituting the solution of the harmonic oscillator found previously (§ 5). In normal form, by (63), this runs as follows:

$$a_n = \sqrt{C(n+1)}\mathrm{e}^{\mathrm{i}\varphi_n}, \qquad C = h/4\pi\omega_0 = h/8\pi^2\nu_0. \tag{82}$$

Thence, using (76), (79), (81) one obtains

$$\left.\begin{aligned} x_n &= -\frac{C}{2\omega_0^2}(2n+1), \\ x'_n &= \frac{C}{3\omega_0^2}\sqrt{(n+1)(n+2)}\mathrm{e}^{\mathrm{i}(\varphi_n+\varphi_{n+1})} \\ y'_n &= \frac{\sqrt{C^3}}{12\omega_0^4}\sqrt{(n+1)(n+2)(n+3)}\mathrm{e}^{\mathrm{i}(\varphi_n+\varphi_{n+1}+\varphi_{n+2})} \end{aligned}\right\} \tag{83}$$

$$\left.\begin{aligned} \omega^{(1)}_{n,n-1} &= 0, \qquad \omega^{(1)}_{n,n-2} = 0, \\ \omega^{(2)}_{n,n-1} &= -\frac{5C}{3\omega_2^3}n; \end{aligned}\right\} \tag{84}$$

that is,

$$\eta_n - \eta_{n-1} = \frac{11C^2}{9\omega_0^4}(2n+1),$$

$$\eta_n = a_n\bar{y}_n + \bar{a}_n y_n = \frac{11C^2}{9\omega_0^4}(n+1)^2.$$

If one sets $y_n = |y_n|\mathrm{e}^{\mathrm{i}\psi_n}$, then

$$|y_n|\cos(\varphi_n - \psi_n) = \frac{\eta_n}{2|a_n|} = \frac{11\sqrt{C^3}}{18\omega_0^4}\sqrt{n+1}^3. \tag{85}$$

In this approximation, y_n cannot be specified any more closely than this.

However, we should like to write out the final equations when one makes the assumption that $\psi_n = \varphi_n$. These are as follows (up to terms of higher than second order in λ):

$$\left.\begin{aligned} \omega(n, n-1) &= \omega_0 - \lambda^2\frac{5C}{3\omega_0^3}n + \ldots, \\ \omega(n, n-2) &= 2\omega_0 + \ldots; \end{aligned}\right\} \tag{86}$$

$$\left.\begin{aligned} q(n, n) &= -\lambda \frac{C}{\omega_0^2}(2n+1)+\ldots, \\ q(n, n-1) &= \sqrt{Cn}\,\mathrm{e}^{\mathrm{i}\varphi_{n-1}}\left(1+\lambda^2 \frac{11Cn}{18\omega_0^4}+\ldots\right), \\ q(n, n-2) &= \lambda \frac{C}{3\omega_0^2}\sqrt{n(n-1)}\,\mathrm{e}^{\mathrm{i}(\varphi_{n-1}+\varphi_{n-2})}+\ldots, \\ q(n, n-3) &= \lambda^2 \frac{\sqrt{C^3}}{12\omega_0^4}\sqrt{n(n-1)(n-2)}\,\mathrm{e}^{\mathrm{i}(\varphi_{n-1}+\varphi_{n-2}+\varphi_{n-3})}+\ldots \end{aligned}\right\} \quad (87)$$

We have also calculated the energy directly and derived the following formula:

$$W_n = h\nu_0(n+\tfrac{1}{2}) - \lambda^2 \frac{5C^2}{3\omega_0^2}\left(n(n+1)+\tfrac{17}{30}\right)+\ldots. \qquad (88)$$

The frequency condition is actually satisfied, since, remembering (82), we have

$$W_n - W_{n-1} = h\nu_0 - \lambda^2 \frac{2C^2}{\omega_0^2} n + \ldots = \frac{h}{2\pi}\,\omega(n, n-1),$$

$$W_n - W_{n-2} = 2h\nu_0 + \ldots \qquad\qquad = \frac{h}{2\pi}\,\omega(n, n-2).$$

With the formula (88) we can associate the observation that already in terms of lowest order there occurs a discrepancy from classical theory which can formally be removed by the introduction of a 'half-integer' quantum number $n'=n+\frac{1}{2}$. This has already been remarked by Heisenberg. Incidentally, our expressions $\omega(n, n-1)$ as given by (86) agree *exactly* with the classical frequencies in all respects. For comparison, we note the classical energy to be[1]

$$W_n^{(\mathrm{cl})} = h\nu_0 n - \lambda^2 \frac{5C^2}{3\omega_0^2} n^2 + \ldots,$$

and thus the classical frequency to be:

[1] See M. Born, Atommechanik (Berlin, 1925), Chapter 4, § 42, p. 294; one has to set $a=\frac{1}{3}$ in the formula (6) in order to obtain agreement with the present treatment.

$$\omega_{\mathrm{cl}} = \frac{1}{h} \frac{\partial W_n^{(\mathrm{cl})}}{\partial n} = h\nu_0 - \lambda^2 \frac{5C^2}{3\omega_0^2} n + \ldots$$

$$= \omega_{\mathrm{qu}}(n, n-1) = \frac{1}{h} (W_n^{(\mathrm{qu})} - W_{n-1}^{(\mathrm{qu})}).$$

We have, finally, checked that the expression (88) can also be derived from the Kramers–Born perturbation formula (up to an additive constant).

Received November 7, 1925

14

THE FUNDAMENTAL EQUATIONS OF QUANTUM MECHANICS

P. A. M. DIRAC

1. Introduction

It is well known that the experimental facts of atomic physics necessitate a departure from the classical theory of electrodynamics in the description of atomic phenomena. This departure takes the form, in Bohr's theory, of the special assumptions of the existence of stationary states of an atom, in which it does not radiate, and of certain rules, called quantum conditions, which fix the stationary states and the frequencies of the radiation emitted during transitions between them. These assumptions are quite foreign to the classical theory, but have been very successful in the interpretation of a restricted region of atomic phenomena. The only way in which the classical theory is used is through the assumption that the classical laws hold for the description of the motion in the stationary states, although they fail completely during transitions, and the assumption, called the Correspondence Principle, that the classical theory gives the right results in the limiting case when the action per cycle of the system is large compared to Planck's constant h, and in certain other special cases.

In a recent paper* Heisenberg puts forward a new theory which suggests that it is not the equations of classical mechanics that are in any way at fault, but that the mathematical operations by which physical results are deduced from them require modification. *All* the information supplied by the classical theory can thus be made use of in the new theory.

Editor's note. This paper was published as Proc. Roy. Soc. A **109** (1926) 642–653. At the time Professor Dirac was 1851 Exhibition Senior Research Student of St. John's College, Cambridge. The paper was communicated by R. H. Fowler, F.R.S.

* Heisenberg, Zs. f. Phys. **33** (1925) 879.

2. Quantum algebra

Consider a multiply periodic non degenerate dynamical system of u degrees of freedom, defined by equations connecting the co-ordinates and their time differential coefficients. We may solve the problem on the classical theory in the following way. Assume that each of the co-ordinates x can be expanded in the form of a multiple Fourier series in the time t, thus,

$$x = \sum_{\alpha_1 \ldots \alpha_u} x(\alpha_1\alpha_2 \ldots \alpha_u) \exp i(\alpha_1\omega_1 + \alpha_2\omega_2 + \ldots + \alpha_u\omega_u)t = \sum_{\alpha} x_\alpha \exp . i(\alpha\omega)t,$$

say, for brevity. Substitute these values in the equations of motion, and equate the coefficients on either side of each harmonic term. The equations obtained in this way (which we shall call the A equations) will determine each of the amplitudes x_α and frequencies $(\alpha\omega)$, (the frequencies being measured in radians per unit time). The solution will not be unique. There will be a u-fold infinity of solutions, which may be labelled by taking the amplitudes and frequencies to be functions of u constants $\kappa_1 \ldots \kappa_u$. Each x_α and $(\alpha\omega)$ is now a function of two sets of numbers, the α's and the κ's, and may be written $x_{\alpha\kappa}, (\alpha\omega)_\kappa$.

In the quantum solution of the problem, according to Heisenberg, we still assume that each co-ordinate can be represented by harmonic components of the form $\exp i\omega t$, the amplitude and frequency of each depending on two sets of numbers $n_1 \ldots n_u$ and $m_1 \ldots m_u$, in this case all integers, and being written $x(nm)$, $\omega(nm)$. The differences $n_r - m_r$ correspond to the previous α_r, but neither the n's nor any functions of the n's and m's play the part of the previous κ's in pointing out to which solution each particular harmonic component belongs. We cannot, for instance, take together all the components for which the n's have a given set of values, and say that these by themselves form a single complete solution of the equations of motion. The quantum solutions are all interlocked, and must be considered as a single whole. The effect of this mathematically is that, while on the classical theory each of the A equations is a relation between amplitudes and frequencies having one particular set of κ's, the amplitudes and frequencies occurring in a quantum A equation do not have one particular set of values for the n's, or for any functions of the n's and m's, but have their n's and m's related in a special way, which will appear later.

On the classical theory we have the obvious relation

$$(\alpha\omega)_\kappa + (\beta\omega)_\kappa = (\alpha + \beta, \omega)_\kappa.$$

Following Heisenberg, we assume that the corresponding relation on the quantum theory is

$$\omega(n, n - \alpha) + \omega(n - \alpha, n - \alpha - \beta) = \omega(n, n - \alpha - \beta)$$

or

$$\omega(nm) + \omega(mk) = \omega(nk). \tag{1}$$

This means that $\omega(nm)$ is of the form $\Omega(n) - \Omega(m)$, the Ω's being frequency levels. On Bohr's theory these would be $2\pi/h$ times the energy levels, but we do not need to assume this.

On the classical theory we can multiply two harmonic components related to the same set of κ's, as follows:

$$[a_{\alpha\kappa} \exp \mathrm{i}(\alpha\omega)_\kappa t][b_{\beta\kappa} \exp \mathrm{i}(\beta\omega)_\kappa t] = (ab)_{\alpha+\beta,\kappa} \exp \mathrm{i}(\alpha + \beta, \omega)_\kappa t$$

where

$$(ab)_{\alpha+\beta,\kappa} = a_{\alpha\kappa} b_{\beta\kappa}.$$

In a corresponding manner on the quantum theory we can multiply an (nm) and an (mk) component

$$[a(nm) \exp \mathrm{i}\omega(nm)t][b(mk) \exp \mathrm{i}\omega(mk)t] = ab(nk) \exp \mathrm{i}\omega(nk)t$$

where

$$ab(nk) = a(nm)b(mk).$$

We are thus led to consider the product of the amplitudes of an (nm) and an (mk) component as an (nk) amplitude. This together with the rule that only amplitudes related to the same pair of sets of numbers can occur added together in an A equation, replaces the classical rule that all amplitudes occurring in an A equation have the same set of κ's.

We are now in a position to perform the ordinary algebraic operations of quantum variables. The sum of x and y is determined by the equations

$$\{x + y\}(nm) = x(nm) + y(nm)$$

and the product by

$$xy(nm) = \sum_k x(nk)\, y(km) \tag{2}$$

similar to the classical product

$$(xy)_{\alpha\kappa} = \sum_r x_{r\kappa} y_{\alpha-r,\kappa}.$$

An important difference now occurs between the two algebras. In general

$$xy(nm) \neq yx(nm)$$

and quantum multiplication is not commutative, although, as is easily verified it is associative and distributive. The quantity with components $xy(nm)$ defined by (2) we shall call the Heisenberg product of x and y, and shall write simply as xy. Whenever two quantum quantities occur multiplied together, the Heisenberg product will be understood. Ordinary multiplication is, of course, implied in the products of amplitudes and frequencies and other quantities that are related to sets of n's which are explicitly stated.

The reciprocal of a quantum quantity x may be defined by either of the relations

$$1/x \cdot x = 1 \quad \text{or} \quad x \cdot 1/x = 1. \tag{3}$$

These two equations are equivalent, since if we multiply both sides of the former by x in front and divide by x behind we get the latter. In a similar way the square root of x may be defined by

$$\sqrt{x} \cdot \sqrt{x} = x. \tag{4}$$

It is not obvious that there always should be solutions to (3) and (4). In particular, one may have to introduce sub-harmonics, i.e., new intermediate frequency levels, in order to express $\sqrt{x}$. One may evade these difficulties by rationalising and multiplying up each equation before interpreting it on the quantum theory and obtaining the A equations from it.

We are now able to take over each of the equations of motion of the system into the quantum theory provided we can decide the correct order of the quantities in each of the products. Any equation deducible from the equations of motion by algebraic processes not involving the interchange of the factors of a product, and by differentiation and integration with respect to t, may also be taken over into the quantum theory. In particular, the energy equation may be thus taken over.

The equations of motion do not suffice to solve the quantum problem. On the classical theory the equations of motion do not determine the $x_{\alpha\kappa}, (\alpha\omega)_\kappa$ as functions of the κ's until we assume something about the κ's which serves to define them. We could, if we liked, complete the solution by choosing the κ's such that $\partial E/\partial \kappa_r = \omega_r/2\pi$, where E is the energy of the system, which would make the κ_r equal the action variables J_r. There must be corresponding equations on the quantum theory, and these constitute the quantum conditions.

3. Quantum differentiation

Up to the present the only differentiation that we have considered on the quantum theory is that with respect to the time t. We shall now determine the form of the most general quantum operation $\mathrm{d}/\mathrm{d}v$ that satisfies the laws

$$\frac{\mathrm{d}}{\mathrm{d}v}(x+y) = \frac{\mathrm{d}}{\mathrm{d}v}x + \frac{\mathrm{d}}{\mathrm{d}v}y, \tag{I}$$

and

$$\frac{\mathrm{d}}{\mathrm{d}v}(xy) = \frac{\mathrm{d}}{\mathrm{d}v}x\cdot y + x\frac{\mathrm{d}}{\mathrm{d}v}y. \tag{II}$$

(Note that the order of x and y is preserved in the last equation).

The first of these laws requires that the amplitudes of the components of $\mathrm{d}x/\mathrm{d}v$ shall be linear functions of those of x, i.e.,

$$\mathrm{d}x/\mathrm{d}v\,(nm) = \sum_{nm} a(nm;n'm')\,x(n'm'). \tag{5}$$

There is one coefficient $a(nm;n'm')$ for any four sets of integral values for the n's, m's, n''s and m''s. The second law imposes conditions on the a's. Substitute for the differential coefficients in II their values according to (5) and equate the (nm) components on either side. The result is

$$\sum_{n'm'k} a(nm;n'm')\,x(n'k)\,y(km') = \sum_{kn'k'} a(nk;n'k')\,x(n'k')\,y(km) + \sum_{kk'm} x(nk)\,a(km;k'm')\,y(k'm').$$

This must be true for all values of the amplitudes of x and y, so that we can equate the coefficients of $x(n'k)\,y(k'm')$ on either side. Using the symbol δ_{mn} to have the value unity when $m=n$ (i.e., when each $m_r=n_r$) and zero when $m\neq n$, we get

$$\delta_{kk'}a(nm;n'm') = \delta_{mm'}a(nk';n'k) + \delta_{nn'}a(km;k'm').$$

To proceed further, we have to consider separate the various cases of equality and inequality between the kk', mm' and nn'.

Take first the case when $k=k'$, $m\neq m'$, $n\neq n'$. This gives

$$a(nm;\,n'm')=0.$$

Hence all the $a(nm;\,n'm')$ vanish except those for which either $n=n'$ or $m=m'$ (or both). The cases $k\neq k'$, $m=m'$, $n\neq n'$ and $k\neq k'$, $m\neq m'$, $n=n'$ do not give us anything new. Now take the case $k=k'$, $m=m'$, $n\neq n'$. This gives

$$a(nm;\,n'm)=a(nk;\,n'k).$$

Hence $a(nm;\,n'm)$ is independent of m provided $n\neq n'$. Similarly, the case $k=k'$, $m\neq m'$, $n=n'$ tells us that $a(nm;\,nm')$ is independent of n provided $m\neq m'$. The case $k\neq k'$, $m=m'$, $n=n'$ now gives

$$a(nk';\,nk)+a(km;\,k'm)=0.$$

We can sum up these results by putting

$$a(nk';\,nk)=a(kk')=-a(km;\,k'm), \tag{6}$$

provided $k\neq k'$. The two-index symbol $a(kk')$ depends, of course, only on the two sets of integers k and k'. The only remaining case is $k=k'$, $m=m'$, $n=n'$, which gives

$$a(nm;\,nm)=a(nk;\,nk)+a(km;\,km).$$

This means we can put

$$a(nm;\,nm)=a(mm)-a(nn). \tag{7}$$

Equation (7) completes equation (6) by defining $a(kk')$ when $k=k'$. Equation (5) now reduces to

$$\begin{aligned}\frac{\mathrm{d}x}{\mathrm{d}v}(nm) &= \sum_{m'\neq m} a(nm;\,nm')\,x(nm') + \sum_{n'\neq n} a(nm;\,n'm)\,x(n'm) \\ &\qquad\qquad + a(nm;\,nm)\,x(nm) \\ &= \sum_{m'\neq m} a(m'm)\,x(nm') - \sum_{n'\neq n} a(nn')\,x(n'm) \\ &\qquad\qquad + \{a(mm)-a(nn)\}\,x(nm) \\ &= \sum_k \{x(nk)\,a(km)-a(nk)\,x(km)\}.\end{aligned}$$

Hence

$$\mathrm{d}x/\mathrm{d}v = xa - ax. \tag{8}$$

Thus the most general operation satisfying the laws I and II that one can perform upon a quantum variable is that of taking the difference of its Heisenberg products with some other quantum variable. It is easily seen that one cannot in general change the order of differentiations, i.e.,

$$\frac{d^2x}{du dv} \neq \frac{d^2x}{dv du}.$$

As an example in quantum differentiation we may take the case when (a) is a constant, so that $a(nm)=0$ except when $n=m$. We get

$$dx/dv(nm) = x(nm)\,a(mm) - a(nn)\,x(nm).$$

In particular, if $ia(mm)=\Omega(m)$, the frequency level previously introduced, we have

$$dx/dv(nm) = i\omega(nm)\,x(nm),$$

and our differentiation with respect to v becomes ordinary differentiation with respect to t.

4. The quantum conditions

We shall now consider to what the expression $(xy-yx)$ corresponds on the classical theory. To do this we suppose that $x(n, n-\alpha)$ varies only slowly with the n's, the n's being large numbers and the α's small ones, so that we can put

$$x(n, n-\alpha) = x_{\alpha\kappa}$$

where $\kappa_r=n_rh$ or $(n_r+\alpha_r)h$, these being practically equivalent. We now have

$$\begin{aligned} x(n, n-\alpha)\,y(n-\alpha, n-\alpha-\beta) &- y(n, n-\beta)\,x(n-\beta, n-\alpha-\beta) \\ &= \{x(n, n-\alpha) - x(n-\beta, n-\beta-\alpha)\}\,y(n-\alpha, n-\alpha-\beta) \\ &- \{y(n, n-\beta) - y(n-\alpha, n-\alpha-\beta)\}\,x(n-\beta, n-\alpha-\beta). \\ &= h\sum_r \left\{\beta_r \frac{\partial x_{\alpha\kappa}}{\partial \kappa_r} y_{\beta\kappa} - \alpha_r \frac{\partial y_{\beta\kappa}}{\partial \kappa_r} x_{\alpha\kappa}\right\}. \qquad (9) \end{aligned}$$

Now

$$2\pi i\beta_r y_\beta \exp i(\beta\omega)t = \frac{\partial}{\partial w_r}\{y_\beta \exp i(\beta\omega)t\},$$

where the w_r are the angle variables, equal to $\omega_r t/2\pi$. Hence the (nm)

component of $(xy-yx)$ corresponds on the classical theory to

$$\frac{ih}{2\pi}\sum_{\alpha+\beta=n-m}\sum_r\left\{\frac{\partial}{\partial\kappa_r}\{x_\alpha\exp i(\alpha\omega)t\}\frac{\partial}{\partial w_r}\{y_\beta\exp i(\beta\omega)t\}\right.$$
$$\left.-\frac{\partial}{\partial\kappa_r}\{y_\beta\exp i(\beta\omega)t\}\frac{\partial}{\partial w_r}\{x_\alpha\exp i(\alpha\omega)t\}\right\}$$

or $(xy-yx)$ itself corresponds to

$$-\frac{ih}{2\pi}\sum_r\left\{\frac{\partial x}{\partial\kappa_r}\frac{\partial y}{\partial w_r}-\frac{\partial y}{\partial\kappa_r}\frac{\partial x}{\partial w_r}\right\}.$$

If we make the κ_r equal the action variables J_r, this becomes $ih/2\pi$ times the Poisson (or Jacobi) bracket expression

$$[x,y]=\sum_r\left\{\frac{\partial x}{\partial w_r}\frac{\partial y}{\partial J_r}-\frac{\partial y}{\partial w_r}\frac{\partial x}{\partial J_r}\right\}=\sum_r\left\{\frac{\partial x}{\partial q_r}\frac{\partial y}{\partial p_r}-\frac{\partial y}{\partial q_r}\frac{\partial x}{\partial p_r}\right\},$$

where the p's and q's are any set of canonical variables of the system.

The elementary Poisson bracket expressions for various combinations of the p's and q's are

$$\begin{aligned}[q_r,q_s]&=0, \qquad [p_r,p_s]=0,\\ [q_r,p_s]&=\delta_{rs}=0 \qquad (r\neq s)\\ &=1. \qquad (r=s)\end{aligned}\tag{10}$$

The general bracket expressions satisfy the laws I and II which now read

$$[x,z]+[y,z]=[x+y,z], \tag{IA}$$
$$[xy,z]=[x,z]y+x[y,z]. \tag{IIA}$$

By means of these laws, together with $[x,y]=-[y,x]$, if x and y are given as algebraic functions of the p_r and q_r, $[x,y]$ can be expressed in terms of the $[q_r,q_s]$, $[p_r,p_s]$ and $[q_r,p_s]$, and thus evaluated, without using the commutative law of multiplication (except in so far as it is used implicitly on account of the proof of IIA requiring it). The bracket expression $[x,y]$ thus has a meaning on the quantum theory when x and y are quantum variables, if we take the elementary bracket expressions to be still given by (10).

We make the fundamental assumption that *the difference between the Heisenberg products of two quantum quantities is equal to* $ih/2\pi$ *times their*

Poisson bracket expression. In symbols,

$$xy - yx = ih/2\pi \cdot [x, y]. \tag{11}$$

We have seen that this is equivalent, in the limiting case of the classical theory, to taking the arbitrary quantities κ_r that label a solution equal to the J_r, and it seems reasonable to take (11) as constituting the general quantum conditions.

It is not obvious that all the information supplied by equation (11) is consistent. Owing to the fact that the quantities on either side of (11) satisfy the same laws I and II or IA and IIA, the only independent conditions given by (11) are those for which x and y are p's or q's, namely

$$\begin{aligned} q_r q_s - q_s q_r &= 0 \\ p_r p_s - p_s p_r &= 0 \\ q_r p_s - p_s q_r &= \delta_{rs} ih/2\pi. \end{aligned} \tag{12}$$

If the only grounds for believing that the equations (12) were consistent with each other and with the equations of motion were that they are known to be consistent in the limit when $h \to 0$, the case would not be very strong, since one might be able to deduce from them the inconsistency that $h=0$, which would not be an inconsistency in the limit. There is much stronger evidence than this, however, owing to the fact that the classical operations obey the same laws as the quantum ones, so that if, by applying the quantum operations, one can get an inconsistency, by applying the classical operations in the same way one must also get an inconsistency. If a series of classical operations leads to the equation $0=0$, the corresponding series of quantum operations must also lead to the equation $0=0$, and not to $h=0$, since there is no way of obtaining a quantity that does not vanish by a quantum operation with quantum variables such that the corresponding classical operation with the corresponding classical variables gives a quantity that does vanish. The possibility mentioned above of deducing by quantum operations the inconsistency $h=0$ thus cannot occur. *The correspondence between the quantum and classical theories lies not so much in the limiting agreement when $h \to 0$ as in the fact that the mathematical operations on the two theories obey in many cases the same laws.*

For a system of one degree of freedom, if we take $p=m\dot{q}$, the only quantum condition is

$$2\pi m(q\dot{q} - \dot{q}q) = ih.$$

Equating the constant part of the left-hand side to ih, we get

$$4\pi m \sum_k q(nk)\, q(kn)\, \omega(kn) = h.$$

This is equivalent to Heisenberg's quantum condition.* By equating the remaining components of the left-hand side to zero we get further relations not given by Heisenberg's theory.

The quantum conditions (12) get over, in many cases, the difficulties concerning the order in which quantities occurring in products in the equations of motion are to be taken. The order does not matter except when a p_r and q_r are multiplied together, and this never occurs in a system describable by a potential energy function that depends only on the q's, and a kinetic energy function that depends only on the p's.

It may be pointed out that the classical theory quantity occurring in Kramers' and Heisenberg's theory of scattering by atoms† has components which are of the form (8) (with $\kappa_r = J_r$), and which are interpreted on the quantum theory in a manner in agreement with the present theory. No classical expression involving differential coefficients can be interpreted on the quantum theory unless it can be put into this form.

5. Properties of the quantum Poisson bracket expressions

In this section we shall deduce certain results that are independent of the assumption of the quantum conditions (11) or (12).

The Poisson bracket expressions satisfy on the classical theory the identity

$$[x, y, z] \equiv [[x, y], z] + [[y, z], x] + [[z, x], y] = 0. \qquad (13)$$

On the quantum theory this result is obviously true when x, y and z are p's or q's. Also, from IA and IIA

$$[x_1 + x_2, y, z] = [x_1, y, z] + [x_2, y, z]$$

and

$$[x_1, x_2, y, z] = x_1[x_2, y, z] + [x_1, y, z]x_2.$$

* Heisenberg, loc. cit. equation (16).

† Kramers and Heisenberg, Zs. f. Phys. **31** (1925) 681, equation (18).

Hence the result must still be true on the quantum theory when x, y and z are expressible in any way as sums and products of p's and q's, so that it must be generally true. Note that the identity corresponding to (13) when the Poisson bracket expressions are replaced by the differences of the Heisenberg products $(xy-yx)$ is obviously true, so that there is no inconsistency with equation (11).

If H is the Hamiltonian function of the system, the equations of motion may be written classically

$$\dot{p}_r = [p_r, H], \qquad \dot{q}_r = [q_r, H].$$

These equations will be true on the quantum theory for systems for which the orders of the factors of products occurring in the equations of motion are unimportant. They may be taken to be true for systems for which these orders are important if one can decide upon the orders of the factors in H. From laws IA and IIA it follows that

$$\dot{x} = [x, H] \tag{14}$$

on the quantum theory for any x.

If A is an integral of the equations of motion on the quantum theory, then

$$[A, H] = 0.$$

The action variables J_r must, of course, satisfy this condition. If A_1 and A_2 are two such integrals, then, by a simple application of (13), it follows that

$$[A_1, A_2] = \text{const.}$$

as on the classical theory.

The conditions on the classical theory that a set of variables P_r, Q_r shall be canonical are

$$[Q_r, Q_s] = 0, \qquad [P_r, P_s] = 0, \qquad [Q_r, P_s] = \delta_{rs}.$$

These equations may be taken over into the quantum theory as the conditions for the quantum variables P_r, Q_r to be canonical.

On the classical theory we can introduce the set of canonical variables ξ_r, η_r, related to the uniformising variables J_r, w_r, by

$$\xi_r = (2\pi)^{-\frac{1}{2}} J_r^{\frac{1}{2}} \exp 2\pi i w_r, \qquad \eta_r = -\,i(2\pi)^{-\frac{1}{2}} J_r^{\frac{1}{2}} \exp -2\pi i w_r.$$

Presumably there will be a corresponding set of canonical variables on the quantum theory, each containing only one kind of component, so that $\xi_r(nm)=0$ except when $m_r=n_r-1$ and $m_s=n_s$ $(s\neq r)$, and $\eta_r(nm)=0$ except when $m_r=n_r+1$ and $m_s=n_s$ $(s\neq r)$. One may consider the existence of such variables as the condition for the system to be multiply periodic on the quantum theory. The components of the Heisenberg products of ξ_r and η_r satisfy the relation

$$\xi_r\eta_r(nn) = \xi_r(nm)\,\eta_r(mn) = \eta_r(mn)\,\xi_r(nm) = \eta_r\xi_r(mm) \qquad (15)$$

where the m's are related to the n's by the formulae $m_r=n_r-1$, $m_s=n_s$ $(s\neq r)$.

The classical ξ's and η's satisfy $\xi_r\eta_r=-(\mathrm{i}/2\pi)J_r$. This relation does not necessarily hold between the quantum ξ's and η's. The quantum relation may, for instance, be $\eta_r\xi_r=-(\mathrm{i}/2\pi)J_r$, or $\frac{1}{2}(\xi_r\eta_r+\eta_r\xi_r)=-(\mathrm{i}/2\pi)J_r$. A detailed investigation of any particular dynamical system is necessary in order to decide what it is. In the event of the last relation being true, we can introduce the set of canonical variables ξ_r', η_r' defined by

$$\xi_r' = (\xi_r + \mathrm{i}\eta_r)/\sqrt{2}, \qquad \eta_r' = (\mathrm{i}\xi_r + \eta_r)/\sqrt{2},$$

and shall then have

$$J_r = \pi(\xi_r'^2 + \eta_r'^2).$$

This is the case that actually occurs for the harmonic oscillator. In general J_r is not necessarily even a rational function of the ξ_r and η_r, an example of this being the rigid rotator considered by Heisenberg.

6. The stationary states

A quantity C, that does not vary with the time, has all its (nm) compenents zero, except those for which $n=m$. It thus becomes convenient to suppose each set of n's to be associated with a definite state of the atom, as on Bohr's theory, so that each $C(nn)$ belongs to a certain state in precisely the same way in which *every* quantity occurring in the classical theory belongs to a certain configuration. The components of a varying quantum quantity are so interlocked, however, that it is impossible to associate the sum of certain of them with a given state.

A relation between quantum quantities reduces, when all the quantities are constants, to a relation between $C(nn)$'s belonging to a definite stationary state n. This relation will be the same as the classical theory relation, on the assumption that the classical laws hold for the description of the stationary states; in particular, the energy will be the same function of the J's as on the classical theory. We have here a justification for Bohr's assumption of the mechanical nature of the stationary states. It should be noted though, that the variable quantities associated with a stationary state on Bohr's theory, the amplitudes and frequencies of orbital motion, have no physical meaning and are of no mathematical importance.

If we apply the fundamental equation (11) to the quantities x and H we get, with the help of (14),

$$x(nm)\,H(mm) - H(nn)\,x(nm) = \mathrm{i}h/2\pi\cdot\dot{x}(nm) = -\,h/2\pi\cdot\omega(nm)\,x(nm),$$

or

$$H(nn) - H(mm) = h/2\pi\cdot\omega(nm).$$

This is just Bohr's relation connecting the frequencies with the energy differences.

The quantum condition (11) applied to the previously introduced canonical variables ξ_r, η_r gives

$$\xi_r\eta_r(nn) - \eta_r\xi_r(nn) = \mathrm{i}h/2\pi\cdot[\xi_r, \eta_r] = \mathrm{i}h/2\pi.$$

This equation combined with (15) shows that

$$\xi_r\eta_r(nn) = -\,n_r\mathrm{i}h/2\pi + \text{const.}$$

It is known physically that an atom has a normal state in which it does not radiate. This is taken account of in the theory by Heisenberg's assumption that all the amplitudes $C(nm)$ having a negative n_r or m_r vanish, or rather do not exist, if we take the normal state to be the one for which every n_r is zero. This makes $\xi_r\eta_r(nn)=0$ when $n_r=0$ on account of equation (15). Hence in general

$$\xi_r\eta_r(nn) = -\,n_r\mathrm{i}h/2\pi.$$

If $\xi_r\eta_r=-(\mathrm{i}/2\pi)J_r$, then $J_r=n_rh$. This is just the ordinary rule for quantising the stationary states, so that in this case the frequencies of the system are the same as those given by Bohr's theory. If $\frac{1}{2}(\xi_r\eta_r+\eta_r\xi_r)=-(\mathrm{i}/2\pi)J_r$, then $J_r=(n_r+\frac{1}{2})h$. Hence in general in this case, half quantum numbers would have to be used to give the correct frequencies by Bohr's theory.*

* In the special case of the Planck oscillator, since the energy is a linear function of J, the frequency would come right in any case.

Up to the present we have considered only multiply periodic systems. There does not seem to be any reason, however, why the fundamental equations (11) and (12) should not apply as well to non-periodic systems, of which none of the constituent particles go off to infinity, such as a general atom. One would not expect the stationary states of such a system to classify, except perhaps when there are pronounced periodic motions, and so one would have to assign a single number n to each stationary state according to an arbitrary plan. Our quantum variables would still have harmonic components, each related to two n's, and Heisenberg multiplication could be carried out exactly as before. There would thus be no ambiguity in the interpretation of equations (12) or of the equations of motion.

I would like to express my thanks to Mr. R. H. Fowler, F.R.S., for many valuable suggestions in the writing of this paper.

Received November 16, 1925

15

ON QUANTUM MECHANICS II

M. BORN, W. HEISENBERG AND P. JORDAN

Göttingen

Abstract: The quantum mechanics developed in Part I of this paper from Heisenberg's approach is here extended to systems having arbitrarily many degrees of freedom. Perturbation theory is carried through for nondegenerate and for a large class of degenerate systems, and its connection with the eigenvalue theory of Hermitian forms is demonstrated. The results so obtained are employed in the derivation of momentum and angular momentum conservation laws, and of selection rules and intensity formulae. Finally, the theory is applied to the statistics of eigenvibrations of a black body cavity.

Introduction

The present paper sets out to develop further a general quantum-theoretical mechanics whose physical and mathematical basis has been treated in two previous papers by the present authors.[1] It was found possible to extend the above theory to systems having several degrees of freedom[2] (Chapter 2), and by the introduction of 'canonical transformations' to reduce the problem of integrating the equations of motion to a known mathematical formulation. From this theory of canonical transformations we were able to derive a perturbation theory (Chapter 1, § 4) which displays close similarity to classical perturbation theory. On the other hand we were able to trace a connection between quantum mechanics and the highly-developed mathematical theory of quadratic forms of infinitely many variables (Chapter 3). Before we go on to

Editor's note. This paper was published as Z. Phys. **35** (1926) 557–615.

[1] W. Heisenberg, Zs. f. Phys. **33** (1925) 879.
M. Born and P. Jordan, Zs. f. Phys. **34** (1925) 858.
Henceforth designated as (Part) I.

[2] *Note added in proof:*
A paper by P. A. M. Dirac (Proc. Roy. Soc. London **109** (1925) 642), which has appeared in the meantime, independently gives some of the results contained in Part I and the present paper, together with further new conclusions to be drawn from the theory.

discuss the presentation of this further development in the theory, we first endeavour to define its physical content more precisely.

The starting point of our theoretical approach was the conviction that the difficulties which have been encountered at every step in quantum theory in the last few years could be surmounted only by establishing a mathematical system for the mechanics of atomic and electronic motions, which would have a unity and simplicity comparable with the system of classical mechanics, and which would entirely consist of relations between quantities that are in principle observable. Admittedly, such a system of quantum-theoretical relations between observable quantities, when compared with the quantum theory employed hitherto, would labour under the disadvantage of not being directly amenable to a geometrically visualizable interpretation, since the motion of electrons cannot be described in terms of the familiar concepts of space and time. A characteristic feature of the new theory lies in the modification it imposes upon kinematics as well as upon mechanics; a notable advantage, however, of this quantum mechanics consists in the fact that the basic postulates of quantum theory form an inherent organic constituent of this mechanics, e.g., that the existence of discrete stationary states is just as natural a feature of the new theory as, say, the existence of discrete vibration frequencies in classical theory (cf. Chapter 3). If one reviews the fundamental differences between classical and quantum theory, differences which stem from the basic quantum theoretical postulates, then the formalism proposed in the two above-mentioned publications and in this paper, if proved to be correct, would appear to represent a system of quantum mechanics as close to that of classical theory as could reasonably be hoped. In this context we merely recall the validity of energy and momentum conservation laws and the form of the equations of motion (Chapter 1, § 2). This similarity of the new theory with classical theory also precludes any question of a separate correspondence principle outside the new theory; rather, the latter can itself be regarded as an exact formulation of Bohr's correspondence considerations. In the further development of the theory, an important task will lie in the closer investigation of the nature of this correspondence and in the description of the manner in which symbolic quantum geometry goes over into visualizable classical geometry. With regard to this question, a particularly important trait in the new theory would seem to us to consist of the way in which both continuous and line spectra arise in it on an equal footing, i.e., as solutions of one and the same equation of motion and closely connected with one another mathematically (cf. Chapter 3, § 3); obviously, in this theory, any distinction between 'quantized' and 'unquantized' motion ceases to be at all meaningful, since the theory contains no mention of a quantization condition which selects only certain types of motion from among a large number of possible types: rather, in place of such a condition one has a basic quantum mechanical equation (Chapter 1, § 1) which is applicable to *all* possible types of motion and which is essential if the dynamic problem is to be given a definite meaning at all.

Now, although we should like to be able to conclude that because of its mathematical simplicity and unity, the proposed theory might reproduce essential characteristics of the actual conditions inherent in problems of atomic structure, we nevertheless have to realize, that the theory is not yet able to furnish a solution to the principal difficulties in quantum theory. The theory has not yet incorporated the forces which in classical theory would be associated with radiation resistance, and in connection with the question of how the coupling problem is to be related to the quantum mechanics postulated here, there exist but a few indistinct indications (cf. Chapter 1, § 5). Nevertheless it would seem that these basic quantum-theoretical difficulties assume an altogether different aspect in the new theory than hitherto and that one might indeed now be more justified in hoping that these problems will in due course be solved. We consider, for instance, the question of collision processes. Recently, Bohr[1] called attention to the basic difficulties which (in the theory as employed hitherto) confronted all attempts to reconcile the fundamental postulates of quantum theory with the law of conservation of energy in fast collisions. In the present theory, however, the fundamental principles of quantum theory and the principle of conservation of energy follow mathematically from the quantum-mechanical equations, and hence the results of the Franck–Hertz collision studies would seem to be natural mathematical consequences of the theory. One may thus hope that a future treatment of collision problems based on the new quantum mechanics may, just because of this organic

[1] N. Bohr, Zs. f. Phys. **34** (1925) 142.

relationship between the basic postulates and this mechanics, avoid difficulties of the type mentioned above.

The question of the anomalous Zeeman effect seems to be hardly different when handled by the theory proposed here than it was before. It is true that the intimate connection between the 'aperiodic' and the 'periodic' orbits inherent in the basic assumptions of this theory entails the fact that we cannot be certain that Larmor's Theorem holds generally (Chapter 4, § 2); the assumptions for the validity of the theorem are satisfied by an oscillator, but not necessarily by a nuclear atom. It is not likely, however, that this standpoint can lead to an interpretation of anomalous Zeeman effects; rather the present quantum mechanics may in the case of Zeeman effects have to content with the same difficulties as the previous theory. Recently, though, the problem of anomalous Zeeman effects has entered a new phase as a result of a Note published by Uhlenbeck and Goudsmit.[1] These authors make the assumption that the electron itself posseses a mechanical and a magnetic moment (whose ratio should be twice as large as for atoms), so that there should actually be no anomalous Zeeman effects. By this assumption, difficulties as to statistical weights are eliminated and a qualitative explanation of various phenomena connected with problems of multiplet structure and Zeeman effects ensues. The question as to whether it can already furnish a quantitative explanation of these phenomena can, of course, be answered only after more rigorous investigations using the methods of quantum mechanics. Some of the results contained in Chapter 4 appear, as regards the Zeeman effects, to substantiate this hope of finding a quantitative interpretation at some later date.

Finally, we have also attempted to treat a well-known statistical problem by means of the methods furnished by the present theory. It is well known that by quantizing the vibrations of a cavity within reflecting walls and using classical methods one can arrive at results which display a certain similarity with the hypotheses in a theory of light quanta and which permit a derivation of Planck's formula. However, as Einstein[2] has always stressed, this semiclassical treatment of cavity radiation yields an erroneous value for the mean square deviation of the energy in a volume element. This result must be

[1] G. Uhlenbeck and S. Goudsmit, Naturwiss. **13** (1925) 953.

[2] A. Einstein, Phys. Zs. **10** (1909) 185, 817.

regarded as a particularly serious objection to earlier methods in quantum theory, since we are concerned here with a breakdown of the theory even for the simple problem of a harmonic oscillator. On the other hand, the above difficulty would arise in the statistical treatment of the eigenvibrations of any mechanical system whatsoever, e.g., a crystal lattice. Now, we have found that with the kinematics and mechanics inherent in the theory presented here, the corresponding calculation leads to a correct value for the mean square deviation and also to Planck's formula, a result which may well be regarded as significant evidence in favour of the quantum mechanics put forward here.

CHAPTER 1. SYSTEMS HAVING ONE DEGREE OF FREEDOM

1. Fundamental principles

I. A quantum-theoretical quantity a, whether representing a coordinate or a momentum or any function of both, is depicted by a set of quantities

$$a(nm)e^{2\pi i\nu(nm)t} \tag{1}$$

or (on leaving off the factor $e^{2\pi i\nu(nm)t}$, which is the same for all quantities belonging to a given system and which depends only upon the indices n and m) by the set of numbers

$$a(nm). \tag{2}$$

We can thus speak of an infinite 'matrix' a.

II. Elementary operations such as addition and multiplication of quantum-theoretical quantities are defined in accordance with the operational rules of matrix calculus.

III. Consider a given function $f(x_1, x_2,\dots,x_s)$ defined through addition and multiplication of given matrices, with $x_1, x_2,\dots,x_s$ denoting quantum-theoretical quantities. We then introduce two types of derivatives of f with respect to one of the quantities x (say, x_1):

(a) Differential coefficient of the first type:

$$\frac{\partial f}{\partial x_1} = \lim_{\alpha\to 0} \frac{f(x_1 + \alpha\mathbf{1}, x_2, \dots, x_s) - f(x_1, x_2, \dots, x_s)}{\alpha}, \tag{3}$$

where α represents a number and $\mathbf{1}$ the unit matrix defined by

$$\mathbf{1} = (\delta_{nm}), \qquad \delta_{nm} = \begin{cases} 1 & \text{for } n = m \\ 0 & \text{,, } n \neq m. \end{cases}$$

(b) Differential coefficient of the second type: Defined through[1]

$$\frac{\partial f}{\partial x_1}(nm) = \frac{\partial D(f)}{\partial x_1(mn)}, \tag{4}$$

where $D(f)$ represents the diagonal sum of the matrix f.

These two forms of differentiation will be distinguished typographically by different fraction strokes [thick stroke for (a), thin for (b)].

The treatment in Part I employed differentiation of the *second* type exclusively since this leads to a simple formulation of the variational principle of quantum mechanics and hence appears to be the more natural. However, for some calculations derivatives of the first type are more convenient to employ. It might be mentioned generally that the introduction of a differential coefficient into quantum mechanics is somewhat of an artifice and that the operations on the left-hand side of the formula (6) which follow represent the natural counterpart to differential coefficients in classical theory. For the formulation of canonical equations it is important to establish the fact that both species of differentiation (3) and (4) become identical in the case of the energy function[2] $H(pq)$.

[1] Cf. Part I [paper **13** in this volume].

[2] For the energy function H of Part I, instead of arbitrary functions such as

$$H^* = \Sigma\, a_{sr} p^s q^r,$$

only those symmetrized functions giving rise to the same Hamilton equations were permitted:

$$H = \Sigma\, a_{sr} \frac{1}{s+1} \sum_{l=0}^{s} p^{s-l} q^r p^l.$$

Now, for these symmetrized functions H the following relations, derived in Part I, apply:

$$\frac{\partial H}{\partial p} = \Sigma\, a_{sr} \frac{1}{s+1} \left\{ \sum_{l=0}^{s-1} (s-l) p^{s-1-l} q^r p^l + \sum_{l=1}^{s} l p^{s-l} q^r p^{l-1} \right\}$$
$$= \Sigma\, a_{sr} \sum_{l=0}^{s-1} p^{s-1-l} q^r p^l = \frac{\partial H}{\partial p}.$$
$$\frac{\partial H}{\partial q} = \Sigma\, a_{sr} \frac{r}{s+1} \sum_{l=0}^{s} p^{s-l} q^{r-1} p^l = \Sigma\, a_{sr} \sum_{j=0}^{r-1} q^{r-1-j} p^s q^j = \frac{\partial H}{\partial q}.$$

IV. Calculations involving quantum-theoretical quantities would yield non-unique results because of the inapplicability of the commutative rule in multiplication unless the value of $pq-qp$ were prescribed.[1] Hence we introduce the following basic quantum-mechanical relation:

$$pq - qp = \frac{h}{2\pi i}\,\mathbf{1}. \tag{5}$$

We shall later discuss the physical significance of this relation according to the correspondence principle. At this stage it would appear important to stress that eq. (5), ch. 1, is the only one of the basic formulae in the quantum mechanics here proposed which contains Planck's constant h. It is satisfying that the constant h already enters into the basic tenets of the theory at this stage in so simple a form. Furthermore, one can see from eq. (5), ch. 1, that in the limit $h=0$, the new theory would converge to classical theory, as is physically required.

A relation which will later prove important can also be derived from eq. (5), ch. 1, namely:
If $f(pq)$ be any function of p and q, then

$$\begin{aligned} fq - qf &= \frac{\partial f}{\partial p}\,\frac{h}{2\pi i}, \\ pf - fp &= \frac{\partial f}{\partial q}\,\frac{h}{2\pi i}, \end{aligned} \tag{6}$$

since, if we assume these formulae to be valid for some given pair of functions, φ and ψ, then they must also hold for $\varphi+\psi$ and $\varphi\cdot\psi$. The former case, $\varphi+\psi$ is trivial; for the latter, $\varphi\cdot\psi$, a simple calculation yields:

$$\begin{aligned} \varphi\cdot\psi q - q\varphi\psi &= \varphi(\psi q - q\psi) + (\varphi q - q\varphi)\psi \\ &= \varphi\left(\frac{\partial\psi}{\partial p} + \frac{\partial\varphi}{\partial p}\,\psi\right)\frac{h}{2\pi i} = \frac{\partial(\varphi\psi)}{\partial p}\,\frac{h}{2\pi i}; \end{aligned}$$

for $p\varphi\psi-\varphi\psi p$, the treatment is similar.

Now, the relations (6) hold for p and q. They must accordingly also apply to every function f which can formally be expressed as a power series in p and q.

[1] The equations of motion merely indicate that this difference has to be a diagonal matrix.

2. The canonical equations, energy conservation and frequency condition

Let an energy function $H(pq)$ be given, together with the associated canonical equations

$$\dot{p} = -\frac{\partial H}{\partial q}; \qquad \dot{q} = \frac{\partial H}{\partial p}. \tag{7}$$

It follows from the frequency combination principle

$$\nu(nm) + \nu(mk) = \nu(nk) \tag{8}$$

that ν can be expressed in the form

$$\nu(nm) = \frac{(W_n - W_m)}{h}. \tag{9}$$

We now introduce a quantum-theoretical quantity W, as 'term', defined through

$$W(nm) = \begin{cases} W_n & \text{for } n = m \\ 0 & \text{for } n \neq m. \end{cases}$$

Thus W is a diagonal matrix.

Then for any quantum-theoretical quantity whatsoever, the following relation holds:

$$\dot{a} = \frac{2\pi i}{h}(Wa - aW). \tag{10}$$

In fact $\dot{a}$ was (cf. Part I) defined through

$$a(\dot{n}m) = 2\pi i\nu(nm)\, a(nm).$$

Among the main tenets of the theory we here seek to build up, we class the law of conservation of energy (H=constant) and the frequency condition

$$\left(\nu(nm) = \frac{H_n - H_m}{h}; \qquad H_n = W_n + \text{const}\right).$$

We carry the proof through for both these conditions by inserting eqs. (6) and (10) into eq. (7), ch. 1. This yields

$$\begin{aligned} Wq - qW &= Hq - qH \\ Wp - qW &= Hp - qH \end{aligned} \tag{11}$$

or, equivalently,

$$(W - H)q - q(W - H) = 0,$$
$$(W - H)p - q(W - H) = 0.$$

The entity $W-H$ commutes with p and q, and hence also with every function of p, q, in particular with H:

$$(W - H)H - H(W - H) = 0.$$

Thence from (10), ch. 1, one has

$$\dot{H} = 0. \tag{12}$$

Thereby the law of conservation of energy is proved, and H is established as a diagonal matrix, $H(nm)=\delta_{nm}H_n$.

The frequency condition now follows directly from (11), ch. 1:

$$q(nm)(H_n - H_m) = q(nm)(W_n - W_m), \tag{13}$$

i.e.,

$$\frac{(H_n - H_m)}{h} = \nu(nm). \tag{14}$$

Thus far, we have proved energy-conservation and the frequency condition from the canonical equations and the basic equation (5), ch. 1. In corollary, we can, however, also invert the proof. We know energy conservation and the frequency condition to be correct. Hence if the energy function H be given as an analytical function of any variables P, Q then, provided that

$$PQ - QP = \frac{h}{2\pi i}\mathbf{1},$$

the following canonical equations always apply:

$$\dot{Q} = \frac{\partial H}{\partial P}, \qquad P = -\frac{\partial H}{\partial Q}. \tag{15}$$

This follows directly from the fact that the quantities $PH-HP$ or $HQ-QH$ can be interpreted in a twofold manner, namely according to (6), ch. 1 and according to (10), ch. 1.

3. Canonical transformations

By a 'canonical transformation' of the variables p, q into new variables

P, Q, we understand a transformation in which

$$pq - qp = PQ - QP = \frac{h}{2\pi i}, \tag{16}$$

as is suggested by the preceding considerations, since then the same canonical equations (7), ch. 1, or (15), ch. 1, apply to P, Q as to p q.

A general transformation which satisfies this condition is

$$\begin{aligned} P &= SpS^{-1} \\ Q &= SqS^{-1}, \end{aligned} \tag{17}$$

wherein S stands for an arbitrary quantum-theoretical quantity. We would surmise that eq. (17), ch. 1, represents in fact the *most general* canonical transformation. The transformation (17), ch. 1, also has the simple property that for any function $f(P, Q)$ it follows that

$$f(P, Q) = Sf(p, q)S^{-1}, \tag{18}$$

wherein $f(p, q)$ is formed from $f(P, Q)$ on replacing P by p and Q by q, retaining the functional form. The proof of this contention for functions in the sense of our above definition follows directly from the observation that the rule holds for sum and product with sum terms or factors p, q.

The importance of the canonical transformation is due to the following theorem: If any pair of values p_0, q_0 be given which satisfy eq. (15), ch. 1, then the problem of integrating the canonical equations for an energy function $H(pq)$ can be reduced to the following: A function S is to be determined, such that when

$$p = Sp_0S^{-1}, \qquad q = Sq_0S^{-1} \tag{19}$$

the function

$$H(pq) = SH(p_0q_0)S^{-1} = W \tag{20}$$

becomes a diagonal matrix. Equation (20), ch. 1, is the analogue to the Hamilton partial differential equation, and in a sense stands for the action function.

4. Perturbation theory

We consider a given mechanical problem defined by the energy function

$$H = H_0(pq) + \lambda H_1(pq) + \lambda^2 H_2(pq) + \dots \tag{21}$$

and assume the mechanical problem defined by the energy function $H_0(pq)$ to be solved. Thus solutions p_0, q_0 of this problem are known; they satisfy the condition $p_0q_0-q_0p_0=(h/2\pi i)\mathbf{1}$ and cause $H_0(p_0q_0)=W_0$ to be a diagonal matrix. We then seek a transformation function S such that

$$p = Sp_0S^{-1}, \qquad q = Sq_0S^{-1}, \tag{22}$$

and that

$$H(pq) = SH(p_0q_0)S^{-1} = W,$$

e.g., that the matrix H becomes diagonalized. To arrive at a solution we try setting

$$S = \mathbf{1} + \lambda S_1 + \lambda S_2 + \dots . \tag{23}$$

Then

$$S^{-1} = \mathbf{1} - \lambda S_1 + \lambda^2(S_1^2 - S_2) + \lambda^3 \dots . \tag{24}$$

If for H we take the expression (21), ch. 1, we can collect together powers of λ to obtain the following equations of approximation:

$$\begin{aligned} H_0(p_0q_0) &= W_0 \\ S_1H_0 - H_0S_1 + H_1 &= W_1 \\ S_2H_0 - H_0S_2 + H_0S_1^2 - S_1H_0S_1 + S_1H_1 - H_1S_1 + H_2 &= W_2 \\ \dots\dots\dots\dots\dots\dots\dots\dots \\ S_rH_0 - H_0S_r + F_r(H_0, \dots, H_r, S_0, \dots, S_{r-1}) &= W_r \end{aligned} \tag{25}$$

where H_0, H_1, ... are throughout to be taken as having arguments p_0, q_0.

The first of the eqs. (25), ch. 1, is already satisfied. The others can be resolved in sequence, actually in just the same manner as in classical theory, namely by first building the mean value in order to determine the energy constant, after which the solution can straightway be written down:

$$W_r = \bar{F}_r, \tag{26}$$

$$S_r(mn) = \frac{F_r(mn)}{h\nu_0(mn)}\,(1 - \delta_{nm}),$$

where $\nu_0(nm)$ are the frequencies of the unperturbed motion. This solution satisfies the condition

$$S\cdot\tilde{S}^* = \mathbf{1}, \tag{27}$$

wherein the tilde represents interchange of rows and columns (transposition) and the star denotes that we take the complex conjugate quantity. Since we shall later return to this condition from a more general standpoint we confine ourselves at this stage merely to verifying it to the first order of approximation, which we shall evaluate right away. To this order, the relation runs

$$S_1 + \tilde{S}_1^* = 0. \tag{28}$$

The significance of eq. (27), ch. 1, lies in the fact that the Hermitian character of the matrices p, q follows from it, since use of (22), ch. 1, shows[1] that

$$q^* = S^* q_0^* S^{*-1} = \tilde{\tilde{S}}^{-1} \tilde{q}_0 \tilde{\tilde{S}} = \tilde{q},$$

and analogously for p.

To first approximation it follows from (26), ch. 1, as also classically, that

$$W_1 = \overline{H}_1, \tag{29}$$

so that

$$S_1(mn) = \frac{H_1(mn)}{h\nu_0(mn)} (1 - \delta_{mn}). \tag{30}$$

This expression indeed satisfies the requirements (28), ch. 1, because H_1 is assumed to be a Hermitian form. We can now evaluate the energy to the second order of approximation and find

$$W_2 = \overline{H}_2 + \frac{1}{h} \sum_l{}' \frac{H_1(nl) H_1(ln)}{\nu_0(nl)}, \tag{31}$$

where the prime on the summation indicates that terms having a vanishing denominator ($l = n$) are to be excluded.

One can progress in this way and successively determine all terms of the W and S series. If we substitute the S series in (22), ch. 1, we obtain the expansions

$$q = q_0 + \lambda q_1 + \lambda^2 q_2 + \ldots,$$
$$p = p_0 + \lambda p_1 + \lambda^2 p_1 + \ldots$$

with known coefficients. Thus, for example, the first-order approximation runs

$$q_1 = S_1 q_0 - q_0 S_1,$$
$$p_1 = S_1 p_0 - p_0 S_1;$$

[1] On noting the rule $\widetilde{(ab)} = \tilde{b}\tilde{a}$.

or, explicitly,

$$\begin{aligned} q_1(mn) &= \frac{1}{h}\sum_k{}' \left(\frac{H_1(mk)q_0(kn)}{\nu_0(mk)} - \frac{q_0(mk)H_1(kn)}{\nu_0(kn)}\right) \\ p_1(mn) &= \frac{1}{h}\sum_k{}' \left(\frac{H_1(mk)q_0(kn)}{\nu_0(mk)} - \frac{q_0(mk)H_1(kn)}{\nu_0(kn)}\right). \end{aligned} \tag{32}$$

The formulae (32), ch. 1, represent the outcome of Kramers' dispersion theory[1] in the limit of an infinitely low-frequency external field; this possibility of attaining a simple derivation of formulae otherwise obtained only on the basis of correspondence considerations seems to provide a strong argument in favour of the theory put forward here. Born[2] has derived eq. (31), ch. 1, on reinterpreting the respective classical formulae. The terms with $m=n$ in eq. (32), ch. 1, correspond to Kramers' formula for normal dispersed light and the remaining terms ($m \neq n$) correspond to the formulae of Kramers and Heisenberg[3] for 'scattered light of combination frequencies'. The latter expressions were used by Pauli[4] to evaluate the intensities of transitions in Hg which take place in presence of external electric fields and which would otherwise be 'forbidden'. In order to derive the general dispersion formulae (if the frequency of the external field does not vanish), one needs more general considerations regarding the action of external fields which change in function of time. We now pass over to such considerations.

5. Systems for which time-variables enter explicitly into the 'energy function'

Treatment of the quantum-mechanical influence of external forces which explicitly depend upon time seems to us to be of especial interest in that therein some characteristic differences crop up between classical and quantum mechanics. The problem of the action of time-dependent external forces can be regarded as a limiting case of the interaction between two systems in which the influence of the inter-

[1] H. A. Kramers, Nature **113** (1924) 673; **114** (1924) 310; cf. also R. Ladenburg, Zs. f. Phys. **4** (1921) 451; R. Ladenburg and F. Reiche, Naturwiss. **11** (1923) 584.
[2] M. Born, Zs. f. Phys. **26** (1924) 379.
[3] H. A. Kramers and W. Heisenberg, Zs. f. Phys. **31** (1925) 681.
[4] W. Pauli, Verh. d. Dän. Akad. d. Wiss. (in press).

action on one of the two systems (termed system A) is so small that the action upon the other system (system B) remains unaffected by this influence. If we now consider the coupling of two systems A, B from the standpoint of quantum mechanics, the Hamilton function decomposes into three parts, H_A, λH_B and $\varepsilon\lambda H_{AB}$ (with λ at this stage an arbitrary parameter and ε a small quantity). We take system A to be known. For calculating the motion of B according to classical theory it suffices to establish the equations of motion [from the Hamilton function $\lambda(H_B+\varepsilon H_{AB})$] for the coordinates of B, whereby for the coordinates of A one substitutes their solutions in function of time (for the definite given values of the constants in A). By this means, apart from the constants of A only the time enters as a new variable into the perturbation problem for B when the reaction is neglected. In the quantum-mechanical calculation the situation is just the same, providing we restrict ourselves to first-order perturbations (i.e., terms proportional to ε in the coordinates and momenta of the system B). It is altogether otherwise, however, for higher-order perturbations, since in the evaluation of higher-order perturbations we encounter products of quantities in which more than one implicitly contains the coordinates of A. But this means that according to the quantum-mechanical rule for building a product it by no means suffices to know the 'external forces in function of time' merely for the *given* values of the constants in A, but these external forces must be known for *all* values of the constants. Thereby, however, the concept of external forces appears in fact to become devoid of meaning. This difficulty seems to us to be overcome on observing that the reaction itself gives rise to terms of order $\lambda\varepsilon^2$ in the coordinates of B, and thus that simultaneous neglect of the reaction and evaluation of terms in B containing ε^2 is meaningful only if λ can also be taken to be very small, i.e., physically, if variation of the quantities in A by amounts of the same order as the associated quantities in B does not bring about any perceptible change in the influence of A upon B. However, in this approximation the quantum-mechanical construction of products and thereby the calculation of the perturbations to higher orders in ε can again be effected. In fact, the rules for this building of products reduce simply to those of classical multiplication, as in this approximation the coordinates, amplitudes and frequencies which enter into H_{AB} do not depend on the constants in A. In this sense one could, for example, treat the action of a strong alternating electromagnetic field on an atom entirely as the influence of an 'external force' with neglect of the reaction, since the field energy can be regarded as infinitely large compared with that of the atom. The action of α-particles upon the electrons of an atom could also be regarded as an 'external force', as in classical theory, because of the relatively large energy of the α-particles, so that in this approximation the Fourier expansion of the force thereby exerted upon the electrons would also be that of classical theory. However, the action of forces due to one atom upon another can never be treated as an operation of external forces – i.e., it can thus be regarded only in the first-order terms, for which such an approach is always possible – since the neglect of the reaction would in the higher-order terms lead to false results.

We can summarize the outcome of our considerations thus: It is meaningful under certain assumptions in quantum as in classical theory to speak of the action of time-dependent forces upon an atom. In such instances, the classical calculation rules can be applied to expressions in which the time parameter figures explicitly: e.g., if the external field of force be periodic with a period ν_0, then the general term of a coordinate q can be written as

$$q(mn, \tau)\ e^{2\pi i[\nu(mn)+\tau\nu_0]t}, \tag{33}$$

and the general term of q^2 as

$$\sum_{k,\tau'} q(mk, \tau - \tau')q(kn, \tau')e^{2\pi i[\nu(mn)+\tau\nu_0]t}. \tag{34}$$

For this reason the case of external forces which vary with time seems in our view to provide a striking illustration of the transition from theoretical quantum kinematics into classical kinematics according to the principle of correspondence.

If one is concerned with the evaluation of the operation of external forces to first order only, the results which ensue from the calculations which follow remain correct even if the assumptions listed at the outset are not obeyed – in exact analogy with the situation in classical theory.

From the preceding considerations it follows that the mathematical treatment of systems in which (provided the assumptions mentioned above are valid) time enters explicitly is simply to be handled in a manner analogous to the corresponding classical procedures. If we again assume the external force to be periodic in time, with period

ν_0, the Hamilton function becomes[1]

$$H = H(p_k, q_k, \cos 2\pi\nu_0 t). \tag{35}$$

We then introduce a new degree of freedom with the variables q', p' and take the following as the Hamiltonian of the new problem, in which time no longer figures explicitly:

$$H' = H(p_k, q_k;\ q') + 2\pi\nu_0\sqrt{1-q'^2}\,p'. \tag{36}$$

Thereby the canonical equations for p_k, q_k remain as hitherto, except that q' is throughout written for $\cos 2\pi\nu_0 t$. The new equations are:

$$\begin{aligned} \dot{q}' &= \frac{\partial H'}{\partial p'} = 2\pi\nu_0\sqrt{1-q'^2}, \\ \dot{p}' &= -\frac{\partial H'}{\partial q'} = -\frac{\partial H}{\partial q'} + 2\pi\nu_0\frac{q'}{\sqrt{(1-q'^2)}}p'. \end{aligned} \tag{37}$$

The first of these equations asserts that q' indeed becomes equal to $\cos 2\pi\nu_0 t$ (up to an arbitrary choice of origin in the time scale), so that the canonical equations for p_k, q_k take on the same form as in the earlier problem; the second equation (37), ch. 1, provides a determination of p'. Thus through (36), ch. 1, the problem (35), ch. 1, is really led back to cases already treated.

Of paramount interest is the question as to the manner in which the perturbation formulae (25), ch. 1, have to be modified if time enters explicitly into $H_1, H_2, \ldots$ but *not* into H_0. Simple considerations show that for this case the perturbation formulae ensue from those cited earlier on replacing every term of the form $H_0S_r - S_rH_0$ by

$$H_0S_r - S_rH_0 + \frac{h}{2\pi i}\frac{\partial S_r}{\partial t}$$

(note that H_0 occurs only in such combinations). Thus the lowest orders of the new perturbation formulae run:

$$\begin{aligned} H_0(p_0q_0) &= W_0, \\ S_1H_0 - H_0S_1 - \frac{h}{2\pi i}\frac{\partial S_1}{\partial t} + H_1 &= W_1, \end{aligned} \tag{38}$$

[1] Here we anticipate for a moment in availing ourselves of results derived in the next chapter for systems having several degrees of freedom.

$$S_2H_0 - H_0S_2 - \frac{h}{2\pi i}\frac{\partial S_2}{\partial t} + \left(H_0S_1 - S_1H_0 + \frac{h}{2\pi i}\frac{\partial S_1}{\partial t}\right)S_1 \qquad (38)$$
$$+ S_1H_1 - H_1S_1 + H_2 = W_2,$$

. . . .

We should like to assume that even if the assumption that the external forces are periodic in time does not apply, these formulae (38), ch. 1, nevertheless remain valid – even though this assumption was incorporated into the derivation of the formulae.

The first-order equations in the formulae (38), ch. 1, which of course remain correct even if the assumptions regarding 'external forces' are no longer valid, taken together with eqs. (22), ch. 1, viz.

$$q = q_0 + \lambda(S_1q_0 - q_0S_1),$$
$$p = p_0 + \lambda(S_1p_0 - p_0S_1),$$

furnish an answer to the problems of dispersion theory in a general sense. In actual fact, if we set:

$$H_1 = Eeq_0 \cos 2\pi\nu_0 t,$$

then

$$H_1(mn, 1) = \frac{Ee}{2}\, q_0(mn), \qquad H_1(mn, -1) = \frac{Ee}{2}\, q_0(mn),$$

$$S_1(mn, 1) = \frac{Ee}{2h}\,\frac{q_0(mn)}{\nu_0(mn) + \nu_0}, \qquad (39)$$

$$S_1(mn, -1) = \frac{Ee}{2h}\,\frac{q_0(mn)}{\nu_0(mn) - \nu_0}.$$

Thence follows (cf. (22), ch. 1):

$$q_1(mn, +1) = \frac{Ee}{2h}\sum_k \left(\frac{q_0(mk)q_0(kn)}{\nu_0(mk) + \nu_0} - \frac{q_0(mk)q_0(kn)}{\nu_0(kn) + \nu_0}\right). \qquad (40)$$

If we assume that we have Cartesian coordinates, i.e., $p = m\dot{q}$, then

$$q_1(mn, 1) = \frac{Ee}{2h \cdot 2\pi i m}\sum_k \frac{q_0(mk)p_0(kn) - p_0(mk)q_0(kn)}{(\nu_0(mk) + \nu_0)(\nu_0(kn) + \nu_0)}; \qquad (41)$$

and similarly

$$q_1(mn, -1) = \frac{Ee}{2h \cdot 2\pi i m}\sum_k \frac{q_0(mk)p_0(kn) - p_0(mk)q_0(kn)}{(\nu_0(mk) - \nu_0)(\nu_0(kn) - \nu_0)}. \qquad (42)$$

The eqs. (40), (41), (42), ch. 1 agree with the formulae obtained from Kramers' dispersion theory.[1] A further particularly interesting case would seem to be that for incident light of very high frequency, $|\nu_0| \gg |\nu_0(mk)|$ or $|\nu_0(kn)|$. Then to first-order approximation one finds

$$q_1 = -\frac{Ee}{h2\pi i\nu_0^2 m}(p_0q_0 - q_0p_0)\cos 2\pi\nu_0 t,$$

or, because of (5), ch. 1,

$$q_1 = +\frac{Ee}{4\pi^2 m\nu_0^2}\cos 2\pi\nu_0 t. \tag{43}$$

This finding indicates that in fact the quantum-mechanical commutation relation (5), ch. 1, ultimately entails the fact that for sufficiently high frequencies the electron behaves on scattering like a free electron. The scattered light of frequency $\nu_0(mn)+\nu_0(m \neq n)$ vanishes, and that of frequency ν_0 has the intensity to be expected for scattering by a free electron.[2]

CHAPTER 2. FUNDAMENTALS OF THE THEORY FOR SYSTEMS HAVING AN ARBITRARY NUMBER OF DEGREES OF FREEDOM

1. The canonical equations of motion; perturbation theory for nondegenerate systems

For several degrees of freedom ($f > 1$) it rather suggests itself that we replace the representation of quantum-theoretical quantities by two-dimensional matrices by one in terms of $2f$-dimensional matrices, corresponding with the $2f$-dimensional manifold of stationary states in the classical J-space:

$$\begin{aligned} q_k &= \big(q_k(n_1 \ldots n_f, m_1 \ldots m_f)\big), \\ p_k &= \big(p_k(n_1 \ldots n_f, m_1 \ldots m_f)\big). \end{aligned} \tag{1}$$

Nevertheless this representation, albeit under certain circumstances

[1] Cf. the discussion at the end of § 4 of results obtained for $\nu_0 = 0$.

[2] Cf. the articles by W. Kuhn, Zs. f. Phys. **33** (1925) 408; W. Thomas, Naturwiss. **13** (1925) 627; F. Reiche and W. Thomas, Zs. f. Phys. **34** (1925) 510.

very convenient and clear, is by no means essential. Even for several degrees of freedom the fundamental dynamical equations assume the form of *matrix equations*, but these matrices can as heretofore also be written in two-dimensional form. It became apparent even for *one* degree of freedom that the *sequence* of the stationary states as given by the ordering of the matrix rows is (in contradistinction to the theory employed hitherto) purely fortuitous and is not governed by any intrinsic property of the system. This observation can now directly be referred to many-dimensional matrices too; one can carry out any arbitrary rearrangements and in particular transform the $2f$-dimensional matrices into two-dimensional ones. This is justified by the fact that the basic definitions of addition and multiplication, as also of differentiation with respect to time, are clearly independent of any *ordering relations* between the basis systems of indices n_1, $n_2, \ldots, n_f$, which taken *singly* specify the *states* and *in pairs* specify the *transitions*.

It is thence also clear that the general rules of matrix analysis, as presented in chapter 1 of Part I and in chapter 1 of the present paper, can be employed in the theory of systems having *several* degrees of freedom also. One can similarly take over the derivation of the equation of motion from the variational principle in I directly, so that we can in like manner write

$$\dot{q}_k = \frac{\partial H}{\partial p_k}; \qquad \dot{p}_k = -\frac{\partial H}{\partial q_k}. \tag{2}$$

The principal new feature distinguishable from those obtaining for systems with just one degree of freedom lies in the general commutation relations for p_k and q_k in the case of several degrees of freedom. Just as in the calculations for but one degree of freedom, so here also calculations with quantum-theoretical quantities would be to some extent indefinite if the 'commutation relations' were not specified.

As a plausible generalization of eqs. (5), ch. 1, the following equations suggest themselves:

$$\begin{aligned} p_k q_l - q_l p_k &= \frac{h}{2\pi i}\,\delta_{kl}, \\ p_k p_l - p_l p_k &= 0, \\ q_k q_l - q_l q_k &= 0, \end{aligned} \tag{3}$$

if H denotes the (symmetrized) energy function, one can in consequence of these relations replace eqs. (2), ch. 2, by

$$\dot{q}_k = \frac{\partial H}{\partial p_k}, \qquad \dot{p}_k = -\frac{\partial H}{\partial q_k}. \tag{2'}$$

Further, it follows from these relations,[1] as in chapter 1 of the present paper, that

$$\begin{aligned} p_k f(q_1 \dots q_f, p_1 \dots p_f) - f p_k &= \frac{h}{2\pi i} \frac{\partial f}{\partial q_k}, \\ f q_k - q_k f &= \frac{h}{2\pi i} \frac{\partial f}{\partial p_k}. \end{aligned} \tag{4}$$

The proof of energy conservation and the frequency condition then follows from (2′) and (4), ch. 2, as shown in ch. 1. Similarly one can show with the aid of (3) and (4) that the canonical equations (2′), ch. 2, apply whenever the relations (3), ch. 2, are satisfied for a system P_k, Q_k and the energy function is given as an analytical function of the P_k and Q_k.

Thus a transformation of the variables p_k, q_k into new variables P_k, Q_k is termed 'canonical' if it leaves the relations (3), ch. 2, unaltered.

A very general class of such transformations is again given by the formulae

$$\begin{aligned} P_k &= S p_k S^{-1}, \\ Q_k &= S q_k S^{-1}. \end{aligned} \tag{5}$$

This transformation again has the property of converting every function $f(PQ)$ into

$$f(P_1, \dots, P_f, Q_1, \dots, Q_f) = S f(p_1, \dots, p_f, q_1, \dots, q_f) S^{-1}. \tag{6}$$

If a system $p_1^0, \dots, p_f^0, q_1^0, \dots, q_f^0$ is known, and satisfies the relations (3), ch. 2, then the problem of integrating eqs. (2), ch. 2, again reduces itself to the simpler problem: A function S is to be sought, such that it satisfies the equations

$$\begin{aligned} p_k &= S p_k^0 S^{-1}, \\ q_k &= S q_k^0 S^{-1} \end{aligned} \tag{5a}$$

[1] The physical significance of these relations for dispersion theory is discussed by H. A. Kramers, Physika, December 1925.

and transforms H into a diagonal matrix,

$$H(pq) = SH(p^0q^0)S^{-1} = W. \tag{7}$$

Equation (7) again represents the counterpart to the Hamilton partial differential equation.

Equations (3), ch. 2 would, together with (2), ch. 2, obviously entail too extensive a set of requirements for the p_k, q_k, if all these equations were *independent* of one another. As an interesting mathematical problem must rank the derivation of eqs. (3) using the least number of independent and mutually consistent assumptions; nevertheless, this question will not be handled here. We shall content ourselves with mentioning that

$$\frac{\mathrm{d}}{\mathrm{d}t} \sum_k (p_k q_k - q_k p_k)^{\cdot} = 0$$

is a general outcome of the equations of motion (1), ch. 2. On the other hand, it will be shown generally that the eqs. (3), ch. 2, together with the equations of motion (2), ch. 2, or the equivalent requirement (7), ch. 2, *can* be satisfied (singular discrepancies apart, of course).

This proof is to be supplied in connection with the generalization of the perturbation theory presented in ch. 1 § 4, when extended to arbitrarily many degrees of freedom. We consider the energy function $H(pq)$ such that it can be written as

$$H = H_0(pq) + \lambda H_1(pq) + \lambda^2 H_2(pq) + \dots, \tag{8}$$

so that

$$H_0(pq) = \sum_{k=1}^{f} H^{(k)}(p_k q_k).$$

Thus for $\lambda=0$ we have f *uncoupled* systems, each having a single degree of freedom; the f cases

$$H = H^{(k)}(p_k q_k)$$

can be solved with

$$q_k = q_k^0, \qquad q_k = p_k^0,$$

wherein q_k^0, p_k^0 are *two-dimensional* matrices,

$$q_k^0 : (q_k^0(nm)); \qquad p_k^0 : (p_k^0(nm)). \tag{10}$$

If we formally regard these f uncoupled systems as a single system having f degrees of freedom, then q_k^0, p_k^0 would be represented as $2f$-dimensional matrices,

$$\left.\begin{aligned} q_k^0 &= \big(q_k^0(n_1 \dots n_f;\quad m_1 \dots m_f)\big), \\ p_k^0 &= \big(p_k^0(n_1 \dots n_f;\quad m_1 \dots m_f)\big), \end{aligned}\right\} \tag{11}$$

for which

$$\begin{aligned} q_k^0(n_1 \dots n_f;\quad m_1 \dots m_f) &= \delta_k q_k^0(n_k m_k), \\ p_k^0(n_1 \dots n_f;\quad m_1 \dots m_f) &= \delta_k p_k^0(n_k m_k), \end{aligned}$$

where $\delta_k=1$ if $n_j=m_j$ for all j except $j=k$, and $\delta_k=0$ if for any $j(j \neq k)$, n_j is not equal to m_j. Thence, however, one sees: firstly, that the equations

$$p_k^0 q_k^0 - q_k^0 p_k^0 = \frac{h}{2\pi \mathrm{i}} \mathbf{1} \tag{12}$$

which originally obtained for the *two-dimensional* matrices (10), ch. 2, also hold for the *2f-dimensional* matrices (11), ch. 2; secondly, that the following relations ensue:

$$\begin{aligned} p_k^0 q_l^0 - q_l^0 p_k^0 &= 0 \quad \text{for} \quad l \neq k, \\ p_k^0 p_l^0 - p_l^0 p_k^0 &= q_k^0 q_l^0 - q_l^0 q_k^0 = 0. \end{aligned} \tag{13}$$

Hence for $\lambda=0$, all the eqs. (13), ch. 2, indeed apply. It is to be shown that p, q can be determined in such a manner that (3), ch. 2, is satisfied simultaneously with $H=W$ for the higher-order approximations also. One again assumes the system H_0 to have been chosen as *non-degenerate*, i.e., that on substituting $q=q^0$, $p=p^0$ no two diagonal elements of H_0 become identical. In this case we again have to set

$$q_k = S q_k^0 S^{-1}; \qquad p_k = S p_k^0 S^{-1} \tag{14}$$

as in eq. (5a), ch. 2, and to determine

$$S = \mathbf{1} + \lambda S_1 + \lambda^2 S_2 + \dots$$

in such a way as to satisfy the relation $H=W$. The eqs. (3), ch. 2, are then jointly also satisfied, since by virtue of (14) they go over into (12), (13). This completes the required proof.

Equations (3) are invariant with respect to a linear orthogonal transformation of the q_k and p_k, for if one sets

$$q'_k = \sum_l a_{kl} q_l, \qquad \sum_l a_{kl} a_{jl} = \delta_{kj},$$
$$p'_k = \sum_l a_{kl} p_l,$$

then

$$p'_k q'_l - q'_l p'_k = \sum_{hj} a_{kh} a_{lj} (p_h q_j - q_j p_h) = \delta_{kl} \frac{h}{2\pi i}$$

and similarly for the other respective relations. If then the conditions (3), ch. 2, hold for a given Cartesian coordinate system, they will also be valid in every other Cartesian coordinate system.

By way of supplement, now that we have established (3), ch. 2, we demonstrate that a well-known law of classical mechanics is also compatible with the new theory.

Let

$$H = E_{\text{kin}} + E_{\text{pot}} = \tfrac{1}{2} \sum_k \frac{p_k^2}{m_k} + E_{\text{pot}}, \tag{15}$$

and let E_{pot} be a homogeneous function of the coordinates of order n. Then from (3), ch. 2,

$$E_{\text{pot}} = \frac{1}{n} \sum_k \frac{\partial E_{\text{pot}}}{\partial q_k} q_k \tag{16}$$

and

$$\frac{d}{dt} \sum_k p_k q_k = \sum_k (\dot{p}_k q_k + p_k \dot{q}_k) = 2E_{\text{kin}} - nE_{\text{pot}},$$

so that for the *mean* values,

$$\bar{E}_{\text{kin}} = \tfrac{1}{2} n \bar{E}_{\text{pot}}. \tag{17}$$

Hence, e.g., for $n=2$ (harmonic oscillations), $\bar{E}_{\text{kin}}=\bar{E}_{\text{pot}}$ and for $n=-1$ (Coulomb force), $\bar{E}_{\text{kin}}=-\tfrac{1}{2}\bar{E}_{\text{pot}}$.

2. Degenerate systems

We now turn to examination of degenerate systems. If we permit some of the frequencies $\nu(nm)$ to vanish (for simplicity, we imagine the matrices to be in two-dimensional representation), then *energy conservation*, $\dot{H}=0$ can still be derived from the considerations employed here and in Part I concerning the equations of motion and the commutation rules (3), ch. 2. But the relation $\dot{H}=0$ no longer necessarily implies that H be a diagonal matrix and in consequence the proof of the frequency condition cannot be carried through. Thus for degenerate systems the equations of motion together with (3), ch. 2, do *not* alone suffice for the unique determination of the properties of a system: we need to strengthen these basic equations. An obvious assumption as to the form of this 'increase in rigour' is:

For basic equations, one should be able generally to choose the commutation relations and the property

$$H = W = \text{diagonal matrix}. \tag{18}$$

This requirement manifestly ensures the validity of the frequency condition for degenerate systems as well. Very probably, the energy W is also thereby uniquely determined (apart from singular instances). On the other hand, the *coordinates* q_k are *not* uniquely determined. Given a solution p_k, q_k of $H(pq)=W$, we can get new solutions from

$$\begin{aligned} p' &= SpS^{-1}, \\ q' &= SqS^{-1}. \end{aligned} \tag{19}$$

Thence

$$H(p'q') = W' = SWS^{-1},$$

and the requirement $W'=W$ yields

$$WS - SW = \dot{S}\frac{h}{2\pi i} = 0,$$

and thus

$$S = \text{const.} \tag{20}$$

Let us at this stage examine this result as regards its implications for nondegenerate systems. From (2), ch. 2, the matrix S has to become a diagonal matrix, and the eqs. (19), ch. 2, imply that

$$\begin{aligned} p'(nm) &= p(nm)S_n S_m^{-1}, \\ q'(nm) &= q(nm)S_n S_m^{-1}, \end{aligned} \tag{19'}$$

writing S_n for $S(nn)$ for the sake of conciseness.

The uncertainty in the solution indicated hereby can significantly be reduced by the requirement that the new solution p', q' should also represent '*real*' motion, expressed in terms of Hermitian matrices, since this yields

$$|S_n S_n^{-1}| = |S_m S_m^{-1}|,$$

or

$$|S_n| = |S_m|. \tag{21}$$

Thus the indeterminacy which has here come to light represents an arbitrariness of the *phase constants*. We namely here find proof of the contention put forward in Part I that in each problem for every state n a phase φ_n always remains undetermined. From (19′) one can perceive the manner in which these phases enter into the elements of the matrices p, q. It was further conjectured in Part I that apart from the above-mentioned arbitrariness of phase for non-degenerate systems, no additional non-uniqueness is to be expected. It is clear that we could still add a constant matrix to each of the 'periodic' matrices S_n in the perturbation calculations of ch. 1 § 4. However, this obviously does not imply that new phases which remain undetermined enter into each approximation. It is easy to see that utilization of this possibility cannot provide any more general solution p, q provided that p^0, q^0 were right from the first taken to have undetermined phases.

If we now go over to degenerate systems, we cannot any longer infer from (20) that S is a diagonal matrix, and accordingly, using (19), we do indeed have the possibility of deriving solutions p', q' which are significantly different from p, q. This indeterminacy seems to lie in the very nature of things. Apparently, degenerate systems possess a *lability* by virtue of which arbitrarily small perturbations can bring about finite changes in coordinates, and this finds its mathematical expression in that in complete absence of perturbations, the solution of the dynamic equations remains partly indeterminate. Naturally, for every actual atom the coordinates which specify the physical properties of the system, in particular the transition probabilities, are always fixed uniquely either by external perturbations or by the previous history of the system.

Now we set out to examine the influence of arbitrary perturbations upon the degenerate system. We set

$$H(\mathsf{pq}) = H_0 + \lambda H_1 + \lambda^2 H_2 + \ldots, \tag{22}$$

and let p^0, q^0 be an arbitrary, but definite, solution of the unperturbed problem:

$$H_0(p^0 q^0) = W_0. \tag{23}$$

Then with

$$p = S p^0 S^{-1},$$
$$q = S q^0 S^{-1},$$

and with

$$S = S_0(1 + \lambda S_1 + \lambda^2 S_2 + \ldots), \tag{24}$$
$$S^{-1} = (1 - \lambda(S_1 + \lambda S_2 \ldots) + \lambda^2 \ldots) S_0^{-1}, \tag{25}$$

we find, on leaving out the arguments p^0, q^0 from $H_0, H_1, \ldots$:

$$S_0 H_0 S_0^{-1} = W_0, \tag{26}$$
$$S_0 S_1 H_0 S_0^{-1} - S_0 H_0 S_1 S_0^{-1} + S_0 H_1 S_0^{-1} = W_1, \tag{27}$$
$$S_0 S_2 H_0 S_0^{-1} - S_0 H_0 S_2 S_0^{-1} + S_0 F_2(H_0 H_1 H_2; S_1) S_0^{-1} = W_2, \tag{28}$$
$$\cdots\cdots\cdots\cdots\cdots\cdots\cdots\cdots$$

$$\begin{aligned} S_0 S_r H_0 S_0^{-1} &- S_0 H_0 S_r S_0^{-1} \\ + S_0 F_r(H_0 H_1 \ldots H_r, S_1 \ldots S_{r-1}) S_0^{-1} &= W_r. \end{aligned} \tag{29}$$

Thus we almost repeat eqs. (26), ch. 1, but with the difference that the left-hand sides are throughout multiplied on the left by S_0 and on the right by S_0^{-1}.

Equation (26), ch. 2, has already been cited above; $S_0(nm)$ becomes zero except for vanishing $\nu_0(nm)$. The remaining arbitrariness in S_0 now has to be used to advantage so far as possible in order to render the next equation soluble. Naturally, one cannot expect that every solution of $H = H_0$, and thus in particular the chosen solution p^0, q^0, will provide the limiting case $\lambda = 0$ of the solution p, q of the problem (22), ch. 2. The function S_0 should serve to obtain from p^0, q^0 that solution of the degenerate problem which possesses this desired property.

We can rewrite eq. (27) as

$$S_1 H_0 - H_0 S_1 + H_1 = S_0^{-1} W_1 S_0. \tag{30}$$

To make this soluble, one has to determine S_0 such that

$$\overline{H}_1 = S_0^{-1} W_1 S_0 \tag{31}$$

for a *diagonal matrix* W_1. An indication as to how one can simultaneously satisfy this eq. (31) and the requirements dictated by (26), ch. 2, can here naturally just as little be given as that for the determination of secular perturbations in classical theory. We shall, however, later use a new algebraic method to arrive at a simple treatment of an extensive class of degeneracies (ch. 3).

If (31), ch. 2, is satisfied, (30), ch. 2, can be solved as in ch. 1. Thereby those terms $S_1(nm)$ of S_1 for which $\nu_0(nm)$ vanishes remain arbitrary, and this indeterminacy has to be utilized in order to solve the next higher order approximation formula, which can be transcribed as

$$S_2H_0 - H_0S_2 + F_2 = S_0^{-1}W_2S_0 \tag{32}$$

in order to fulfil the necessary relation

$$\overline{F_2(H_0, H_1, H_2; S_1)} = S_0^{-1}W_2S_0 \tag{31'}$$

with W_2 a diagonal matrix. This has to be satisfied for the problem to be soluble. The continuation of the procedure is clear.

The difficulty lies in the fact that at each order of approximation equations have to be satisfied by matrices which are already fixed to a large extent, so that it is not perceptible whether or not these equations will really prove soluble. In classical theory there is, though, an altogether analogous difficulty. These difficulties can, at least in the higher orders of approximation, be removed if in some approximation the system becomes nondegenerate.

Suppose, for example, that $p^{(1)}$ and $q^{(1)}$ in

$$q = q^0 + \lambda q^{(1)} + \ldots,$$
$$p = p^0 + \lambda p^{(1)} + \ldots$$

have really been determined, so that with

$$Q = q_0 + \lambda q^{(1)}$$
$$P = p^0 + \lambda p^{(1)}$$

one has

$$H(PQ) = W_0 + \lambda W_1 + \lambda^2 H_2' + \lambda^3 H_3' + \ldots,$$

and suppose

$$\nu_0(nm) + \lambda\nu_1(nm) \neq 0 \quad \text{for} \quad n \neq m.$$

If for brevity we write H_0' for $W_0+\lambda W_1$ and set

$$p = SPS^{-1}$$
$$q = SQS^{-1},$$

then we have to build the following relation,

$$S(H_0' + \lambda^2 H_2' + \lambda^3 H_3' + \ldots)S^{-1} = W,$$

which, with the procedures of ch. 1, can be achieved with

$$S = \mathbf{1} + \lambda^2 S_2 + \lambda^3 S_3 + \ldots.$$

The generalization of these considerations for the case in which only in the rth approximation can one attain a nondegenerate system $W = W_0 + \lambda W_1 + \ldots + \lambda^r W_r$ follows of itself.[1]

In conclusion, we deem it important to point out that the notorious convergence difficulties encountered in the classical perturbation series, which play so decisive a rôle in the discussion of the three-body problem, do not arise here in quantum-mechanical perturbation theory; rather, one would here in general expect finite orbits to be periodic also.

CHAPTER 3. CONNECTION WITH THE THEORY OF EIGENVALUES OF HERMITIAN FORMS

1. General method

The treatment in the preceding sections has aimed at solving the basic quantum-theoretical equations in a manner as closely parallel to classical theory as possible. But behind the formalism of this perturbation theory there lurks a very simple, purely algebraic connection and it is well worth while to bring this into the limelight. Apart from the deeper insight into the mathematical structure of the theory, we thereby gain the advantage of being able to use the methods and results developed earlier in mathematics. We shall thus arrive at a new definition of the energy constants ('terms') which remains valid in the case of aperiodic motion also, i.e., of continuously-varying indices. Thereby we attain the prospect of finding methods for direct

[1] Analogous cases in classical mechanics have been discussed by M. Born and W. Heisenberg, Ann. d. Phys. **74** (1924) 1.

calculation of the energy without explicitly solving the problem of motion: methods which correspond to Sommerfeld's method of complex integration. We shall then be able to treat perturbations of an extensive class of degenerate systems completely, which the above-mentioned perturbation methods were not yet able to handle.

In considering a problem of f degrees of freedom specified by the energy function $H(pq)$, we can first select any system of matrices p_k^0, q_k^0 whatsoever such that at all events the commutation relations (3), ch. 2, are satisfied: for example, we can take the p_k, q_k for a system of noncoupled harmonic oscillators.

Then, as mentioned in ch. 2 § 1, the dynamic problem, e.g., the determination of the p_k, q_k can be formulated as: A transformation $(p_k^0 q_k^0) \to (p_k q_k)$ is to be found which leaves eqs. (3), ch. 2, invariant and at the same time reduces the energy to a diagonal matrix.

The transformation of matrices can most easily be grasped if one regards them as a system of coefficients for linear transformations or bilinear forms. We therefore premise some known results of the algebra of such forms.

To every matrix $a = (a(nm))$ there belongs a *bilinear form*

$$A(xy) = \sum_{nm} a(nm) x_n y_m \tag{1}$$

of two series of variables $x_1, x_2, \ldots$ and $y_1, y_2 \ldots$. If the matrix be Hermitian, i.e., if the *transposed matrix* $\tilde{a} = (a(mn))$ be equal to the complex conjugate of the original matrix,

$$\tilde{a} = a^*, \qquad a(mn) = a^*(nm), \tag{2}$$

then the form A assumes real values if in place of the variables y_n one substitutes the complex conjugate values x_n:

$$A(xx^*) = \sum_{nm} a(nm) x_n x_m^*. \tag{1a}$$

We recall the readily demonstrable transposition rule

$$(\widetilde{ab}) = \tilde{b}\,\tilde{a} \tag{3}$$

and now subject the x_n to a linear transformation

$$x_n = \sum_l v(ln) y_l \tag{4}$$

with the aid of the (complex) matrix $v = (v(ln))$.

Then the form A goes over into

$$A(xx^*) = B(yy^*) = \sum_{nm} b(nm) y_n y_m^*, \tag{5}$$

with

$$b(nm) = \sum_{kl} v(nk) a(kl) v^*(ml),$$

or, in matrix notation,

$$\mathsf{b} = \mathsf{vav}^*. \tag{6}$$

This is termed the generation of a matrix b by the transformation v applied to a.

The matrix b is again of Hermitian type, for, with (3), ch. 3,

$$\tilde{\mathsf{b}} = \mathsf{v}^* \tilde{\mathsf{a}} \tilde{\mathsf{v}} = \mathsf{v}^* \mathsf{a}^* \tilde{\mathsf{v}} = \mathsf{b}^*. \tag{7}$$

The matrix v is called *orthogonal* if the respective transformation leaves the Hermitian unit form

$$E(xx^*) = \sum_n x_n x_n^*$$

invariant; from the result derived above, this is the case if and only if

$$\mathsf{v}\tilde{\mathsf{v}}^* = \mathbf{1}, \quad \text{or} \quad \tilde{\mathsf{v}}^* = \mathsf{v}^{-1}. \tag{8}$$

Thus, for instance, the permutation matrices mentioned in ch. 1 § 2 are real orthogonal matrices.

As is known, it is always possible for a finite number of variables to effect an orthogonal transformation of a form into a sum of squares (transformation to principal axes).[1]

$$A(xx^*) = \sum_n W_n y_n y_n^*. \tag{9}$$

For matrices, this means: a matrix exists for which

$$\mathsf{v}\tilde{\mathsf{v}}^* = \mathbf{1} \quad \text{and} \quad \mathsf{va}\tilde{\mathsf{v}}^* = \mathsf{vav}^{-1} = \mathsf{W}, \tag{10}$$

where $\mathsf{W} = (W_n \delta_{nm})$ is a diagonal matrix.

For infinite matrices, all the cases investigated so far have been found to obey an analogous rule; it can however occur that the index n on the right-hand side runs not only through a set of discrete numbers

[1] We write the coefficients of the transformed form W_n because in quantum mechanics they stand for the 'energy'.

but also through a continuous range of values; this would correspond[1] to an integral constituent of (9) and the transformation (4).

The quantities W_n are termed 'eigenvalues', their ensemble is the 'mathematical spectrum' of the form, made up of 'point-' and 'continuous' spectrum. As we shall see, this is identical with the 'term-spectrum' in physics, whereas the 'frequency spectrum' is obtained from this by forming differences.

This transformation to principal axes now directly presents us with the solution of our dynamic problem which consists in seeking a transformation $(p^0q^0) \to (pq)$ such that the eqs. (3), ch. 2 are left invariant and at the same time the energy is brought into diagonal matrix form.

By the above rules of algebra, there exists an orthogonal matrix S for which

$$S\tilde{S}^* = \mathbf{1}, \qquad \tilde{S}^*S = \mathbf{1} \tag{11}$$

and for which the transformations

$$\begin{aligned} p_k &= Sp_k^0\tilde{S}^* = Sp_k^0S^{-1}, \\ q_k &= Sq_k^0\tilde{S}^* = Sq_k^0S^{-1} \end{aligned} \tag{12}$$

leave

(i) the Hermitian character of p_k^0, q_k^0 conserved also for the p_k, q_k;

(ii) the eqs. (3), ch. 2, invariant;

(iii) the energy

$$H(pq) = SH(p^0q^0)S^{-1} = W \tag{13}$$

converted into diagonal matrix form.

We wish to discuss the question of the uniqueness of this solution and in particular whether one could not generate other energy values through another orthogonal transformation T. Let us assume that W', as given by

$$TH(p^0q^0)T^{-1} = W',$$

is a diagonal matrix which differs from W. One would then have

$$TS^{-1}SHS^{-1}ST^{-1} = TS^{-1}W(TS^{-1})^{-1},$$

[1] Up till now, the theory of quadratic (or Hermitian) forms of infinitely many variables has been developed mainly for a special class ('bounded' forms) (D. Hilbert, Grundzüge einer allgemeinen Theorie der linearen Integralgleichungen; E. Hellinger, Crelles Journ. **136** (1910) 1). But here we are concerned just with non-bounded forms. We may nevertheless assume that in the main the rules run likewise.

and our question is equivalent to asking whether it is possible, starting from a diagonal matrix W to build another, W', through te transformation

$$W' = MWM^{-1}, \qquad M\tilde{M}^* = 1 \tag{14}$$

such that W' can *not* be derived from W by a permutation of the diagonal elements.

However, eq. (14), ch. 3, can be written

$$W'M - MW = 0.$$

and thus implies

$$M(nm)(W'_n - W_m) = 0. \tag{14a}$$

From the orthogonality of M, it follows in particular for $m=n$ that

$$\sum_k |M(nk)|^2 = 1, \qquad \sum_k |M(kn)|^2 = 1;$$

and consequently for a fixed n neither all the $M(nk)$ nor all the $M(kn)$ can vanish. But then (14a), ch. 3, shows that for every n there is certainly an m for which $W'_n=W_m$, i.e., all the W'_n appear among the W_m. The same holds inversely.

Thus all solutions derived from (12), ch. 3, lead (for given p^0_k, q^0_k) to the same values for the energies of the stationary states, in accord with the conjecture stated in ch. 2 that the energies are always uniquely determined by the fundamental dynamic equations.

Degenerate systems will be characterized by the fact that multiple eigenvalues occur. The multiplicity of the eigenvalue W_n, i.e., the number of linear independent solutions $v(ln)$ of eq. (4), ch. 3, yields the statistical weight of the respective state.

The importance of eq. (9), ch. 3, for our physical theory lies in the fact that various methods[1] exist in the algebra of finite or bounded infinite forms for determining the eigenvalues of a form without actually carrying the transformation through. It is to be hoped that such methods will prove of much avail in the future treatment of certain physical systems.

[1] For finite forms, the eigenvalues are the roots of an algebraic equation. Here, and also for bounded infinite matrices, they can be determined, e.g., by the method of Graeffe and Bernoulli; see, for example, R. Courant and D. Hilbert, Methoden der mathematischen Physik **1** (Springer, Berlin, 1924) § 3, pp. 14, 15.

2. Application to perturbation theory

In the following, we show that our present algebraic conception of the dynamic problem not only leads to exactly those formulae which were previously derived in ch. 1 § 4 in connection with perturbation theory in classical mechanics, but that when applied to degenerate systems it is considerably superior to the theory used hitherto.

We thus again assume that H has the form

$$H = H_0 + \lambda H_1 + \lambda^2 H_2 + \ldots,$$

and that the dynamic problem specified by H_0 has the solution p_k^0, q_k^0. We take these quantities as our starting coordinates from which the p_k, q_k are to be found, using an orthogonal transformation S. Naturally, the form assumed for H does not basically represent any limitation in generality, inasmuch as one can obviously separate off from H a component H_0 of any desired form; however, the convergence of the power series in λ will depend essentially upon an apposite choice of H_0.

To undertake a principal-axes transformation of the Hermitian form

$$\sum_{mn} H_{mn} x_m x_n^*$$

we can, as is known, proceed as follows:
We attempt to find a solution of the linear equations

$$W x_k - \sum_l H(kl) x_l = 0; \tag{15}$$

this is possible only for certain values of the parameter W, namely $W = W_n$, when W_n again denotes the eigenvalues (energy values). We first assume that no degeneracy is present, so that all W_n are different. Then to each W_n there corresponds a solution $x_k = x_{kn}$ (determined except for a multiplicative factor), and hence the identities

$$W_n x_{kn} - \sum_l H(kl) x_{ln} = 0,$$

$$W_m x_{km}^* - \sum_l H^*(kl) x_{lm}^* = 0$$

obtain. On multiplying the former by x_{km}^*, the latter by x_{kn} and summing over k, it follows on subtraction (because of the Hermitian character of H) that

$$(W_n - W_m) \sum_k x_{kn} x_{km}^* = 0.$$

By choosing the proportionality factor suitably, one can normalize to

$$\sum_k x_{kn}x_{kn}^* = 1.$$

Hence the x_{kn} form an orthogonal matrix

$$S = (x_{kn}).$$

It is precisely this which transforms the given form to a sum of squares, since if we substitute

$$x_k = \sum_n x_{kn}y_n$$

into the form, we obtain

$$\begin{aligned}\sum_{kl} H(kl)x_k x_l^* &= \sum_{kl}\sum_{mn} H(kl)x_{km}x_{ln}^*y_m y_n^* \\ &= \sum_{mn}\sum_{l} W_m x_{lm}x_{ln}^*y_m y_n^* \\ &= \sum_m W_m y_m y_m^*.\end{aligned}$$

From our assumption as to the form of H, the coefficients of eq. (15), ch. 3, now have the form

$$H(kl) = \delta_{kl}W_l^0 + \lambda H_1(kl) + \lambda^2 H_2(kl) + \ldots.$$

We thus seek to find the solution of (15), ch. 3, through expansions of the form

$$\begin{aligned}W &= W^0 + \lambda W^{(1)} + \lambda^2 W^{(2)} + \ldots \\ x_k &= x_k^0 + \lambda x_k^{(1)} + \lambda^2 x_k^{(2)} + \ldots.\end{aligned} \tag{16}$$

If we substitute the above in (15), ch. 3, we obtain the approximation equations

$$\begin{aligned}&\text{(a)}\quad x_k^0(W^0 - W_k^0) = 0, \\ &\text{(b)}\quad x_k^{(1)}(W^0 - W_k^0) = -x_k^0 W^{(1)} + \sum_l H^{(1)}(kl)x_l^0, \\ &\text{(c)}\quad x_k^{(2)}(W^0 - W_k^0) = -(x_k^{(1)}W^{(1)} + x_k^{(0)}W^{(2)}) \\ &\qquad\qquad + \sum_l (H^{(1)}(kl)x_l^{(1)} + H^{(2)}(kl)x_l^0).\end{aligned} \tag{17}$$

It follows from (17a), ch. 3, that W has to become equal to one of the W_k, since otherwise all x_k^0 would vanish and we could then also infer the vanishing of $x_k^{(1)}, x_k^{(2)}, \ldots$ in sequence from the subsequent approximation equations.

If, then, we take our starting system as nondegenerate, and thus all the W_k^0 as different from one another, the solution of (17a), ch. 3, is

$$W = W_n^0; \qquad x_{nn}^0 = y_n^0; \qquad x_{kn}^0 = 0 \qquad \text{for} \qquad k \neq n. \tag{18}$$

Herein, y_n^0 is an arbitrary number.

If we substitute this in (17b), ch. 3, we find, depending upon whether $k=n$ or $k \neq n$,

$$0 = y_n^0(-W^{(1)} + H^{(1)}(nn)),$$
$$x_k^{(1)}(W_n^0 - W_k^0) = H^{(1)}(kn)y_n^0, \qquad k \neq n.$$

Thus the solution runs

$$\begin{aligned} W^{(1)} &= H^{(1)}(nn); \qquad x_{nn}^{(1)} = y_n^{(1)}; \\ x_{kn}^{(1)} &= -\frac{H^{(1)}(kn)}{h\nu_0(kn)}\, y_n^0 \qquad \text{for} \qquad k \neq n, \end{aligned} \tag{19}$$

where again $y_n^{(1)}$ is an arbitrary number.

Hence it similarly follows from (17c), ch. 3, that

$$\begin{aligned} W^{(2)} &= H^{(2)}(nn) - \frac{1}{h}\sum_l{}' \frac{H^{(1)}(nl)H^{(1)}(ln)}{\nu_0(ln)}, \\ x_{nn}^{(2)} &= y_n^{(2)} \\ x_{kn}^{(2)} &= \left(\frac{1}{h^2}\sum_l{}' \frac{H^{(1)}(kl)H^{(1)}(ln)}{\nu_0(kn)\nu_0(ln)} - \frac{H^{(1)}(nn)H^{(1)}(kn)}{h^2\nu_0(kn)^2} \right. \\ &\qquad \left. - \frac{H^{(2)}(kn)}{h\nu_0(kn)}\right) y_n^0 - \frac{H^{(1)}(kn)}{h\nu_0(kn)}\, y_n^{(1)}. \end{aligned} \tag{20}$$

The solution of the third-order approximation can be derived just as easily; we cite only the energy value:

$$W^{(3)} = H^{(3)}(nn) - \frac{1}{h}\sum_l{}' \frac{H^{(1)}(nl)H^{(2)}(ln) + H^{(2)}(nl)H^{(1)}(ln)}{\nu_0(ln)}$$
$$+ \frac{1}{h^2}\left(\sum_{kl}{}' \frac{H^{(1)}(nl)H^{(1)}(lk)H^{(1)}(kn)}{\nu_0(ln)\nu_0(kn)} - H^{(1)}(nn)\sum_l{}' \frac{H^{(1)}(nl)H^{(1)}(ln)}{\nu_0(ln)^2}\right).$$

The quantities $y_n^{(0)}, y_n^{(1)}, \ldots$, which for the present are arbitrary, serve to normalize the solution (it is orthogonal of itself); the condition

$$\sum_k x_{kn} x^*_{kn} = 1$$

yields, for

$$x_{kn} = x^0_{kn} + \lambda x^{(1)}_{kn} + \lambda^2 x^{(2)}_{kn} + \dots,$$

the equations

$$\sum_k x^0_{kn} x^{*0}_{kn} = 1$$

$$\sum_k (x^0_{kn} x^{*(1)}_{kn} + x^{(1)}_{kn} x^{*0}_{kn}) = 0$$

$$\dots\dots\dots\dots\dots$$

On substituting the solution just obtained, it follows successively that

$$|y^0_n|^2 = 1$$

$$y^0_n y^{*(1)}_n + y^{*0}_n y^{(1)}_n = 0$$

$$\dots\dots\dots\dots$$

If we now set

$$y^{(p)}_n = a^{(p)}_n e^{i\varphi_n(p)}, \tag{21}$$

we obtain

$$a^0_n = 1$$

$$2a^{(1)}_n \cos(\varphi^0_n - \varphi^{(1)}_n) = 0$$

$$\dots\dots\dots\dots\dots\dots\dots\dots$$

$$2a^{(r)}_n \cos(\varphi^0_n - \varphi^{(r)}_n) = F^{(r)}(a^{(r-1)}, \varphi^{(r-1)}, \dots).$$

Thus the phase constants $\varphi^0_n, \varphi^{(1)}_n, \dots$ can be chosen arbitrarily; the $a^0_n, a^{(1)}_n, \dots$ can be evaluated in sequence and determined uniquely. This stands in agreement with the result we found earlier (§ 3), namely that the phases of the diagonal terms of S remain undetermined.

On substituting the values $a^0_n = 1, \dots$ obtained above into (21), ch. 3, and this in turn into (18), (19), (20), ch. 3, we see that the 'perturbation procedure' carried through earlier yielded just the solution for which the phases $\varphi^{(p)}_n$ vanish, i.e., for which the diagonal terms of S are real.

We now turn to consideration of the case in which the starting system is degenerate and in which W^0_n is an r-fold eigenvalue. This means that eq. (17a), ch. 3, has the solution

$$W = W_n; \qquad x^0_{nn} = y^0_{1,n}, \qquad x^0_{n,n+1} = y_{2,n} \qquad \dots$$

$$x_{n,n+r-1} = y_{r,n}, \tag{23}$$

$$x^0_{kn} = 0 \qquad \text{for} \qquad k \neq n, n+1, \dots, n+r-1.$$

The left-hand side of (17b), ch. 3, then vanishes for

$$k = n, n + 1, \ldots, n + r - 1;$$

this yields (r) equations:

$$W^{(1)} y^0_{kn} - \sum_{l=1}^{r} H^{(1)}(n + k, n + l) y^0_{ln} = 0; \qquad k = 1, 2, \ldots, r, \tag{24}$$

whose array of coefficients is again of Hermitian type.

On setting the determinant to zero, one obtains a secular equation of the rth order for $W^{(1)}$:

$$\det \left(W^{(1)} \delta_{kl} - H^{(1)}(n + k, n + l)\right) = 0, \tag{25}$$

whose roots are certainly real. To each root there belong one or more independent solutions of eqs. (24), ch. 3.

If one selects one of these solutions, the perturbation procedure can be pursued: we shall, however, not go into this further here.

It suffices to have recognized that our algebraic method is able to handle all degeneracies of finite multiplicity, i.e., that it can reduce the problem to the solution of algebraic equations. If, for example, each eigenvalue occurs twice, so that to each there belongs a vanishing frequency $\nu_0(nm)$, the perturbation problem leads to a quadratic equation:

$$\begin{vmatrix} W^{(1)} - H^{(1)}(n, n) & -H^{(1)}(n, n + 1) \\ -H^{(1)}(n + 1, n) & W^{(1)} - H^{(1)}(n + 1, n + 1) \end{vmatrix} = 0.$$

This case obtains when two originally identical nondegenerate systems (in which all frequencies in each of the respective systems are to be different) are coupled through some force.

Further, the orthogonality relation

$$\sum_k x^0_{kn} x^{*0}_{kn} = 1$$

has an interesting meaning in the case of degenerate systems. Because of (23), this relation goes over into

$$\sum_{l=1}^{r} y^0_{ln} y^{*0}_{ln} = 1.$$

From this it follows that, if m denotes any number in the series $n, n+1, \ldots, n+r-1$, and k denotes any number outside this set, the sums

$$\sum_{m=n}^{n+r-1} p^0(mk)p^{*0}(mk),$$

$$\sum_{m=n}^{n+r-1} q^0(mk)q^{*0}(mk)$$

are uniquely determined, even for degenerate systems, e.g., the summations are invariant with respect to those transformations which, by (19), ch. 2, allow new and altogether different solutions p', q' to arise from certain solutions p, q in the case of degeneracy. This result provides a mathematical representation of the so-called spectroscopic stability, which has played an important part in the more recent theories of fine-structure intensities (cf. ch. 4).

3. Continuous spectra

The simultaneous appearance of both continuous and line spectra as solutions of the same equations of motion and the same commutation relations seemed to us to represent a particularly significant feature of the new theory. In spite of this close connection between the two kinds of spectra, there nevertheless are characteristic distinctions, both mathematically and physically, between continuous and discrete spectra, corresponding to the difference between Fourier series and Fourier integrals in classical theory; it therefore strikes us as desirable to indicate the rough outlines of the treatment of continuous spectra here. The mathematical theory of continuous spectra which occur for infinite quadratic forms has, starting from the fundamental investigations of Hilbert, explicitly been developed by Hellinger (*loc.cit.*) for the case of bounded quadratic forms. If we here permit ourselves to take over Hellinger's results to the unbounded forms which appear in our case, we feel ourselves to be justified by the fact that Hellinger's methods obviously conform exactly to the physical content of the problem posed.

Let us first briefly examine the classical analogue to our problem, namely aperiodic motion and its Fourier integral. Whereas in a Fourier series a certain amplitude $a(\nu)$ always belongs to an oscillation $\exp(2\pi i\nu t)$, in the case of a Fourier integral one has a quantity of the form $\varphi(\nu)d\nu$ in place of $a(\nu)$, where $\varphi(\nu)$ might in a sense be conceived as an amplitude-density per frequency interval $d\nu$. In a similar and physically immediately obvious manner, one can always relate all quantities such as intensity, polarization, etc. to a frequency interval $d\nu$ between ν and $\nu+d\nu$, but never to a definite frequency itself. We shall have to expect quite similar conditions to apply in quantum mechanics. Instead of quantities $q(kl)$ we shall have quantities of the form $q(k, W)dW$ or $q(W, W')dWdW'$, depending upon whether one or both of the two indices lie in the continuous region. Indeed, in place of the energy W itself, there will have to be a 'total energy' per interval dW, since the probability for an atom to have an absolutely definite energy W in the continuous region is zero. To elucidate these questions we shall in the following briefly sketch Hellinger's mathematical theory.

For infinite quadratic forms, the case may arise that the form

$$\sum_{mn} H(mn)x_m x_n^*$$

cannot be converted into the expression $\sum_n W_n y_n y_n^*$ by an orthogonal substitution. We may then assume, in analogy with the results for bounded forms, that a representation with a continuous spectrum exists,

$$\sum_{mn} H(mn)x_m x_n^* = \sum_n W_n y_n y_n^* + \int W(\varphi)y(\varphi)y^*(\varphi)d\varphi, \tag{26}$$

in which the original variables are connected with new variables y_n, $y(\varphi)$ through an 'orthogonal transformation'; one only has to specify more clearly what is here understood by an orthogonal transformation.

If we again consider the linear equations (15), ch. 3,

$$Wx_k - \sum_l H(kl)x_l = 0, \tag{27}$$

the case under review in which (26), ch. 3, contains an integral component will occur when there are not only discrete values W_n, for which these equations can be solved, but also a continuum of such values comprising one or more 'segments' on the W-axis (continuous spectrum). For any given point W of this continuum, there exists a solution $x_l(W)$ (or several, which we for simplicity wish to exclude); for two such W-values, W' and W'', the equations

$$\begin{aligned} W'x_k(W') - \sum_l H(kl)x_l(W') &= 0, \\ W''x_k^*(W'') - \sum H^*(kl)x_l^*(W'') &= 0 \end{aligned} \tag{28}$$

obtain, from which, as above, we conclude that

$$(W' - W'') \sum_k x_k(W')x_k(W'') = 0. \tag{29}$$

If one tries imposing the normalization condition

$$\sum_k |x_k(W)|^2 = 1$$

on top of these orthogonality relations, one observes that the function of two variables

$$\sum_k x_k(W')x_k(W'')$$

becomes wildly irregular, if it exists at all. The above sum does not in fact converge and therefore does not represent a function.

Accordingly, a different type of normalization is required. With Hellinger, we set

$$\sum_k |\int x_k(W)\mathrm{d}W|^2 = \varphi(W). \tag{30}$$

The series on the left-hand side is in general convergent and represents a monotonous function $\varphi(W)$, which apart from certain restrictions can be chosen arbitrarily, since the $x_k(W)$ are of course determined only up to a factor which is independent of k. We shall later discuss the physical significance of this function $\varphi(W)$, by which the solutions $x_k(W)$ are defined. Hellinger has termed $\varphi(W)$ the 'basis function' and has shown that the orthogonality conditions can be derived in the following form: If Δ_1 and Δ_2 be any two intervals of the continuous spectrum and Δ_{12} the interval common to them both (which may also be absent), then

$$\begin{aligned}\sum_k \int_{\Delta_1} x_k(W')\,\mathrm{d}W' \int_{\Delta_2} x_k(W'')\,\mathrm{d}W'' &= \int_{\Delta_{12}} \mathrm{d}\varphi(W) \\ &= \varphi(W^{(2)}) - \varphi(W^{(1)}),\end{aligned} \tag{31}$$

where $W^{(1)}$, $W^{(2)}$ are the end-points of Δ_{12}. Hence if there is no overlap between the intervals Δ_1, Δ_2, a zero stands on the right-hand side.

If one conceives the intervals Δ_1, Δ_2, Δ_{12} to be very small, one can symbolically write

$$\sum_k x_k(W')\,\mathrm{d}W' \cdot x_k(W'')\,\mathrm{d}W'' = \mathrm{d}\varphi(W). \tag{32}$$

This relation prompts the suggestion that one operate generally with the quantities $x_k(W)\,\mathrm{d}W$ as 'differential solutions' of (27), ch. 3, whereby one has to note that the respective equations are always to be interpreted in the sense of (31), ch. 3. These differential solutions are orthogonal in the usual way, but instead of being normalized to unity, are normalized to the differential of the basis function $\varphi(W)$.

The totality of discrete values x_{kn}, and of values $x_k(W)$ which are discrete in one index and have a continuous distribution in the other, comprises the elements of the 'orthogonal' matrix

$$S = (x_{kn}, x_k(W)\,\mathrm{d}W),$$

which can schematically be represented as:

$S =$ 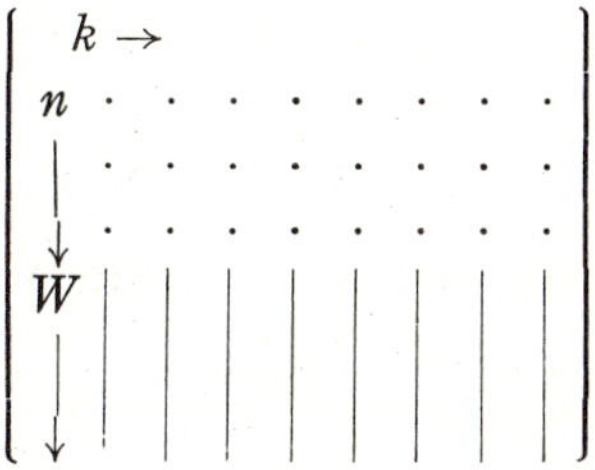

. (33)

The orthogonality and normalization equations for the entire matrix split into four different groups:

$$\sum_k x_{km}x^*_{kn} = \delta_{mn};$$

$$\sum_k x_{kn}x^*_k(W)\,\mathrm{d}W = 0; \qquad \sum_k x_k(W)\,\mathrm{d}W\cdot x^*_{kn} = 0; \tag{34}$$

$$\sum_k x_k(W')\,\mathrm{d}W'\cdot x^*_k(W'')\,\mathrm{d}W'' = \mathrm{d}\varphi.$$

We can also write the orthogonality relations for the columns, which read

$$\sum_n x_{kn}x^*_{ln} + \int \frac{x_k(W)\,\mathrm{d}W\cdot x^*_l(W)\,\mathrm{d}W}{\mathrm{d}\varphi}$$

$$= \sum_n x_{kn}x^*_{ln} + \int \frac{\mathrm{d}W}{\varphi'}\,x_k(W)x^*_l(W) = \delta_{kl}, \tag{35}$$

where the prime denotes differentiation, $\varphi' = \mathrm{d}\varphi/\mathrm{d}W$.

With the aid of this matrix, we have to transform the variables x_k into new ones, y_n, $y(\varphi)\,\mathrm{d}\varphi$. We set:

$$y_n = \sum_k x_{kn}\cdot x_k,$$
$$y(\varphi)\,\mathrm{d}\varphi = \sum_k x_k(W)\,\mathrm{d}W\cdot x_k. \tag{36}$$

A simple calculation then yields

$$\sum_n W_n y_n y_n^* + \int W(\varphi) y(\varphi) y^*(\varphi) \, \mathrm{d}\varphi = \sum_{kl} H(kl) x_k x_l^*. \tag{37}$$

The principal-axes transformation has thereby been carried through.

Let us now investigate which representation of coordinate and momentum matrices is obtained with the aid of this orthogonal transformation, e.g., what is meant here by the equations

$$\begin{aligned} p &= S p_0 S^{-1}, \\ q &= S q_0 S^{-1}, \end{aligned} \tag{38}$$

or, generally, by

$$f(pq) = S f(p_0 q_0) S^{-1}. \tag{39}$$

We find, for example, four types of elements for p:

$$\begin{aligned} p(mn) &= \sum_{kl} x_{km}^* p^0(kl) x_{ln} \\ p(m, W) \, \mathrm{d}W &= \sum_{kl} x_{km}^* p^0(kl) x_l(W) \, \mathrm{d}W, \\ p(W, n) \, \mathrm{d}W &= \sum_{kl} x_k^*(W) \, \mathrm{d}W \cdot p^0(kl) x_{ln}, \\ p(W', W'') \, \mathrm{d}W' \, \mathrm{d}W'' &= \sum_{kl} x_k^*(W') \, \mathrm{d}W' p^0(kl) x_l(W'') \, \mathrm{d}W''. \end{aligned} \tag{40}$$

In a similar manner, instead of the amplitudes $p(mn)$, 'amplitude densities' $p(mW) \, \mathrm{d}W$ (which refer to an interval $\mathrm{d}W$) occur generally in the case of a continuously variable index. This accords with our previously declared expectation. It is, however, not necessary to take just the energy as the continuously variable index. In place of the energy, one could, for example, introduce the quantity $\varphi(W)$. Then in place of $p(mW) \, \mathrm{d}W$ one would have $p(m\varphi)(\mathrm{d}W/\mathrm{d}\varphi) \, \mathrm{d}\varphi$. Finally, in the continuous case the energy W_n is replaced by the quantity $W(\varphi) \, \mathrm{d}\varphi$. In place of the energy of the individual atom, we get a sort of total energy per interval $\mathrm{d}W$. Thence $\mathrm{d}\varphi$ essentially represents the number of atoms having an energy which lies between W and $W + \mathrm{d}W$, or the *a priori* probability that the energy of the atom lies between W and $W + \mathrm{d}W$. We here most clearly observe the difference between the cases with discrete stationary states on the one hand and those with a continuous manifold of states on the other hand, and we can see a simple connection between the problem of statistical weights and the question of the normalization of the solution of (27), ch. 3. In the case of discrete states when there are no multiple eigenvalues, we make the simple physical contention that each state should have the statistical weight 1. This was ensured by the fact that we normalized the x_{kn} on the basis of the requirement

$$\sum_k x_{kn}x_{kn}^* = 1.$$

In the case of continuous manifolds of states, it was not possible to fix the *a priori* probabilities so simply; more detailed investigations of the problem in question are necessary for their determination and hence also for the evaluation of the function φ. Hence the connection between transition probabilities and the amplitudes might also assume a somewhat more complicated aspect in the case of continuous spectra than for line spectra.

The matrices of p, q or $f(p, q)$ represented by (40), ch. 3, and corresponding forms, can for the general case be made clear by the adjoining scheme:

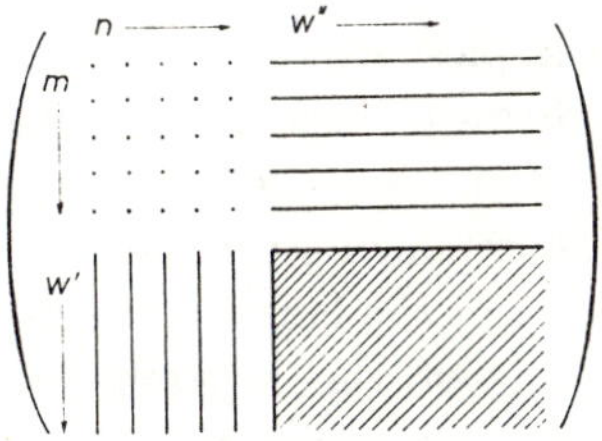

The physical meaning of this scheme is self-evident.

There are four types of 'transitions' which to some extent furnish a simple analogue to the 'transitions' postulated hitherto in the theory of the hydrogen atom, viz. (1) from ellipse to ellipse; (2) from ellipse to hyperbola; (3) from hyperbola to ellipse; (4) from hyperbola to hyperbola.

One can still raise the objection against the formulae (38) and (40), ch. 3, that manifestly in some instances the infinite sums on the right-hand sides do not converge, and hence do not represent a function, since of course in classical theory also, the representation of a function $f(p, q)$ by Fourier integrals is sometimes impossible, as for instance if the respective functions f increase linearly with time at large times (as is in general the case with coordinates). To this objection, one may, however, rejoin that the observable effects of the atom (such as radiation, the force upon another atom, etc.) do not in general belong to this type of function, and thus that the appropriate sums of the same type as the formulae (40), ch. 3, might indeed converge.

CHAPTER 4. PHYSICAL APPLICATIONS OF THE THEORY

1. Laws of conservation of momentum and angular momentum; intensity formulae and selection rules

By way of applying the general theory as established in the aforegoing sections, we now derive the known features concerning 'quantization' of angular momentum and some associated principles.

We shall thereby at the same time become acquainted with some characteristic examples involving *integration* of the quantum-mechanical equations of motion. The previously-discussed perturbation methods can, of course, be applied successfully only when a set of particularly simple examples, which can be selected as unperturbed systems H_0, has been integrated in some other way. Now, the quantum-mechanical equations of motion coming from the decomposition of matrix equations into components present the special difficulty that – apart from the instance of the harmonic oscillator – infinitely many unknowns occur in each of the separate equations. A technique frequently employed in overcoming this difficulty in the following and, as it seems, of wide applicability, consists of the following procedure: By analogy with classical theory, one first seeks integrals of the equations of motion, i.e., functions $A(p, q)$ which on the basis of the equations of motion and the commutation rules are constant in function of time and consequently become diagonal matrices in the case of nondegenerate periodic systems. Now if $\varphi(p, q)$ be any function whatsoever, the difference

$$\varphi A - A\varphi = \psi$$

can be evaluated with the help of the commutation rules; if A is a diagonal matrix, a system of equations results, each of which contains only a finite number of unknowns, namely a single component of the matrices φ and ψ (and two diagonal terms of A) in each.

If in Cartesian coordinates, $H = H'(p) + H''(q)$, which includes the case of relativistic mechanics, then one can see immediately that the components of the angular momentum $\mathfrak{M}$, viz.

$$\begin{aligned} M_x &= \sum_{k=1}^{f/3} (p_{ky}q_{kz} - q_{ky}p_{kz}), \\ M_y &= \sum_{k=1}^{f/3} (p_{kz}q_{kx} - q_{kz}p_{kx}), \\ M_z &= \sum_{k=1}^{f/3} (p_{kx}q_{ky} - q_{kx}p_{ky}) \end{aligned} \tag{1}$$

become constant under the same general conditions as in classical theory. This is because a sum,

$$\dot{M}_z = \varphi(q) + \psi(p),$$

ensues for the derivative of, say, M_z with respect to time, and since all the p commute with one another, as do all the q, the quantities φ, ψ vanish under the same conditions as in classical theory.

The same remarks are to be applied to the linear momentum

$$\mathfrak{p} = \sum_{k=1}^{f/3} \mathfrak{p}_k; \qquad \text{i.e.,} \qquad p_x = \sum_{k=1}^{f/3} p_{kx}, \ldots, \tag{2}$$

which likewise becomes constant. Thus the centre-of-mass theorem holds just as in classical theory.

We immediately note here a formula which will be used later and which can be derived from the commutation relations (3), ch. 2. We find

$$\begin{aligned} M_xM_y - M_yM_x &= \sum_{kl} \{(p_{ky}q_{kz} - q_{ky}p_{kz})(p_{lz}q_{lx} - q_{lz}p_{lx}) \\ &\qquad - (p_{kz}q_{kx} - q_{kz}p_{kx})(p_{ly}q_{lz} - q_{ly}p_{lz})\}, \\ &= \sum_{kl} \{p_{ky}q_{lx}(q_{kz}p_{lz} - p_{lz}q_{kz}) \\ &\qquad + q_{ky}p_{lx}(p_{kz}q_{lz} - q_{lz}p_{kz})\}, \\ &= \frac{h}{2\pi i} \sum_k (p_{kx}q_{ky} - q_{kx}p_{ky}), \end{aligned}$$

i.e.,

$$M_xM_y - M_yM_x = \varepsilon M_z, \qquad (\text{where } \varepsilon = h/2\pi i). \tag{3}$$

Incidentally, one can directly see from this formula that the theorem of conservation of angular momentum invariably holds for at most one or alternatively for all three axes, as in classical theory.

In the following we shall assume that on treating the problem with which we are confronted by the methods developed in the preceding chapter we are led to obtain *discrete* energy values (point spectrum). If then $\dot{\mathsf{M}}_z=0$ for a *nondegenerate* system – this will for instance be the case if forces which are symmetrical about the z-axis act upon the atom – M_z has to become a *diagonal matrix*: the separate diagonal terms are to be regarded as the angular moments of the atom about the z-axis for the individual *states* of the atom. For the investigation of the motions of the electrons in this case, we first note that the relation

$$\mathsf{q}_{lz}\mathsf{M}_z - \mathsf{M}_z\mathsf{q}_{lz} = 0 \tag{4}$$

follows from (1), ch. 4, and since $\mathsf{M}_z(nm)=\delta_{nm}M_{zn}$, this means that

$$q_{lz}(nm)(M_{zn} - M_{zm}) = 0. \tag{5}$$

One sees that: *For a quantum jump in which there is a change in the angular momentum M_z, the 'plane of vibration' of the generated 'spherical wave' lies perpendicular to the z-axis.*

Furthermore, one has

$$\begin{aligned} \mathsf{q}_{lx}\mathsf{M}_z - \mathsf{M}_z\mathsf{q}_{lx} &= -\varepsilon\mathsf{q}_{ly}, \\ \mathsf{q}_{ly}\mathsf{M}_z - \mathsf{M}_z\mathsf{q}_{ly} &= \varepsilon\mathsf{q}_{lx}, \end{aligned} \tag{6}$$

i.e.,

$$\begin{aligned} q_{lx}(nm)(M_{zn} - M_{zm}) &= -\varepsilon q_{ly}(nm), \\ q_{ly}(nm)(M_{zn} - M_{zm}) &= \varepsilon q_{lx}(nm). \end{aligned} \tag{7}$$

Thus for jumps in which no change in M_z occurs, the emitted light is linearly polarized parallel to the z-axis.

Further, from (7), ch. 4, it follows that

$$\{(M_{zn} - M_{zm})^2 - (h^2/4\pi^2)\}q_{l\eta}(nm) = 0; \qquad \eta = x, y. \tag{8}$$

One finally concludes: *For every quantum jump M_{zn} changes by* 0, *or by $\pm h/2\pi$. The light emitted in the latter case is circularly polarized, as follows from* (7), *ch.* 4.

In accordance with the above finding concerning the possible changes in M_z, the quantity M_{zn} can be represented in the form

$$M_{zn} = \frac{h}{2\pi}(n_1 + C), \qquad n_1 = \ldots, -2, -1, 0, 1, 2, \ldots. \tag{9}$$

If there were states whose angular momentum did not fit into this set, no transitions and no interactions whatsoever could occur between these and the states depicted by (9), ch. 4. Equation (9), ch. 4, can be taken as a motive for splitting n into two components, one of which is the number n_1, introduced in (9), ch. 4, whereas the other, n_2, counts off the various n with the same n_1. Our matrices then become four-dimensional, and the results we found for the motions of electrons may be summarized as:

$$q_{lz}(nm) = \delta_{n_1, m_1} q_{lz}(nm); \tag{10}$$

$$\begin{aligned} q_{lx}(nm) &= \delta_{1, |n_1 - m_1|} q_{lx}(nm), \\ q_{ly}(nm) &= \delta_{1, |n_1 - m_1|} q_{ly}(nm); \end{aligned} \tag{10'}$$

$$q_{lx}(n_1, n_2; n_1 \pm 1, m_2) \mp i q_{ly}(n_1, n_2; n_1 \pm 1, m_2) = 0. \tag{10''}$$

Further, from (4) and (6), ch. 4, it follows that if we set

$$\mathfrak{q}_l^2 = q_l^2 = q_{lx}^2 + q_{ly}^2 + q_{lz}^2,$$

then

$$q_l^2 M_z - M_z q_l^2 = 0. \tag{11}$$

This relation means that q_l^2 is a diagonal matrix with respect to the 'quantum number' n_1.

The relations (4) to (7), ch. 4 and (10), (11), ch. 4, also hold if in place of the q_{lx}, q_{ly}, q_{lz} we insert p_{lx}, p_{ly}, p_{lz} or alternatively M_x, M_y, M_z. Thus in particular we have:

$$\begin{aligned} &M_x(nm) = \delta_{1, |n_1 - m_1|} M_x(nm); \qquad M_y(nm) = \delta_{1, |n_1 - m_1|} M_y(nm), \\ &M_x(n_1, n_2; n_1 \pm 1, m_2) \pm i M_y(n_1, n_2; n_1 \pm 1, m_2) = 0. \end{aligned} \tag{12}$$

Further (cf. eq. (1), ch. 4), $\mathfrak{M}^2 = M^2 = M_x^2 + M_y^2 + M_z^2$ is a diagonal matrix with respect to n_1, since

$$M^2 M_z - M_z M^2 = 0. \tag{13}$$

For a system in which all three angular momentum conservation theorems apply, the constant components of $\mathfrak{M}$ certainly cannot collectively be diagonal matrices, since otherwise the above considerations for M_z to be a diagonal matrix could be applied to each of these components, which would lead to discrepancies. Hence such a system is necessarily *degenerate*.

We now set out to consider a system $H = H_0 + \lambda H_1 + \ldots$ of the following type: *All three angular momentum theorems are to apply for* $\lambda = 0$. For

$\lambda \neq 0$ *the system is to be nondegenerate; the constancy of* M_z *is to remain undisturbed. The energy* H_0 *is to be independent of* n_1. The results we shall obtain from this investigation of the case $\lambda \neq 0$ can in part also be carried over to the degenerate system H_0, namely insofar as *they are independent firstly of* λ *and secondly of the distinguished direction z.*

The assumed degeneracy of the system for $\lambda = 0$ is expressed by the fact that $\dot{\mathsf{M}}_x$, $\dot{\mathsf{M}}_y$, $(d/dt)(\mathsf{M}^2)$ contain no terms of zeroth order in λ. Thus

$$\begin{aligned} \nu_0(nm) M_\eta(nm) &= 0, \qquad \eta = x, y; \\ \nu_0(nm) M^2(nm) &= 0. \end{aligned} \tag{14}$$

Since W_0 is independent of the quantum number n_1 introduced earlier, whence $\nu_0(n_1, n_2; m_1, n_2) = 0$, whereas $\nu_0(n_1, n_2; m_1, m_2) \neq 0$ is invariably non-zero for $n_2 \neq m_2$, it follows from (14), ch. 4, that

$$\begin{aligned} M^0_\eta(nm) &= \delta_{n_2 m_2} M^0_\eta(nm), \\ M^{0^2}(nm) &= \delta_{n_2 m_2} M^{0^2}(nm). \end{aligned} \tag{15}$$

The square of the total momentum $(\mathsf{M}^0)^2$ is a diagonal matrix in consequence of (13), (15), ch. 4. The double sum representing an element of the matrix M^0_x, M^0_y reduces to a simple sum

$$\begin{aligned} &\sum_{k_1 k_2} M^0_x(n_1 n_2; k_1 k_2) M^0_y(k_1 k_2; m_1 m_2) \\ &\qquad = \delta_{n_2 m_2} \sum_{k_1} M^0_x(n_1 n_2; k_1 n_2) M^0_x(k_1 n_2; m_1 n_2), \end{aligned} \tag{16}$$

which contains only a finite number of summation terms because of the finite number of possible n_1 at fixed n_2 (the terms of

$$\mathsf{M}^{0^2} = \mathsf{M}^{0^2}_x + \mathsf{M}^{0^2}_y + \mathsf{M}^{0^2}_z \geqslant \mathsf{M}^2_z$$

do not depend on n_1). In (3), ch. 4, applied to M^0_x, M^0_y, M^0_z, we can at any given time sum the equations which belong to a given n_2 over n_1 and obtain,[1] for fixed n_2:

$$\sum_{n_1} M_z(n_1 n_2; m_1 n_2) = \sum_{n_1} (n_1 + C) \frac{h}{2\pi} = 0. \tag{17}$$

On noting additionally that, by (12) and (16), ch. 4, the sum (17), ch. 4 vanishes for *every* single uninterrupted sequence of the n_1 it follows that at fixed n_2 the possible values of $n_1 + C$ form an unbroken series and lie symmetrically with respect to zero. Hence they must

[1] In I we already noted that in the case of a finite diagonal sum $D(ab)$ we always have $D(ab) = D(ba)$.

necessarily constitute either *integer* or *half-integer* numbers, the latter being numbers in the series ..., $-\frac{3}{2}$, $-\frac{1}{2}$, $\frac{1}{2}$, $\frac{3}{2}$,.... If for the moment M_z about the z-axis we now introduce the notation usually used in the literature, namely $m(h/2\pi)$ in place of $(n_1+C)(h/2\pi)$, this result accordingly shows that the selection rule $m\to(m+1, m, m-1)$ applies to m and that m is either 'integer' or 'half-integer'.

Our result demonstrates further that exclusion of individual states, such as was, for example, necessary in the past theory of the hydrogen atom in order to prevent collisions between the electron and the nucleus, has no place in the theory proposed here.

We now attempt to derive the selection principle for the 'total momentum quantum number', as also the intensities for the Zeeman effect, from our theory, proceeding from (5) and (8), ch. 4.

Let us recall the derivation of these selection rules in classical theory: There it is only necessary to introduce a coordinate system whose z-axis coincides with the direction of the total angular momentum; in the new coordinates the same results can be derived for $\mathfrak{M}$ as were previously obtained for M_z. Let us accordingly set up such a coordinate system x', y', z'. The relation

$$z' = x\frac{M_x}{M} + y\frac{M_y}{M} + z\frac{M_z}{M}$$

has to hold anyway in order that the z'-axis lie in the direction of the total momentum. (In the following, we shall again drop the index ^{0} for simplicity in all momenta and coordinates: the calculations throughout refer to the limiting case $\lambda=0$). Further, we can so arrange it that the x'-axis lies in the x, y-plane. Everything is thereby fixed, and we have

$$x' = y\frac{M_x}{\sqrt{(M_x^2+M_y^2)}} - x\frac{M_y}{\sqrt{(M_x^2+M_y^2)}}$$

$$y' = \frac{z(M_x^2+M_y^2) - xM_zM_x - yM_zM_y}{M\sqrt{(M_x^2+M_y^2)}}.$$

Now let us try a similar procedure in quantum mechanics. We introduce the three quantities

$$\begin{aligned} \mathsf{Z}_l &= \mathsf{q}_{lx}\mathsf{M}_x + \mathsf{q}_{ly}\mathsf{M}_y + \mathsf{q}_{lz}\mathsf{M}_z, \\ \mathsf{X}_l &= \mathsf{q}_{ly}\mathsf{M}_x - \mathsf{M}_y\mathsf{q}_{lx}, \\ \mathsf{Y}_l &= \mathsf{M}_x\mathsf{q}_{lz}\mathsf{M}_x + \mathsf{M}_y\mathsf{q}_{lz}\mathsf{M}_y - \mathsf{q}_{lx}\mathsf{M}_z\mathsf{M}_x - \mathsf{M}_y\mathsf{M}_z\mathsf{q}_{ly}. \end{aligned} \tag{18}$$

In order to derive the desired selection rules, we still need some commutation relations, which result from (4) and (6), ch. 4 ($\varepsilon=h/2\pi i$):

$$q_{lx}M^2 - M^2 q_{lx} = 2\varepsilon(q_{lz}M_y - M_z q_{ly}) \tag{19}$$

and the equations for q_{ly}, q_{lz} which ensue from this on cyclic permutation. It then follows[1] from (3), (4), (6) and (19), ch. 4, that

$$\begin{aligned} X_l M^2 - M^2 X_l &= -2\varepsilon Y_l, \\ Y_l M^2 - M^2 Y_l &= \varepsilon(X_l M^2 + M^2 X_l), \\ Z_l M^2 - M^2 Z_l &= 0. \end{aligned} \tag{20}$$

These equations are fully analogous to the relations (4) and (6), ch. 4, which determine the selection rules for M_z; since we shall later show that the q_{lx}, q_{ly}, q_{lz} really can be expressed as linear functions of the X_l, Y_l, Z_l, with coefficients which for $\lambda=0$ are constant with time, we can determine the selection rules for M directly from (20), ch. 4. As M^2 is a

[1] The first and third formulae in eq. (20), ch. 4, result from a quite simple calculation. The second of eqs. (20), ch. 4, can be derived in the following way: From (18), ch. 4,

$$Y_l = M_x q_{lz} M_x + M_y q_{ly} M_y - q_{lx} M_z M_x - M_y M_z q_{ly},$$

and because of (6), ch. 4,

$$\begin{aligned} Y_l &= q_{lz}(M_x^2 + M_y^2) - \varepsilon q_{ly} M_x + \varepsilon M_y q_{lx} + \varepsilon^2 q_{lz} \\ &\qquad - q_{lx} M_z M_x - M_y M_z q_{ly} \\ &= q_{lz}(M^2 - M_z^2) - \varepsilon X_l + \varepsilon^2 q_{lz} - q_{lx} M_z M_x - M_y M_z q_{ly}. \end{aligned}$$

In the evaluation of $Y_l M^2 - M^2 Y_l$ we now have to note that M^2 commutes with M_x, M_y, M_z. Hence for the second part of the formula for Y_l written above, it follows that (cf. (19), ch. 4)

$$\begin{aligned} &(q_{lx} M_z M_x + M_y M_z q_{ly})M^2 - M^2(q_{lx} M_z M_x + M_y M_z q_{ly}) \\ &\qquad = 2\varepsilon(q_{lz} M_y M_z M_x - M_z q_{ly} M_z M_x + M_y M_z q_{lx} M_z - M_y M_z M_x q_{lz}). \end{aligned}$$

On noting that (eq. (19), ch. 4) $q_{lz}M^2 - M^2 q_{lz} = 2\varepsilon X_l$, it follows from the commutation relations that

$$\begin{aligned} q_{lz} M_y M_z M_x - M_y M_z M_x q_{lz} &= \varepsilon(M_y M_z q_{ly} - q_{lx} M_z M_x), \\ M_y M_z q_{lx} M_z - M_z q_{ly} M_z M_x &= -X_l . M_z^2 - \varepsilon(M_z q_{ly} M_y - q_{lx} M_x M_z), \end{aligned}$$

and finally we obtain the desired formula (20), ch. 4:

$$\begin{aligned} Y_l M^2 - M^2 Y_l &= 2\varepsilon X_l(M^2 - M_z^2 + \varepsilon^2) - \varepsilon(X_l M^2 - M^2 X_l) + 2\varepsilon X_l M_z^2 \\ &\qquad - 2\varepsilon^2(q_{lx} M_x M_z - q_{lx} M_z M_x + M_y M_z q_{ly} - M_z M_y q_{ly}) \\ &= 2\varepsilon X_l(M^2 - M_z^2 + \varepsilon^2) - \varepsilon(X_l M^2 - M^2 X_l) + 2\varepsilon X_l M_z^2 - 2\varepsilon^3 X_l \\ &= \varepsilon(X_l M^2 + M^2 X_l). \end{aligned}$$

diagonal matrix, it follows from (20), ch. 4, that

$$\begin{aligned} X_l(nm)(M_m^2 - M_n^2) &= -\,2\varepsilon Y_l(nm), \\ Y_l(nm)(M_m^2 - M_n^2) &= \varepsilon X_l(nm)(M_m^2 + M_n^2), \\ Z_l(nm)(M_m^2 - M_n^2) &= 0. \end{aligned} \tag{21}$$

The last of the eqs. (21), ch. 4, states that no vibrations take place in Z which could entail a change in M^2. It follows from the first two equations that

$$X_l(nm)\left\{(M_m^2 - M_n^2)^2 - \frac{h^2}{2\pi^2}(M_m^2 + M_n^2)\right\} = 0. \tag{22}$$

If we now set $M_m^2 = (h/2\pi)^2(a_m^2 - \frac{1}{4})$, where a_m denotes any function of the quantum numbers, eq. (22), ch. 4 yields

$$X_l(nm)\big((a_n - a_m)^2 - 1\big)\big((a_n + a_m)^2 - 1\big) = 0,$$

or, if $X_l(nm)$ does not vanish,

$$a_n = \pm a_m \pm 1. \tag{23}$$

There is no sacrifice of generality in taking a_m as positive and $\geqq \frac{1}{2}$ throughout. The a_m thus constitute a series of the form $C, 1+C, 2+C, \ldots$ where C denotes a constant which is $\geqq \frac{1}{2}$. Setting $a_m = j + \frac{1}{2}$ yields

$$M^2 = j(j+1)(h/2\pi)^2, \tag{24}$$

and the following selection rule holds for j:

$$j \to \begin{cases} j+1 \\ j \\ j-1 \end{cases}.$$

This result is formally reminiscent of the values of M^2 which enter the Landé g-formula.

If for M_z we now again introduce the designation $m(h/2\pi)$, we find from (12), ch. 4, and the relations

$$M^2 = M_x^2 + M_y^2 + M_z^2$$

and

$$(M_x + iM_y)(M_x - iM_y) = M_x^2 + M_y^2 - i\varepsilon M_z = M^2 - M_z^2 - i\varepsilon M_z$$

that

$$M_x(j, m-1; j, m) + iM_y(j, m-1; j, m) = \frac{h}{2\pi}\sqrt{(j(j+1) - m(m-1))},$$

$$M_x(j, m; j, m-1) - iM_y(j, m; j, m-1) = \frac{h}{2\pi}\sqrt{(j(j+1) - m(m-1))}. \tag{25}$$

For a given value of j, the maximum value m_{max} of m is characterized by the absence of the jumps $m_{\text{max}} \rightarrow m_{\text{max}}+1$, i.e., the right-hand side of (24), ch. 4, for example vanishes for such jumps. This gives

$$j = m_{\text{max}}.$$

Hence j also can be 'integer' or 'half-integer' only.

The calculation of the intensity formulae for the Zeeman effect, e.g., the dependence of q_{lx}, q_{ly}, q_{lz} upon m, now appears very simple. From (18), ch. 4, we derive the relations

$$\begin{aligned} q_{lz} &= (Z_l M_z + \varepsilon X_l + Y_l)M^{-2}, \\ q_{lx} + iq_{ly} &= [Z_l - q_{lz}(M_z + i\varepsilon) + iX_l](M_x - iM_y)^{-1}, \\ q_{lx} - iq_{ly} &= [Z_l - q_{lz}(M_z - i\varepsilon) - iX_l](M_x + iM_y)^{-1}, \end{aligned} \tag{26}$$

by solving for q_{lx}, q_{ly}, q_{lz}. These equations also furnish the previously postponed proof that the q_{lx}, q_{ly}, q_{lz} can be represented as linear functions of the X_l, Y_l, Z_l with coefficients which for $\lambda=0$ are constant with time. At the same time, eqs. (26) ch. 4, include the desired intensity formulae. This can be seen by first noting that the X_l, Y_l, Z_l are diagonal matrices with respect to m, since

$$\begin{aligned} X_l M_z - M_z X_l &= 0, \\ Y_l M_z - M_z Y_l &= 0, \\ Z_l M_z - M_z Z_l &= 0. \end{aligned} \tag{27}$$

Our problem now resolves itself into two parts, namely discussion of intensities for jumps $j \rightarrow j$ and $j \rightarrow j-1$ (the jumps $j \rightarrow j+1$ then do not provide anything new). We first consider the transitions $j \rightarrow j$. For these, equation (20), ch. 4 shows that only terms in Z_l are present. We shall call these terms $Z_l(j, m)$. Then, on setting $M_z = m(h/2\pi)$ and taking note of (24), ch. 4, eqs. (26), ch. 4 yield:

$$q_{lz}(j, m) = \frac{2\pi}{h} Z_l(j, m) \frac{m}{j(j+1)},$$

$$(q_{lx} + \mathrm{i}q_{ly})(j, m-1; j, m)$$
$$= \frac{2\pi}{h} Z_l(j, m-1) \sqrt{\frac{j(j+1) - m(m-1)}{j(j+1)}}, \tag{28}$$

$$(q_{lx} - \mathrm{i}q_{ly})(j, m; j, m-1)$$
$$= \frac{2\pi}{h} Z_l(j, m) \sqrt{\frac{j(j+1) - m(m-1)}{j(j+1)}}.$$

Finally, to establish the dependence of the quantity $Z_l(j, m)$ upon m, we might use the relation

$$\mathsf{M}_x q_{ly} - q_{ly}\mathsf{M}_x = \varepsilon q_{lz}; \tag{29}$$

it demonstrates in our case that $Z_l(j, m)$ does not depend on m. For the transitions $j \to j$ we thus obtain:

$$\begin{aligned} q_{lz}(j, m) &: (q_{lx}+\mathrm{i}q_{ly})(j, m-1; j, m) : (q_{lx}-\mathrm{i}q_{ly})(j, m; j, m-1) \\ &= m : \sqrt{\{j(j+1) - m(m-1)\}} : \sqrt{\{j(j+1) - m(m-1)\}}. \end{aligned} \tag{30}$$

We treat the jumps $j \to j-1$ analogously. For these, according to (21), ch. 4, we have $X_l(j, m; j-1, m) = (\varepsilon/j) Y_l(j, m; j-1, m)$. If, using (26), ch. 4, we express the intensities in terms of $X_l(j, m; j-1, m)$, we obtain:

$$q_{lz}(j, m; j-1, m) = \mathrm{i}\frac{2\pi}{h} X_l(j, m; j-1, m)\frac{1}{j},$$

$$(q_{lx} + \mathrm{i}q_{ly})(j, m-1; j-1, m)$$
$$= \mathrm{i}\frac{2\pi}{h} X_l(j, m-1; j-1, m-1) \frac{\sqrt{(j-m)}}{j\sqrt{(j+m-1)}}, \tag{31}$$

$$(q_{lx} - \mathrm{i}q_{ly})(j, m; j-1, m-1)$$
$$= -\mathrm{i}\frac{2\pi}{h} X_l(j, m; j-1, m) \frac{\sqrt{(j+m-1)}}{j\sqrt{(j-m)}}.$$

In conclusion, to establish the dependence of the quantity $X_l(j, m; j-1, m)$ upon m, we again use the relation (29), ch. 4, which by way of a simple calculation here yields:

$$X_l(j, m; j-1, m) = A(j, j-1)\sqrt{(j^2 - m^2)}. \tag{32}$$

We thus find that

$$\begin{aligned} &q_{lz}(j, m; j-1, m) : (q_{lx} + iq_{ly})(j, m-1; j-1, m) \\ &: (q_{lx} - iq_{ly})(j, m; j-1, m-1) = \sqrt{(j^2 - m^2)} : \sqrt{((j-m)(j-m+1))} \\ &\qquad : -\sqrt{((j+m)(j+m-1))}. \end{aligned} \tag{33}$$

The jumps $j \to j+1$ essentially give the same intensities; we here find that

$$\begin{aligned} &q_{lz}(j, m; j+1, m) : (q_{lx} + iq_{ly})(j, m; j+1, m+1) \\ &: (q_{lx} - iq_{ly})(j, m+1; j+1, m) = \sqrt{((j+1)^2 - m^2)} \\ &: \sqrt{((j+m+2)(j+m+1))} : -\sqrt{((j-m+1)(j-m))}. \end{aligned} \tag{34}$$

The formulae (30), (33), (34), ch. 4 agree with the intensity formulae derived from correspondence considerations.[1]

We wish just to draw attention to a simple deduction from (21), ch. 4: *The jumps $\Delta j = 0$ occur only in the 'Z_l-direction'*. If we consider the motion of a *single* electron about a nucleus, that is, examine the hydrogen atom, it follows directly from (1), ch. 4, that Z vanishes. Hence in this case the jumps $\Delta j = 0$ never take place.

2. The Zeeman effect

If one carries the Lorentz force $(e/c)[\mathfrak{v}\mathfrak{H}]$ exerted by a magnetic field $\mathfrak{H}$ upon an electron over into quantum mechanics, it seems obvious at first that the normal Zeeman Effect ensues for atoms, since under exactly the same assumptions as are introduced to derive Larmor's Theorem classically for the nuclear atom – namely, neglect of terms with $\mathfrak{H}^2$ – one can derive this theorem here. There is, nevertheless, a certain difference between classical theory and quantum mechanics insofar as the justification for dropping terms in $\mathfrak{H}^2$ is concerned. The neglect of $\mathfrak{H}^2$ in classical theory is certainly permissible for orbits of small dimensions and certainly *impermissible* for very large orbits or indeed, hyperbolic orbits. In quantum mechanics all these orbits, be they the innermost or the outermost, are so closely connected with one another as a result of the kinematics specific to quantum me-

[1] S. Goudsmit and R. de L. Kronig, Naturwiss. **13** (1925) 90; H. Hönl, Zs. f. Phys. **32** (1925) 340.

chanics, that indication of the neglect of the quantity $\mathfrak{H}^2$ is not immediately apparent. The probabilities of transitions to free electrons are indeed considerable, even from the ground state.

For an oscillator, we are thus sure of the normal Zeeman effect; on the other hand for the nuclear atom it does not seem to be entirely excluded that the intimate connection between innermost and outermost orbits leads to findings which differ somewhat from the normal Zeeman effect. However, we must emphasize that a whole set of weighty reasons speak against the possibility of explaining the anomalous Zeeman effects on this basis. Rather, one might perhaps hope that the hypothesis of Uhlenbeck and Goudsmit might later provide a quantitative description of the above-mentioned phenomena.

3. Coupled harmonic resonators. Statistics of wave fields

A system of coupled harmonic oscillators given by

$$H = \frac{1}{2} \sum_{k=1}^{f} \frac{p_k^2}{m_k} + Q(q), \tag{35}$$

with a quadratic form $Q(q)$ of the coordinates (with numerical coefficients) represents the simplest conceivable system having several degrees of freedom. As was established in ch. 2 § 1, the commutation rules remain invariant on simultaneous orthogonal transformation of coordinates and momenta. Therefore, as in classical theory, the system (35), ch. 4, can be transformed into a system of *uncoupled* oscillators. In particular, the vibrations of a crystal lattice can be analyzed into *eigenvibrations*, just as in classical theory. Each individual eigenvibration is to be treated as a simple linear oscillator according to the manner discussed previously in detail, and the synthesis of the various uncoupled oscillators to a single system is to be undertaken in the way explained in ch. 2 § 1. The same will also apply if we go over to the limiting case of a system with infinitely many degrees of freedom and for instance consider the vibrations of an elastic body idealized to a continuum or finally of an electromagnetic cavity.

In the previous quantum theory also, vibrations of an electromagnetic cavity constituted the subject of many detailed investigations, since on the one hand the problem of the harmonic oscillator represents just about the simplest problem which can be treated with the methods used hitherto, and on the other hand the familiar result that the energy of an eigenvibration should be an integer multiple of $h\nu$ exhibits a formal similarity to the fundamental assumptions of the theory of light quanta, so that one might hence expect to gain insight into the nature of light quanta through the consideration of black-body radiation. To be sure, it is clear from the very outset that attacking the problem of light quanta from the above standpoint cannot by any means account for the most important aspect of this problem, namely the phenomenon of coupling of distant atoms, for this problem does not enter at all into the formulation of our questions regarding the vibrations of a cavity. So strong an association between the eigenvibrations of a cavity and the light quanta postulated formerly can nonetheless be drawn that every statistics of cavity eigenvibrations corresponds to a definite statistics of light quanta, and conversely.

Debye[1] has attempted to arrive at such a form of statistics, starting from a distribution of individual light quanta among the eigenvibrations of the cavity. In this manner he was able to derive Planck's formula. However such a mixture of theoretical wave and light-quantum considerations would seem to us hardly to accord with the real nature of the problem. Rather, we believe it to be consistent to separate the theoretical wave-aspect of the problem completely from the theory of light quanta, that is to say, to treat the wave-statistics of black-body radiation throughout by the more general statistical rules applying e.g., to the quantum theory of atomic systems. The statistics applicable to light quanta is then, as we shall show, Bose statistics.[2] This finding hardly seems unnatural, since this statistics has nothing to do with the hypothesis of independent light-corpuscles, but rather is to be regarded as carried over from the statistics of eigenvibrations – which just shows that the assumption of statistically independent light-corpuscles would not meet the case correctly.

However, in each such treatment of cavity radiation by quantum theory hitherto, one encountered the fundamental difficulty that although it led to Planck's law of radiation, it did not yield the correct mean square deviation of energy in an element of volume. One thus finds that a consistent treatment of the natural vibrations of a me-

[1] P. Debye, Ann. d. Phys. **33** (1910) 1427; cf. also P. Ehrenfest, Phys. Zs. **7** (1906) 528.

[2] S. N. Bose, Zs. f. Phys. **26** (1924) 178.

chanical system or an electromagnetic cavity in accordance with past theory leads to most serious contradictions. This caused us to hope that the modified kinematics which forms an inherent feature of the theory proposed here would yield the correct value for the interference fluctuations, thus precluding the above contradictions and opening the possibility of setting up a consistent system of statistics for black-body radiation.

The states of the system of oscillators can be characterized by 'quantum numbers' $n_1, n_2, n_3, \ldots$ of the individual oscillators, so that apart from an additive constant the energies of the individual states are given by

$$E_n = h \sum_k \nu_k n_k. \tag{36}$$

The additive constant, the so-called *zero-point energy* is

$$C = \tfrac{1}{2} h \sum_k \nu_k \tag{36'}$$

(in particular, for the limiting case of infinitely many degrees of freedom, it would be infinitely large). From now on, let us simply call the quantity E_n in (36), ch. 4, the *thermal energy*. In accordance with what was stated in Part I, the same statistical weight is to be attached to each of the states of the system characterized by a certain set of values $n_1, n_2, n_3, \ldots$. The consequences of this can immediately be perceived on the basis of the following remark: If waves are propagated with a phase velocity v in an s-dimensional isotropic part of space $V = l^s$ the number of *eigenvibrations* for the frequency range $d\nu$ is equal to the number of '*cells*' for $d\nu$ (in the Bose–Einstein sense), and this in fact holds for arbitrary s, hence e.g. also for *vibrating membranes* or *strings*. This follows from the fact that, if we omit consideration of polarization properties, etc., the number of eigenvibrations for the range $d\nu$ is furnished by the solution of the problem of determining the number of ways in which one can choose a set of positive integers $m_1, \ldots m_s$ such that the ν determined by the relation

$$\frac{2l}{v} \nu = \sqrt{(m_1^2 + \ldots + m_s^2)}$$

falls within the interval $d\nu$. If $K_s(a)$ be the volume of an s-dimensional sphere of radius a, there are $(V/v^s) K_s(\nu)$ eigenvibrations which have a frequency less than ν. On the other hand, the number of cells for the range $d\nu$ can be determined as follows: The momentum components $p_1, \ldots, p_s$ of the quantum satisfy the equation

$$h\nu/v = \sqrt{(p_1^2 + \ldots + p_s^2)},$$

and the size of the cells is h^s in the $2s$-dimensional phase space. One can see from this that the number of cells belonging to a frequency lower than ν is also equal to $(V/v^s)K_s(\nu)$.

Hence, as mentioned above, one can effect a one-to-one correspondence of cells to eigenvibrations such that the individual pairs always belong the same $d\nu$. This correspondence can, incidentally, be so carried out that the *directions* of an eigenvibration and those of the light quanta in the respective cell fall within the same infinitesimal angular range. From (36), ch. 4, the quantum number of an *oscillator* is then to be set equal to the *number of quanta in the appropriate cell.* Every system of light-quanta statistics yields an associated statistics of natural vibrations and conversely. It can be seen that the statement made above concerning the weighting of the states of the system of oscillators goes directly over into the basic postulate of Bose–Einstein statistics because of this association. The equally probable complexions are defined through a declaration of the number of quanta sitting in each cell.[1]

In Debye statistics, the number of oscillators involving r quanta is (except for a factor which depends on ν only) equal to

$$\frac{1}{r}\,\mathrm{e}^{-r(h\nu/kT)}, \tag{37}$$

and Planck's law arises from

$$\sum_{r=1}^{\infty} \mathrm{e}^{-r(h\nu/kT)} = \frac{1}{\mathrm{e}^{h\nu/kT} - 1}.$$

It is unsatisfactory that eq. (37), ch. 4, holds only for $r>0$ and does not also give the number of oscillators involving no quanta. From the new point of view, we have to replace (37), ch. 4, according to

[1] A. Einstein, Sitzungsber. d. Preuss. Akad. d. Wiss. (1925) p. 3. Our considerations naturally cannot yield any fresh viewpoint for the valuation of Einstein's hypothesis that this form of statistics is also applicable to an ideal gas.

Bose,[1] by

$$(1 - e^{-h\nu/kT})\, e^{-r(h\nu/kT)}, \tag{38}$$

which (to use the terminology of the theory of light quanta) gives the number of 'r-fold occupied cells', and Planck's formula results from

$$\sum_{r=0}^{\infty} r(1 - e^{h\nu/kT})\, e^{-r(h\nu/kT)} = \frac{1}{e^{h\nu/kT} - 1}.$$

The light-quanta statistics corresponding to Debye's vibration statistics is represented by the theory developed by Wolfke[2] and Bothe.[3] To be sure, these authors do not speak of r-fold occupied cells, but designate (37), ch. 4, as the number of 'r-quantal light-quanta molecules'.

As is known, the above-mentioned shortcomings of classical wave theory become evident in the study of energy deviations in the radiation field as follows: If there is communication between a volume V and a very large volume such that waves having frequencies which lie within a small range ν to $\nu + d\nu$ can pass unhindered from one to the other, whereas for all other waves the volumes remain detached, and if E be the energy of the waves with frequency ν in V, then according to Einstein the mean square deviation $\overline{\Delta^2} = \overline{(E - \bar{E})^2}$ can be calculated by an inversion of the Boltzmann Principle. If $z_\nu\, d\nu$ be the number of eigenvibrations (cells) in the range $d\nu$ per unit volume, so that

$$\bar{E} = \frac{z_\nu h\nu}{e^{h\nu/kT} - 1} \cdot V, \tag{39}$$

then it follows that

$$\overline{\Delta^2} = h\nu\bar{E} + \frac{\bar{E}^2}{z_\nu V}. \tag{40}$$

If, however, one calculates the energy deviations from *interferences* in the wave field, classical theory yields only the second summation

[1] This expression naturally has to be assumed for example also in the case of elastic waves in a continuum, which necessitates a certain modification to considerations by Schrödinger (Phys. Zs. **25** (1924) 89) concerning the thermal equilibrium between light- and sound-beams. This modification can easily be carried out in analogy with the probability theorem for the Compton effect on assuming Einstein's gas theory to be valid, as has earlier been pointed out (P. Jordan, Zs. f. Phys. **33** (1925) 649).

[2] M. Wolfke, Phys. Zs. **22** (1921) 375.

[3] W. Bothe, Zs. f. Phys. **20** (1923) 145; **23** (1924) 214.

term in (40), ch. 4, as has explicitly been shown by Lorentz.[1] This discrepancy naturally also exists quite generally for such waves as those in a crystal lattice or in an elastic continuum. According to Ehrenfest,[2] its origins are to be sought in the fact that in the Einstein treatment, *additivity of the entropies* of V and of the large volume was assumed. However, this additivity of entropies applies, according to the classical theory of natural vibrations, only in the region of validity of the Rayleigh–Jeans Law. Precisely the nonexistence of statistical independence of the volume elements in the general case is so unnatural a result of the theory of cavity radiation to date that one is obliged to conclude that this theory breaks down even in the simple problem of the harmonic oscillator.

We now calculate the mean square deviation $\overline{\Delta^2}$ from the interferences using quantum mechanics. To avoid calculational complications which have no bearing upon the nature of the case, we base ourselves on the simplest conceivable case, namely a *vibrating string* fastened at its ends. Incidentally, all essential points of the calculation can immediately be taken over in more general instances. We first cite the classical approach.

Let the length of the string be l and its lateral displacement be $u(x, t)$. On introducing the Fourier coefficients $q_k(t)$ as given by

$$u(x, t) = \sum_{k=1}^{\infty} q_k(t) \sin k \frac{\pi}{l} x, \tag{41}$$

or

$$q_k(t) = \frac{2}{l} \int_0^l u(x, t) \sin k \frac{\pi}{l} x \, dx \tag{41'}$$

as coordinates, the energy of the string goes over into a sum of squares. Namely, for suitable choice of units,

$$H = \frac{1}{2} \int_0^l \left\{ u^2 + \left(\frac{\partial u}{\partial x} \right)^2 \right\} dx = \frac{l}{4} \sum_{k=1}^{\infty} \left\{ \dot{q}_k(t)^2 + \left(k \frac{\pi}{l} \right)^2 q_k(t)^2 \right\}. \tag{42}$$

[1] H. A. Lorentz, Les Théories Statistiques en Thermodynamique (Leipzig, 1916), p. 59.

[2] P. Ehrenfest, Lecture in the Göttingen seminar on the Structure of Matter, Summer 1925. The contents of this lecture were of great assistance to our present considerations. Meanwhile published in Zs. f. Phys. **34** (1925) 362.

More generally, for the energy E in a segment $(0, a)$ of the string, we obtain

$$E = \frac{1}{2}\int_0^a \sum_{j,k=1}^{\infty} \left\{\dot{q}_j\dot{q}_k \sin j\frac{\pi}{l}x \sin k\frac{\pi}{l}x + q_jq_kjk\left(\frac{\pi}{l}\right)^2 \cos j\frac{\pi}{l}x \cos k\frac{\pi}{l}x\right\} dx. \tag{43}$$

If in (43), ch. 4, we take only the terms with $j=k$, we find (under the explicit assumption that all wavelengths which come into consideration are small with respect to a) just the value $(a/l)H$. From this one sees: The difference

$$\Delta = E - \bar{E},$$

wherein the bar represents an average over the phases φ_k in

$$q_k = a_k \cos(\omega_k t + \varphi_k); \qquad \omega_k = k\frac{\pi}{l}, \tag{44}$$

can be derived from (43), ch. 4, by omitting terms of the sum which have $j=k$. This phase average is identical with the time average. On carrying out the integration, one then finds

$$\Delta = \frac{1}{4}\sum_{\substack{j,k=1 \\ j\neq k}}^{\infty} \left\{\dot{q}_j\dot{q}_k K_{jk} + jkq_jq_k\left(\frac{\pi}{l}\right)^2 K'_{jk}\right\}, \tag{45}$$

with

$$\left.\begin{aligned} K_{jk} &= \frac{\sin(j-k)\frac{\pi}{l}a}{(j-k)\frac{\pi}{l}} - \frac{\sin(j+k)\frac{\pi}{l}a}{(j+k)\frac{\pi}{l}} \\ &= \frac{\sin(\omega_j-\omega_k)a}{\omega_j-\omega_k} - \frac{\sin(\omega_j+\omega_k)a}{\omega_j+\omega_k}, \\ K'_{jk} &= \frac{\sin(j-k)\frac{\pi}{l}a}{(j-k)\frac{\pi}{l}} + \frac{\sin(j+k)\frac{\pi}{l}a}{(j+k)\frac{\pi}{l}} \\ &= \frac{\sin(\omega_j-\omega_k)a}{\omega_j-\omega_k} + \frac{\sin(\omega_j+\omega_k)a}{\omega_j+\omega_k}. \end{aligned}\right\} \tag{45'}$$

In consideration of later quantum-mechanical calculations, we write out the mean square deviation $\overline{\Delta^2}$ explicitly. It is

$$\Delta^2 = (\Delta_1 + \Delta_2)^2 = \Delta_1^2 + \Delta_2^2 + \Delta_1\Delta_2 + \Delta_2\Delta_1, \tag{46}$$

with

$$\Delta_1^2 + \Delta_2^2 = \frac{1}{16} \sum_{\substack{j,k=1 \\ j \neq k}}^{\infty} \sum_{\substack{\iota,\varkappa=1 \\ \iota \neq \varkappa}}^{\infty} \Big\{ \dot{q}_j \dot{q}_k \dot{q}_\iota \dot{q}_\varkappa K_{jk} K_{\iota\varkappa} + jk\iota\varkappa \left(\frac{\pi}{l}\right)^4 q_j q_k q_\iota q_\varkappa K'_{jk} K'_{\iota\varkappa} \Big\}; \tag{46'}$$

$$\Delta_1\Delta_2 + \Delta_2\Delta_1 = \frac{1}{16} \sum_{\substack{j,k=1 \\ j \neq k}}^{\infty} \sum_{\substack{\iota,\varkappa=1 \\ \iota \neq \varkappa}}^{\infty} \left(\frac{\pi}{l}\right)^2 \{ jk q_j q_k \dot{q}_\iota \dot{q}_\varkappa K'_{jk} K_{\iota\varkappa} + \iota\varkappa \dot{q}_j \dot{q}_k q_\iota q_\varkappa K_{jk} K'_{\iota\varkappa} \}. \tag{46''}$$

Equation (44), ch. 4, implies $\overline{\Delta_1\Delta_2 + \Delta_2\Delta_1} = 0$ and

$$\overline{\Delta^2} = \overline{\Delta_1^2} + \overline{\Delta_2^2} = \frac{1}{8} \sum_{j,k=1}^{\infty} \left\{ \overline{\dot{q}_j^2}\,\overline{\dot{q}_k^2} K_{jk}^2 + j^2k^2 \left(\frac{\pi}{l}\right)^4 \overline{q_j^2}\,\overline{q_k^2} K'^2_{jk} \right\}. \tag{47}$$

If we now let the string's length l become very large, the ω_k get ever closer together, according to (44), ch. 4, so that the sum (47), ch. 4, goes over into an integral:

$$\overline{\Delta^2} = \overline{\Delta_1^2} + \overline{\Delta_2^2} = \frac{1}{8} \int_0^\infty \int_0^\infty d\omega_j\, d\omega_k \frac{l^2}{\pi^2} \left\{ \overline{\dot{q}_j^2}\,\overline{\dot{q}_k^2} K_{jk}^2 + j^2k^2 \left(\frac{\pi}{l}\right)^4 \overline{q_j^2}\,\overline{q_k^2} K'^2_{jk} \right\}. \tag{47'}$$

Finally, we also assume the 'volume' a to become very large and employ the relation

$$\lim_{a\to\infty} \frac{1}{a} \int_{-\Omega}^{\Omega'} \frac{\sin^2 \omega\alpha}{\omega^2} f(\omega)\, d\omega = \pi f(0) \qquad \text{for} \qquad \Omega, \Omega' > 0. \tag{48}$$

We then see that only the first sum terms $(\sin(\omega_j - \omega_k)a)/(\omega_j - \omega_k)$ in (45), ch. 4, provide an appreciable contribution, and we find for (47′), ch. 4,

$$\overline{\Delta^2} = \frac{al}{8\pi} \int_0^\infty d\omega \{ (\overline{\dot{q}_\omega^2})^2 + (\omega^2 \overline{q_\omega^2})^2 \}. \tag{49}$$

On the other hand, by (42), ch. 4, the mean energy in the volume a becomes equal to

$$\bar{E} = \frac{a}{l}\cdot\frac{l}{4}\int_0^\infty d\omega\,\frac{l}{\pi}\cdot\{\overline{\dot{q}_\omega^2} + \omega^2\overline{q_\omega^2}\} = \frac{al}{4\pi}\int_0^\infty d\omega\{\overline{\dot{q}_\omega^2} + \omega^2\overline{q_\omega^2}\}. \quad (50)$$

Therein we have

$$\overline{\dot{q}_\omega^2} = \omega^2\overline{q_\omega^2}, \quad (51)$$

a relation which, as we would recall, remains valid in quantum theory too, according to ch. 1. In order to obtain the quantities $\overline{\Delta^2}$, $\bar{E}$ employed in (39), (40), ch. 4, we have merely to extract those parts referring to $d\nu = d\omega/2\pi$ from (49), (50), ch. 4, and to divide these by $d\nu$. With $v = a$ we then obtain

$$\overline{\Delta^2} = \frac{\overline{E}^2}{2v}. \quad (52)$$

We see from (44), ch. 4, that in our case $z_\nu = 2$, since

$$d\omega_k = 2\pi\, d\nu_k = \frac{\pi}{l}\, dk.$$

Thence (52), ch. 4, in fact gives precisely the second term in (40), ch. 4.

On going over to quantum mechanics, we have to regard (41, (41'), (42), (43), ch. 4, as matrix equations for u, H, q, E. The quantity x, however, remains a number, since if in place of the continuous string we consider an elastic *series of points*, x would denote the *number* (multiplied by the lattice constant) of any given point.

The matrix q_k has $2f$ dimensions if f be the number of eigenvibrations, i.e., infinitely many in the case of an elastic string. Each of the components $q_k(nm)$ of q_k vanishes except for those with

$$\left.\begin{aligned} n_j - m_j &= 0 \qquad \text{for} \qquad j \neq k, \\ n_k - m_k &= \pm 1. \end{aligned}\right\} \quad (53)$$

The phase average of a matrix is that diagonal matrix which coincides with the diagonal of the respective matrix. From (53), ch. 4, in part similar conclusions can be drawn to those derivable from (44), ch. 4. The considerations which formerly led to (46), (46'), (46"), ch. 4, remain valid in quantum theory. The formulae (47), (47'), ch. 4 with matrices

q_k also hold for the diagonal matrix $\overline{\Delta_1^2+\Delta_2^2}$ and finally, according to (52), ch. 4, if we denote those parts of $\overline{\Delta^2}$ which belong to a given frequency ν as $\overline{\Delta^2}$, we find

$$\overline{\Delta_1^2 + \Delta_2^2} = \frac{\overline{E^{*2}}}{2v}. \tag{52'}$$

The quantity E^* in (52′), ch. 4, is, by (49), (50), (51), ch. 4, no longer the mean *thermal* energy, but rather the sum of this and the *zero-point energy*: from the elementary oscillator formulae, we have

$$\overline{E^*} = h\nu \cdot V + \bar{E},$$

$$\overline{\Delta_1^2 + \Delta_2^2} = \tfrac{1}{2}(h\nu)^2\, V + h\nu\bar{E} + \frac{\overline{E^2}}{2V}, \tag{54}$$

since for dν the zero-point energy becomes equal to

$$\frac{v}{l}\cdot\frac{h\nu}{2}\cdot l z_\nu\, \mathrm{d}\nu = h\nu\cdot V\, \mathrm{d}\nu.$$

We now still have to consider $\Delta_1\Delta_2+\Delta_2\Delta_1$. In treating this quantity in just the same way as $\overline{\Delta_1^2+\Delta_2^2}$ we obtain, in accordance with (49), ch. 4, the expression:

$$\overline{\Delta_1\Delta_2 + \Delta_2\Delta_1} = \frac{al^2}{8\pi}\int_0^\infty \mathrm{d}\omega\cdot\omega^2\{(q_\omega\dot{q}_\omega)^2 + (\dot{q}_\omega q_\omega)^2\}.$$

However, since the quantity $\frac{1}{2}l$ is, from (42), ch. 4, to be regarded as the 'mass' of the resonators, the commutation rules give us

$$-q_j\dot{q}_j(nn) = \dot{q}_j q_j(nn) = \frac{1}{2}\cdot\frac{2}{l}\cdot\frac{h}{2\pi \mathrm{i}} = \frac{h}{2l\pi \mathrm{i}}.$$

Hence the part $\overline{\Delta_1\Delta_2+\Delta_2\Delta_1}$ of $\overline{\Delta_1\Delta_2+\Delta_2\Delta_1}$ which belongs to dν is, after division by dν, equal to

$$\overline{\Delta_1\Delta_2 + \Delta_2\Delta_1} = -\tfrac{1}{2}(h\nu)^2 V,$$

and, with (54), ch. 4, we have in fact

$$\overline{\Delta^2} = h\nu\bar{E} + \frac{\overline{E^2}}{z_\nu V}, \tag{55}$$

which agrees with (40), ch. 4.

If one bears in mind that the question considered here is actually somewhat remote from the problems whose investigation led to the growth of quantum mechanics, the result (55), ch. 4, can be regarded as particularly encouraging for the further development of the theory.

From Ehrenfest's finding mentioned above, one could save oneself calculation of energy deviations involving interference considerations and at the same time acquire the assurance that also for other similar problems no contradictions are possible – if the additivity of the entropies of volume elements could directly be proved in the quantum mechanics of wave fields. Our above findings lead us to expect this additivity to hold generally.

The reasons leading to the appearance in (55), ch. 4, of a term which is not provided by the classical theory are obviously closely connected with the reasons for occurrence of a zero-point energy. The basic difference between the theory proposed here and that used hitherto in both instances lies in the characteristic kinematics and not in a disparity of the mechanical laws. One could indeed perceive one of the most evident examples of the difference between quantum-theoretical kinematics and that existing hitherto on examining the formula (55), ch. 4, which actually involves no mechanical principles whatsoever.

If the quantum mechanics proposed here should prove to be correct even in its essential features, we might quite generally designate the following as constituting the most important advance of this as against the past theory: that in our theory, kinematics and mechanics have again been brought into as close a relationship as that prevailing in classical theory, and that the new fundamental viewpoints, stemming as they do from the basic postulates of quantum theory for the mechanical concepts together with the concepts of space and time, find adequate expression in kinematics just as in mechanics and in the connection between kinematics and mechanics.

Received January 17, 1926

16

ON THE HYDROGEN SPECTRUM FROM THE STANDPOINT OF THE NEW QUANTUM MECHANICS

W. PAULI JR

It is shown that the Balmer terms of an atom with a single electron are yielded correctly by the new quantum mechanics and that the difficulties (particularly evident in the case of crossed fields) which arose in the earlier theory through the extra prohibition of singularities in the motion, disappear in the new theory. The influence of external electric and magnetic fields of force, too, on the hydrogen spectrum is discussed from the standpoint of the new quantum mechanics. However, relativistic corrections have not been taken into account and the calculation of transition probabilities (intensities) has for the present been omitted from consideration.

1. The fundamentals of the new quantum mechanics

Heisenberg[1] has recently published a formulation of the principles of quantum theory which represents a considerable advance over the previous theory of multiply-periodic systems. Heisenberg's form of quantum theory completely avoids a mechanical-kinematic visualization of the motion of electrons in the stationary states of an atom. Apart from time averages of classical kinematic quantities, only harmonic partial vibrations are introduced, which are associated with each transition between two stationary states and which are directly related to the spontaneous transition probabilities of the system. If

$$x^n_m = a^n_m \exp\,[2\pi i(\nu^n_m t + \delta^n_m)]$$

is the partial vibration of the Cartesian coordinate x of a given electron in an atom, associated with the transition from a state n to another state m, then it contributes an amount

$$\frac{1}{h\nu^n_m}\,\frac{2}{3}\,\frac{e^2}{c^3}\,(2\pi\nu^n_m)^4\,|x^n_m|^2\cdot 2$$

Editor's note. This paper was published as Zs. f. Phys. **36** (1926) 336–363.

[1] W. Heisenberg, Zs. f. Phys. **33** (1925) 879.

to the value of the coefficient of the spontaneous emission probability A^n_m belonging to this transition. Whereas in the earlier theory this relation could, according to the correspondence principle, be regarded only as asymptotically valid in the limit of large quantum numbers, it can now be taken as a universally valid definition of the amplitudes x^n_m. More generally, the partial vibrations assigned to each of the transition processes are physically defined by the intensity and polarization of the emitted radiation. These partial vibrations however can no longer be combined into definite 'orbits' of the atomic electrons, since they are assigned to transition processes, and not to stationary states.

Heisenberg's formulation of the quantum theory was further extended by Born and Jordan,[1] Dirac,[2] and Born, Heisenberg and Jordan.[3] A consistent mathematical system thus ensued, in which all relations formerly taken from classical mechanics were replaced by analogously constructed quantum-theoretical relations between the time averages x^n_n and the partial vibrations x^n_m of the coordinates of each of the atomic particles. In order to formulate these relations, it proved convenient to assign a matrix to each classical kinematic quantity x. The diagonal terms of such a matrix are the time averages x^n_n belonging to the individual stationary states, and the (n, m) and (m, n) elements (nth row, mth column, and mth row, nth column, respectively) are complex conjugate vibrations

$$x^n_m = a^n_m \exp\,[2\pi i(\nu^m_n t + \delta^n_m)] \quad \text{and} \quad x^m_n = a^m_n \exp\,[2\pi i(\nu^m_n t + \delta^m_n)], \quad (1)$$

with a^n_m (equal to a^n_m) positive and real, and

$$\nu^m_n = -\,\nu^n_m, \qquad \delta^m_n = -\,\delta^n_m. \qquad (2)$$

The harmonic vibration x^m_n belongs to the transition from m to n and the harmonic vibration x^n_m to the reverse transition from n to m, so that one of these transitions betokens an emission, and the other an absorption.

To the time derivative $\dot{x}$ we assign the matrix whose individual elements are time derivatives of the corresponding elements of the matrix $\mathbf{x}$, i.e.,

$$\dot{x}^n_m = 2\pi i \nu^n_m x^n_m. \qquad (3)$$

[1] M. Born and P. Jordan, Zs. f. Phys. **34** (1925) 858.

[2] P. A. M. Dirac, Proc. Roy. Soc. **109** (1925) 642.

[3] M. Born, W. Heisenberg and P. Jordan, Zs. f. Phys. **35** (1926) 557; hereafter quoted as 'Quantenmechanik II'.

In particular, we have $\dot{x}^n_n=0$, i.e., the diagonal terms of $\dot{x}$ vanish. Since $\nu^m_n=-\nu^n_m$, it also follows that $\dot{x}^n_m$ and $\dot{x}^m_n$ are conjugate complex (the matrices are Hermitian in character). To the energy E we have to assign a diagonal matrix, i.e., one whose off-diagonal terms vanish. The energy value of the quantum state characterized by the index n is given by $E_n=E^n_n$, whence follows the frequency condition,

$$h\nu^n_m = E^n_n - E^m_m, \tag{I}$$

in agreement with the above prescription $\nu^m_n=-\nu^n_m$, $\nu^n_n=0$.

The essential point, as stressed by Heisenberg, lies in the fact that the multiplication of two matrices $\mathbf{x}$ and $\mathbf{y}$ has now acquired a proper meaning, in view of the frequency condition. The product $\mathbf{xy}$ of the two matrices $\mathbf{x}$ and $\mathbf{y}$ is defined by

$$(xy)^n_m = \sum_l x^n_l y^l_m. \tag{4}$$

From (I) follows the combination rule

$$\nu^n_l + \nu^l_m = \nu^n_m, \tag{5}$$

and therefore the quantity $(xy)^n_m$ indeed again represents a harmonic vibration of frequency ν^n_m, if x^n_l and y^l_m are harmonic vibrations having frequencies ν^n_l and ν^l_m, respectively. For the phases δ^n_m, too, a combination rule

$$\delta^n_l + \delta^l_m = \delta^n_m \tag{6}$$

has to be assumed.

All the normal calculational rules apply to the multiplication of two matrices, with the exception of the commutation law: in general $\mathbf{xy}$ differs from $\mathbf{yx}$. Thus, e.g., the difference $\mathbf{Ex}-\mathbf{xE}$ (where $\mathbf{E}$ denotes the diagonal energy matrix and products are to be formed according to the general prescription (4)) can be simply related to the matrix $\dot{\mathbf{x}}$, the time derivative of $\mathbf{x}$:

$$\mathbf{Ex} - \mathbf{xE} = \frac{h}{2\pi i}\dot{\mathbf{x}}, \tag{7}$$

using (3) and the frequency condition (I). This relation holds for any arbitrary matrix $\mathbf{x}$.

The requisite relations for calculating the matrices $\mathbf{x}$ for a given mechanical system, i.e., the basic physical laws of the new quantum mechanics, have been brought by Born and Jordan into the following form (which we immediately write down for systems having arbitrarily many degrees of freedom). We denote the Cartesian coordinates and the respective momenta of the atomic particles by q_ϱ and p_ϱ, $(\varrho=1 \dots f)$ with $p_x = m\dot{x}$, etc. Then, in addition to the frequency condition (I), we have the 'quantum conditions'

$$p_\varrho p_\sigma - p_\sigma p_\varrho = 0, \qquad q_\varrho q_\sigma - q_\sigma q_\varrho = 0,$$
$$p_\varrho q_\sigma - q_\sigma p_\varrho = \begin{cases} 0 & \text{for} \quad \varrho \neq \sigma, \\ \dfrac{h}{2\pi i} \cdot \mathbf{1} & \text{for} \quad \varrho = \sigma. \end{cases} \tag{II}$$

Here the symbol **1** denotes the unit matrix (whose off-diagonal terms vanish and whose diagonal terms are each equal to 1). As Kramers[1] has shown, these relations can be interpreted from the standpoint of the quantum-theoretical dispersion formulae of Ladenburg, Kramers, and Kramers and Heisenberg, if one postulates that the individual atomic particles behave as free particles with respect to external short-period forces. Finally, as the last of the quantum laws, one has the energy conservation law:

$$H(p, q) = E \text{ (diagonal matrix).} \tag{III}$$

The matrix function $H(p, q)$ characterizes a given mechanical system and the most obvious assumption to make is to expect this function to coincide formally with the classical function when Cartesian coordinates are used. It suffices to consider the case in which it comprises two parts, corresponding to kinetic and potential energy, of which the one depends only on p and the other only on q. According to the multiplication rule (4), only those matrix functions are defined in the first instance which can be written in form of a power series in p and q (with positive and negative powers). Born, Heisenberg and Jordan have shown that in this case the basic laws (I), (II) and (III) lead to matrix relations that are completely analogous to the equations of motion in classical mechanics. They can be written as

$$\dot{q}_\varrho = \frac{\partial H(p, q)}{\partial p_\varrho}, \qquad \dot{p}_\varrho = -\frac{\partial H(p, q)}{\partial q_\varrho}, \tag{8}$$

by appropriately defining the partial differential coefficients which occur on the right-hand side.

[1] H. A. Kramers, Physica **5** (1925) 369.

It may further be remarked that the sequence in which the stationary states of the system under consideration are arranged within the matrices is immaterial and that in the new theory the concept of 'quantum number' does not enter into the basic laws. Furthermore, in the new theory, by contrast with the treatment used hitherto, the values of the transition probabilities are in principle quantitatively determined even for small quantum numbers.

2. General survey of methods and results of subsequent calculations

The present paper sets out to apply the new theory to an atom having a single electron. However, up to the present we have not, in the case of such a hydrogen-like atom, succeeded in developing all the consequences of the basic laws of the new theory, and in particular we have not yet attempted to evaluate transition probabilities for hydrogen-like spectra. We have confined ourselves to calculation of the energy values of stationary states of the hydrogen atom for the unperturbed case, and for the case where external electric and magnetic fields are present (with elimination of the transition probabilities). The relativistic correction terms have for the present not been taken into consideration. As a result, the terms of the Balmer series and the Stark effect are obtained in agreement with observation. Furthermore, difficulties disappear which had arisen in the old theory through the additional exclusion of singular motions in which the electron comes arbitrarily near the nucleus, and which became particularly evident for the case of crossed electric and magnetic fields. In view of this, we explain these difficulties in some detail.

We start by considering the case of parallel electric and magnetic fields. With e and m_0 as the electron charge and mass, Ze the nuclear charge, a the semi-axis of the electron orbit and F and H the field strengths of the electric and magnetic fields, respectively, the Larmor frequency is given by

$$o_H = \frac{eH}{4\pi m_0 c}, \tag{9}$$

and the secular Stark effect frequency, o_F, by

$$o_F = \frac{3}{4\pi}\sqrt{\frac{a}{Zm_0}}\,F = \frac{3}{2}eF\frac{a_1}{h}n, \tag{10}$$

where the quantity a_1, given by

$$a_1 = \frac{h^2}{4\pi^2 Z e^2 m_0},$$

represents the radius of the circular first quantum orbit in the atom. In the presence of external fields, we have two additional quantum conditions which fix the projection z of the distance of the orbit's electric midpoint from the nucleus along the field direction, viz.

$$z = \tfrac{3}{2}a(s/n), \tag{11}$$

and which fix the moment of momentum P_z parallel to the field, viz.

$$P_z = m(h/2\pi). \tag{12}$$

For given n and $|m| \leqq n$, the Stark-effect quantum number s here ranges over the sequence of values[1]

$$s = -(n-|m|), \quad -(n-|m|-2), \ldots, \quad (n-|m|-2), \quad n-|m|, \tag{13}$$

where $|m| \leqq n$, which lie symmetrically with respect to zero and differ by *two* units from one another (for the moment, we have ignored additional restrictions). The extra energy in presence of a field is then given by

$$E_1 = (s o_F + m o_H)h. \tag{14}$$

In generalizing this to the case of crossed fields, it is appropriate to introduce frequencies

$$\omega_1 = o_H + o_F, \qquad \omega_2 = |o_H - o_F|$$

in place of o_F and o_H. The relations (13) and (14) are then equivalent to

$$E_1 = (\tfrac{1}{2}n - n_1)\omega_1 h + (\tfrac{1}{2}n - n_2)\omega_2 h \tag{15}$$

with

$$0 \leqq n_1 \leqq n, \qquad 0 \leqq n_2 \leqq n. \tag{16}$$

Since ω_1 and ω_2 are always defined as positive quantities, i.e.,

$$\omega_2 = o_H - o_F \quad \text{for} \quad o_H > o_F,$$
$$\omega_2 = o_F - o_H \quad \text{for} \quad o_H < o_F,$$

[1] This follows from the connection between s and n with the quantum numbers n_ξ, n_η of the parabolic coordinates ξ, η, viz.

$$n = n_\xi + n_\eta + |m|, \qquad s = n_\xi - n_\eta,$$
$$0 \leqq n_\xi \leqq n, \qquad 0 \leqq n_\eta \leqq n.$$

the connection between the numbers s and m, and the numbers n_1 and n_2, is given by

$$\begin{aligned} m &= n-(n_1+n_2), & s &= n_2-n_1 \quad \text{for} \quad o_H > o_F, \\ m &= n_2-n_1, & s &= n-(n_1+n_2) \quad \text{for} \quad o_H < o_F \end{aligned} \tag{17}$$

(for $o_H = o_F$ it follows that $\omega_2 = 0$, and the system is degenerate).

Now, in the general case of crossed electric and magnetic fields, the results of Klein[1] and Lenz[2] imply that the expression (15) for the perturbation energy in the quantum states of the system remains valid provided the frequencies ω_1 and ω_2 are defined as follows. We take $\mathfrak{o}_F$ and $\mathfrak{o}_H$ to be *vectors* parallel to the directions of the applied electric and magnetic fields, respectively, whose magnitudes coincide with the secular frequencies (10) or (9) which would be produced by either of these fields alone. We then form the *vectorial* sum and difference of $\mathfrak{o}_F$ and $\mathfrak{o}_H$, and take the respective moduli, to obtain

$$\omega_1 = |\mathfrak{o}_H + \mathfrak{o}_F|, \qquad \omega_2 = |\mathfrak{o}_H - \mathfrak{o}_F|. \tag{18}$$

For parallel electric and magnetic fields, this agrees with the earlier prescription.

This result leads to considerable difficulties if one relates it to the exclusion of such orbits as would cause the electron either to fall into the nucleus or to come arbitrarily near it in the course of its motion. The first of such additional exclusion rules already appeared in Sommerfeld's relativistic theory of fine structure. There, states with vanishing momentum quantum number k, in which the electron would oscillate forwards and backwards along a rectilinear path through the centre of the nucleus, had to be excluded as unsuitable for stationary states:

$$k \neq 0. \tag{19}$$

Correspondingly, in the case of the Stark effect, the value $|s| = n$ of the Stark-effect quantum number represents such a rectilinear oscil-

[1] O. Klein, Zs. f. Phys. **22** (1924) 109.

[2] W. Lenz, Zs. f. Phys. **24** (1924) 197. The numbers designated as n_1 and n_2 in that paper are non-integers for odd n; with a common difference of one unit, they run from $-\frac{1}{2}n$ to $\frac{1}{2}n$, including the extreme values (if one adheres to the quantum rules for periodic systems).

latory motion, and empirically it is certain that it can never occur in reality:

$$|s| \neq n. \tag{20}$$

By comparing the number of relativistic fine-structure stationary states under the influence of weak axially-symmetric fields of force with that for the Stark effect, Bohr was able to demonstrate quite generally that all orbits with $m=0$ had to be excluded as well, because of the additional restriction (19) when applied to axially-symmetric fields. Incidentally, for such orbits the electron would in the case of the Stark effect approach the nucleus arbitrarily closely. Thus

$$m \neq 0. \tag{20'}$$

Condition (2) is contained in (20′) as a special case, since for $s=n$ the number m can only take the value zero, according to (13). Now, for crossed fields it is possible to carry orbits allowed as stationary states over continuously into orbits excluded by (20) or (20′). For this, one need merely carry out the following adiabatic process: Assume both fields to be parallel in the first instance and o_H to differ from o_F, e.g., $o_H > o_F$. Then, after having slowly rotated the field directions with respect to one another, reduce the magnetic field intensity until a point is reached when $|o_H| < |o_F|$; finally, re-align the fields parallel to one another. In this process, ω_1 and ω_2 always remain nonzero according to (18) and the quantum numbers n_1 and n_2 therefore retain the same values throughout. Since, however, initially $o_H > o_F$ and finally $o_H < o_F$, it follows from (17) that the process has the effect of transforming into each other states in which *the electric quantum number s and the magnetic quantum number m are interchanged.* In particular, the oscillatory orbit $s=n$, $m=0$ is converted into the circular orbit $s=0$, $m=n$, whose plane lies perpendicular to the field direction. It thus becomes apparent that the additional exclusion rules which have the effect of forbidding rectilinear oscillatory orbits cannot be consistently applied within the framework of a quantum theory of multiply-periodic systems.

The calculation carried out below (§ 5) now shows that special additional exclusion rules become superfluous in the new quantum mechanics, in which we have not conceived the stationary states as represented by particular electron orbits; hence the difficulties indicated above disappear automatically. Thus, for the nth quantum state of the unperturbed atom, with energy

$$E_n = -\,RhZ^2/n^2 \tag{21}$$

(R = Rydberg constant), we obtain once more the values (14) and (15) for the extra energy in presence of external parallel and of crossed electric and magnetic fields, respectively. The quantities o_H and o_F are again given by (9) and (10), and ω_1, ω_2 by (18). However, according to the new mechanics, n has to be replaced throughout by

$$n^* = n - 1 \tag{22}$$

in equations (13), (16) and (17). This value now acts as a maximum limit for the values of s, m and n_1, n_2, so that we now have

$$s = -(n^*-|m|),\quad -(n^*-|m|-2),\ \ldots\quad (n^*-|m|-2),\quad n^*-|m|, \qquad \text{with}\quad |m| \leqq n^*, \tag{13*}$$

$$0 \leqq n_1 \leqq n^*,\quad 0 \leqq n_2 \leqq n^*, \tag{16*}$$

$$\begin{aligned} m &= n^* - (n_1+n_2), & s &= n_2 - n_1, & &\text{for}\quad o_H > o_F,\\ m &= n_2 - n_1, & s &= n^* - (n_1+n_2), & &\text{for}\quad o_H < o_F. \end{aligned} \tag{17*}$$

In particular, we have for the Stark effect, according to (10) and (14),

$$E_1 = \tfrac{3}{2}eFa_1ns, \quad \text{with}\quad 0 \leqq s \leqq n^*, \tag{23}$$

as demanded by experiment. Further, one sees that the set of values taken on by m and s is now entirely symmetric, as required by the above-mentioned adiabatic process for crossed fields.

When the degeneracy of the unperturbed atom is removed by an additional central field of force (e.g., the field arising from the relativistic corrections) and an external magnetic field, the nth quantum state of the atom, whose energy is given by (21), decomposes according to the new quantum mechanics into states which can be characterized by quantum numbers k and m, satisfying the familiar selection rules

$$\Delta k = \pm 1, \qquad \Delta m = 0, \pm 1.$$

The integer m again specifies, according to (12), the atomic momentum component parallel to the field, whereas no such direct dynamical meaning can be ascribed to the number k which determines the value of the perturbation energy of the central field. For a quantized state of order n, the number k takes on $n-1$ consecutive values whose common difference is unity. Thus the number of fine-structure levels comes out correctly in any event, without any additional restrictions being introduced (though we cannot as yet make any predictions about their energy values). We want to normalize the quantum number k in just such a way that for each state characterized by n and k in presence of an external magnetic field, the quantum number m assumes an integral value within the range

$$-k \leqq m \leqq k. \tag{24}$$

In an nth-order quantum state, the number k normalized in this way can take the values

$$k = 0, 1, 2, \ldots, n^*. \tag{25}$$

It is possible to set up a unique correspondence between the states classified by (24) and (25) and those classified by (13*). The weight of the nth-order quantum state is (in each case) equal to n^2.

In particular, it follows from the above set of terms for a hydrogen-like atom in external fields, as furnished by the new theory, that for the ground state of such an atom, where $n=1$, $n^*=0$, the quantum number m can have no value other than $m=0$, and hence that this state is non-magnetic. This conclusion may appear surprising, especially in analogy with the behaviour of alkali atoms. In this connection, it should be stressed that the present version of the new quantum mechanics apparently cannot, as yet, account for the anomalous Zeeman effect (breakdown of Larmor's theorem) and that it might accordingly still require some modification. It may not be impossible that such modifications of the theory could become apparent even in the case of atoms having but a single electron. We shall revert to this point toward the end of this paper (§ 6).

As for the method employed below to solve the matrix equations of the new theory in the case of an atom having only one electron, we must first (in § 3) develop the requisite rules for simultaneously operating with matrices x, y, z of the Cartesian coordinates of the electron (combined into a vector matrix $\mathfrak{r}$), the matrix r of the magnitude of the radius vector, and their time derivatives. The present version of the laws of the new quantum mechanics requires that we avoid the introduction of a polar angle φ. Since this is not confined within finite limits, it cannot, namely, be formally represented as a matrix in the same way as the above-mentioned coordinates, which execute librations in classical mechanics.

For just this reason, the following special integration method, applicable to Coulomb forces in classical mechanics, and previously utilized by Lenz,[1] proves to be particularly suitable for going over to the new quantum mechanics. If

$$\mathfrak{P} = m_0[\mathfrak{r}\mathfrak{v}] \tag{26}$$

denotes the time-independent angular momentum of the electron about the nucleus, and

$$\mathfrak{p} = m_0\mathfrak{v}$$

the linear momentum, then it can be shown directly from the equations of motion in classical mechanics that the vector

$$\mathfrak{A} = \frac{1}{Ze^2m_0}\,[\mathfrak{P}\mathfrak{p}] + \frac{\mathfrak{r}}{r} \tag{27}$$

is constant in time. Scalar multiplication with $\mathfrak{r}$ then gives

$$(\mathfrak{A}\mathfrak{r}) = -\frac{1}{Ze^2m_0}\,\mathfrak{P}^2 + r. \tag{28}$$

This is the equation of a conic section, and it can from this be seen that $\mathfrak{A}$ lies along the direction from the nucleus to the aphelion of the ellipse and that its magnitude is equal to the numerical eccentricity of the ellipse. Squaring (27), one obtains

$$1 - \mathfrak{A}^2 = -\frac{2E}{m_0Z^2e^4}\,\mathfrak{P}^2, \tag{29}$$

where E represents the energy.

In § 4 it will be shown that in the new mechanics, too, a time-independent vector matrix $\boldsymbol{\mathfrak{A}}$ analogous to (27) can be introduced for which, together with the vector matrix of angular momentum $\boldsymbol{\mathfrak{P}}$ (also constant in time), relations analogous to (28) and (29) hold. If, in addition, one uses the quantum conditions (II) which are characteristic of the new mechanics, together with the relations derived in § 3, one obtains a system of matrix equations that comprise only the time-independent matrices $\boldsymbol{\mathfrak{A}}$, $\boldsymbol{\mathfrak{P}}$ and E; the coordinates (i.e., the transition probabilities) are eliminated. The solution of these latter equations which can be effected by elementary methods (§ 5), then leads to the results already discussed in the present section.

[1] W. Lenz, Zs. f. Phys. **24** (1924) 197.

3. Calculational rules for the radius-vector matrix. Momentum conservation law for central forces

We start by setting up the rules of calculation for the matrices x, y, z of the Cartesian coordinates of the electron, which make up the components of the vector matrix $\mathfrak{r}$, and for the matrix r which represents the magnitude of the radius vector. Obviously, they must satisfy the relation

$$r^2 = x^2 + y + z^2. \tag{30}$$

The quantum conditions (II) not only express the commutability of x and y, of x and $\dot{y}$, and of $\dot{x}$ and $\dot{y}$ (and correspondingly for the remaining coordinates),

$$xy = yx, \ldots; \qquad x\dot{y} = \dot{y}x, \ldots; \qquad \dot{x}\dot{y} = \dot{y}\dot{x}, \ldots. \tag{31a}$$

They also contain relations for the product of x with the momentum component p_x (and similarly for the other canonical quantities):

$$p_x x - x p_x = \frac{h}{2\pi i} \mathbf{1}, \ldots \tag{31b}$$

($p_x = m\dot{x}$ denotes the x-component of the linear momentum $\mathfrak{p} = m\mathfrak{v}$, a vector matrix with the components p_x, p_y, p_z). Here and in what follows, we shall denote by '...' the presence of analogous equations for the remaining coordinates, obtained by cyclic permutation of the coordinates in the expressions cited.

These rules can be extended by the following additional relations, on making use of the matrix r. Firstly, r also commutes with x, y, z. This can be written in the form of a vector equation:

$$r\mathfrak{r} = \mathfrak{r}r. \tag{32}$$

Secondly, for any arbitrary rational function f of r, x, y, z, the relation

$$p_x f - f p_x = \frac{h}{2\pi i} \frac{\partial f}{\partial x}, \ldots \tag{33}$$

holds, and in particular for $f = r$:

$$\mathfrak{p}r - r\mathfrak{p} = \frac{h}{2\pi i} \frac{\mathfrak{r}}{r}. \tag{34}$$

Conversely, (33) follows generally from (31) and (34) for every function which can be expressed as a series of positive and negative powers of

x, y, z and r, as can easily be shown by induction. Relation (34) is also in accord with (30). For this reason, the existence of relations (32) and (33) constitutes a necessary requirement for the energy conservation law,

$$\tfrac{1}{2}m\mathfrak{v}^2 + F(x, y, z, r) = E \quad \text{(diagonal matrix)} \tag{35}$$

(for simplicity, we assume here that there is only a single particle), together with the frequency condition (leading to the equation

$$E\Phi - \Phi E = \frac{h}{2\pi i}\dot{\Phi}$$

for any quantity Φ) to yield the equations of motion

$$\frac{dp_x}{dt} = -\frac{\partial F}{\partial x}, \ldots. \tag{36}$$

We therefore postulate the existence of a matrix r which satisfies relations (30), (32), (34).

We now introduce a vector matrix $\mathfrak{P}$, which represents the angular momentum of the particle about the origin. Let us first of all remark that we shall define, as in ordinary vector algebra, the scalar product of two vector matrices $\mathfrak{A}$ and $\mathfrak{B}$ as the expression

$$(\mathfrak{A}\mathfrak{B}) = \mathfrak{A}_x\mathfrak{B}_x + \mathfrak{A}_y\mathfrak{B}_y + \mathfrak{A}_z\mathfrak{B}_z,$$

and the vector product $[\mathfrak{A}\mathfrak{B}]$, as a new vector matrix having components

$$[\mathfrak{A}\mathfrak{B}]_x = \mathfrak{A}_y\mathfrak{B}_z - \mathfrak{A}_z\mathfrak{B}_y, \ldots . \tag{37}$$

In general, the order of the factors for $\mathfrak{A}$ and $\mathfrak{B}$ is important: the expressions $(\mathfrak{A}\mathfrak{B})-(\mathfrak{B}\mathfrak{A})$ and $[\mathfrak{A}\mathfrak{B}]+[\mathfrak{B}\mathfrak{A}]$ do not in general vanish here, since the commutation law for multiplication does not apply. Also, the components of the vector product $[\mathfrak{A}\mathfrak{A}]$ of a matrix $\mathfrak{A}$ with itself are in general different from zero:

$$[\mathfrak{A}\mathfrak{A}]_x = \mathfrak{A}_y\mathfrak{A}_z - \mathfrak{A}_z\mathfrak{A}_y, \ldots . \tag{37'}$$

However, a special case is obtained if we form the vector product $[\mathfrak{r}\mathfrak{v}]$, since this is equal to $-[\mathfrak{v}\mathfrak{r}]$ because of the commutability of x with $\dot{y}$. We can thus define the vector matrix

$$\mathfrak{P} = m[\mathfrak{r}\mathfrak{v}] = -m[\mathfrak{v}\mathfrak{r}] \tag{38}$$

to be the angular momentum of the particle.

This matrix satisfies the following commutation rules, which are a direct consequence of (31a) and (31b):

$$xP_x = P_x x, \ldots; \qquad xP_y - P_y x = P_x y - yP_x = -\frac{h}{2\pi i} z, \ldots; \qquad (\mathfrak{r}\mathfrak{P}) = (\mathfrak{P}\mathfrak{r}) = 0, \tag{39}$$

and also

$$p_x P_x = P_x p_x, \ldots; \qquad p_x P_y - P_y p_x = P_x p_y - p_y P_x = -\frac{h}{2\pi i} p_z, \ldots; \qquad (\mathfrak{P}\mathfrak{p}) = (\mathfrak{p}\mathfrak{P}) = 0. \tag{40}$$

From this it follows that $v^2 = \dot{x}^2 + \dot{y}^2 + \dot{z}^2$ commutes[1] with P_x, P_y, P_z:

$$v^2 \mathfrak{P} = \mathfrak{P} v^2. \tag{41}$$

Further, we deduce from (34) the commutability of r with P_x, P_y, P_z,

$$r\mathfrak{P} = \mathfrak{P} r; \tag{42}$$

therefore every function $F(r)$ of r alone commutes with $\mathfrak{P}$. If we are dealing with a central force, for which the potential energy depends only on r,

$$\tfrac{1}{2} m \mathfrak{v}^2 + F(r) = E \qquad \text{(diagonal matrix)}, \tag{35'}$$

it follows that

$$E\mathfrak{P} = \mathfrak{P} E,$$

and hence the vector matrix $\mathfrak{P}$ remains constant with time (angular momentum integral).

For the vector product of $\mathfrak{P}$ with itself (cf. (37′)), it is easy to obtain a relation which will be of use[2] later on,

$$[\mathfrak{P}\mathfrak{P}] = -\frac{h}{2\pi i} \mathfrak{P}. \tag{43}$$

For example, from (39) and (40) we obtain for the z-component of $[\mathfrak{P}\mathfrak{P}]$:

$$\begin{aligned} P_x P_y - P_y P_x &= P_x(zp_x - xp_z) - (zp_x - xp_z)P_x \\ &= (P_x z - zP_x)p_x - x(P_x p_z - p_z P_x) \\ &= \frac{h}{2\pi i}(yp_x - xp_y) = -\frac{h}{2\pi i} P_z. \end{aligned}$$

[1] Note the identity $a^2b - ba^2 \equiv a(ab - ba) + (ab - ba)a$.

[2] Cf. 'Quantenmechanik II', p. 597, eq. (3). The quantities denoted by M_x, M_y, M_z there, represent the *negative* angular momentum components.

We are now in a position to evaluate the radial momentum $p_r = m\dot{r}$, for this is in fact equal to

$$p_r = \frac{2\pi i}{h} m(Er - rE) = \frac{2\pi i}{h} \tfrac{1}{2}[(p_x^2 + p_y^2 + p_z^2)r - r(p_x^2 + p_y^2 + p_z^2)]$$

$$= \frac{2\pi i}{h} \tfrac{1}{2}[p_x(p_x r - r p_x) + p_y(p_y r - r p_y) + p_z(p_z r - r p_z)$$

$$+ (p_x r - r p_x)p_x + (p_y r - r p_y)p_y + (p_z r - r p_z)p_z],$$

.nd hence, because of (34),

$$p_r = \frac{1}{2}\left[\left(\mathfrak{p}\,\frac{\mathfrak{r}}{r}\right) + \left(\frac{\mathfrak{r}}{r}\,\mathfrak{p}\right)\right]. \tag{44}$$

Now, from (33)

$$\left(\mathfrak{p}\,\frac{\mathfrak{r}}{r}\right) - \left(\frac{\mathfrak{r}}{r}\,\mathfrak{p}\right) = \frac{h}{2\pi i} \operatorname{div} \frac{\mathfrak{r}}{r} = \frac{h}{2\pi i}\,\frac{2}{r},$$

so that (44) can also be written

$$p_r = (\mathfrak{p}\mathfrak{r})\,\frac{1}{r} - \frac{h}{2\pi i}\,\frac{1}{r} = \frac{1}{r}\,(\mathfrak{r}\mathfrak{p}) + \frac{h}{2\pi i}\,\frac{1}{r}. \tag{44'}$$

Multiplying this by r, we obtain, first, the relation

$$p_r r + r p_r = (\mathfrak{p}\mathfrak{r}) + (\mathfrak{r}\mathfrak{p}), \tag{45}$$

which is also obtained directly by differentiating (30) with respect to time. Also

$$p_r r - r p_r = (\mathfrak{p}\mathfrak{r}) - (\mathfrak{r}\mathfrak{p}) - \frac{h}{2\pi i}\,\mathbf{2}.$$

The meaning of $(\mathfrak{p}\mathfrak{r}) - (\mathfrak{r}\mathfrak{p})$ is now $(p_x x - x p_x) + (p_y y - y p_y) + (p_z z - z p_z)$. From (31b), each of the bracketed terms has the value

$$\frac{h}{2\pi i}\,\mathbf{1}.$$

We thus obtain the overall result

$$p_r r - r p_r = \frac{h}{2\pi i}\,\mathbf{1}. \tag{46}$$

Finally, in preparation for the application which follows, we evaluate the time-derivative of $\mathfrak{r}/r$. For example, for the x-component we make use of (34) to obtain

$$\frac{d}{dt}\frac{x}{r} = \frac{2\pi i}{h}\left(E\frac{x}{r} - \frac{x}{r}E\right) = \frac{2\pi i}{h}\frac{1}{2m}\left(\mathfrak{p}^2\frac{x}{r} - \frac{x}{r}\mathfrak{p}^2\right)$$

$$= \frac{2\pi i}{h}\frac{1}{2m}\left\{\mathfrak{p}\left(\mathfrak{p}\frac{x}{r} - \frac{x}{r}\mathfrak{p}\right) + \left(\mathfrak{p}\frac{x}{r} - \frac{x}{r}\mathfrak{p}\right)\mathfrak{p}\right\}$$

$$= \frac{1}{2m}\left\{\left(p_x\frac{y^2+z^2}{r^3} - p_y\frac{xy}{r^3} - p_z\frac{xz}{r^3}\right)\right.$$

$$\left. + \left(\frac{y^2+z^2}{r^3}p_x - \frac{xy}{r^3}p_y - \frac{xz}{r^3}p_z\right)\right\}$$

$$= \frac{1}{2m}\left\{\left(P_y\frac{z}{r^3} - P_z\frac{y}{r^3}\right) + \left(\frac{z}{r^3}P_y - \frac{y}{r^3}P_z\right)\right\}.$$

Thus, generally,

$$\frac{d}{dt}\frac{\mathfrak{r}}{r} = \frac{1}{2m}\left\{\left[\mathfrak{P}\frac{\mathfrak{r}}{r^3}\right] - \left[\frac{\mathfrak{r}}{r^3}\mathfrak{P}\right]\right\}. \tag{47}$$

4. Introduction of the vector matrix $\mathfrak{A}$, constant in time, for Coulomb forces. Elimination of the coordinates

Let us now consider an atom consisting of a fixed nucleus having charge $+Ze$ and exerting a Coulomb attraction on a single orbital electron of mass m_0 and charge $-e$. For the Hamiltonian, we have to set

$$\frac{1}{2m_0}\mathfrak{p}^2 - \frac{Ze^2}{r} = E \quad \text{(diagonal matrix)}, \tag{48}$$

i.e., for this special case we set

$$F(r) = -Ze^2/r$$

in (35′). The equations of motion (36) derived from energy-conservation with the aid of the quantum rules assume the same form here as in classical mechanics:

$$\dot{\mathfrak{p}} = m_0\ddot{\mathfrak{r}} = -\frac{Ze^2}{r^3}\mathfrak{r}. \tag{49}$$

Analogously to classical mechanics (cf. (27)), it follows from (47) that the vector matrix $\mathfrak{A}$ defined by

$$\mathfrak{A} = \frac{1}{Ze^2m_0}\,\frac{1}{2}\,\{[\mathfrak{P}\mathfrak{p}] - [\mathfrak{p}\mathfrak{P}]\} + \frac{\mathfrak{r}}{r} \tag{50}$$

is constant with time in the special case of a Coulomb field of force. Using (40), we can also write

$$\mathfrak{A} = \frac{1}{Ze^2m_0}\left\{[\mathfrak{P}\mathfrak{p}] + \frac{h}{2\pi\mathrm{i}}\,\mathfrak{p}\right\} + \frac{\mathfrak{r}}{r}$$

$$= -\frac{1}{Ze^2m_0}\left\{[\mathfrak{p}\mathfrak{P}] + \frac{h}{2\pi i}\,\mathfrak{p}\right\} + \frac{\mathfrak{r}}{r}. \tag{51'}$$

The remaining calculations are quite elementary when use is made of the rules collected in the preceding section. In the first instance, in analogy with equation (28) of a conic in classical mechanics, we obtain the relation

$$\tfrac{1}{2}[(\mathfrak{A}\mathfrak{r}) + (\mathfrak{r}\mathfrak{A})] = -\frac{1}{Ze^2m_0}\left[\mathfrak{P}^2 + \frac{3}{2}\,\frac{h^2}{4\pi^2}\right] + r, \tag{51}$$

and also the commutation rule

$$\tfrac{1}{2}([\mathfrak{A}\mathfrak{r}] + [\mathfrak{r}\mathfrak{A}]) = -\frac{h}{2\pi\mathrm{i}}\,\frac{3}{2}\,\frac{1}{Ze^2m_0}\,\mathfrak{P}. \tag{52}$$

Furthermore, the following relations are found to be valid, in which the coordinates x, y, z, r have been entirely eliminated, leaving only the matrices $\mathfrak{A}$, $\mathfrak{P}$ and E which are constant with time:

$$[\mathfrak{P}\mathfrak{P}] = -\frac{h}{2\pi\mathrm{i}}\,\mathfrak{P}. \tag{I}$$

$$\left.\begin{aligned} &A_xP_x = P_xA_x, \ldots, \\ &A_xP_y - P_yA_x = P_xA_y - A_yP_x = -\frac{h}{2\pi\mathrm{i}}\,A_z, \ldots, \\ &(\mathfrak{A}\mathfrak{P}) = (\mathfrak{P}\mathfrak{A}) = 0 \end{aligned}\right\} \tag{II}$$

$$[\mathfrak{A}\mathfrak{A}] = \frac{h}{2\pi\mathrm{i}}\,\frac{2}{m_0Z^2e^4}\,E\mathfrak{P}. \tag{III}$$

$$1 - \mathfrak{A}^2 = -\frac{2}{m_0Z^2e^4}\,E\left(\mathfrak{P}^2 + \frac{h^2}{4\pi^2}\right). \tag{IV}$$

Equation (I) is identical with equation (43) of the previous section, (II) is analogous with (39) in form, (IV) is analogous to the classical equation (29), though the occurrence of the additional term $h^2/4\pi^2$ (as also of the additional term $\frac{3}{2}h^2/4\pi^2$ in (51)) is characteristic of the new mechanics.

From the existence of the vector matrix $\mathfrak{A}$, constant in time, we can infer that an atom with a single electron constitutes a degenerate system, even apart from the spatial orientation of the atom (a situation similar to Kepler motion in classical mechanics). Namely, we can readily conclude from the relations derived above that in general $\mathfrak{A}\mathfrak{P}^2-\mathfrak{P}^2\mathfrak{A}$ cannot vanish. Since, on the other hand, $\mathfrak{A}E-E\mathfrak{A}$ vanishes, it is obvious that for every value of the energy E there is not just a single value of $\mathfrak{P}^2$; the system is thus in fact degenerate.

As discussed in detail by Born, Heisenberg and Jordan,[1] when such a system is treated from the standpoint of the new quantum mechanics the amplitudes of the various partial vibrations belonging to transitions between states of predetermined energy are not uniquely established by the quantum-mechanical equations. Furthermore, matrices which are constant with time need not in general be diagonal, inasmuch as non-zero elements can occupy positions (n, m) which correspond to a vanishing frequency $\nu^n_m=(E_n-E_m)/h=0$. In our case, to each energy value (to each value of the principal quantum number) there belongs a matrix which contains the time-independent parts of a quantity (e.g., x or r) and whose number of rows or columns is equal to the weight of that particular energy state. This matrix is obtained by setting equal to zero all those elements in the original matrix that are located in positions relating to a transition process associated with a change in the energy value. It is called the time-average of the corresponding quantity and is denoted by a bar above that quantity, (e.g., $\bar{x}$ or $\bar{r}$).

Even though in the case of a degenerate system the individual partial vibrations of a kinematic quantity which belong to the same frequency ν^n_m are not uniquely determined, nevertheless the energy values and statistical weights of these states are.[2] Thus in principle it should be possible to derive the Balmer terms and the corresponding statistical weights from equations (I) to (IV) without any further specifying assumptions as to the type of solution. This we have, regrettably, not succeeded in doing and in the following we sidestep

[1] Quantenmechanik II, Chapter 2, § 2.

[2] Quantenmechanik II, Chapter 2, § 2 and Chapter 3.

this difficulty by introducing additional requirements (in various ways) which render the solution of equations (I) to (IV) unique.

If the degeneracy is removed by an additional perturbing field whose Hamiltonian is H_1, then the time-average of the perturbation function $\overline{H}_1$, taken over the unperturbed motion, must be a diagonal matrix, as can be shown by carrying through the perturbation calculation according to the method of Born, Heisenberg and Jordan.[1] In our case, this mean value depends in general not only on the energy E of the unperturbed motion but also on $\mathfrak{P}$ and $\mathfrak{A}$.

If, in particular, the perturbing field is due to an additional non-Coulomb central force, the above time average depends only on $\mathfrak{P}^2$ (apart from its E-dependence), since there exists no preferred direction in space here. Further, the perturbation energy of a magnetic field in the z-direction depends only on the momentum component P_z, which is parallel to the field. The requirement that $\mathfrak{P}^2$ *and* P_z are to be *diagonal* matrices therefore leads to a special solution of equations (I) to (IV), adapted to the relativistic fine structure and to an additional weak magnetic field. We shall treat this case in the next section.

A second case of particular interest is furnished by the Stark effect. Here, we are concerned with the existence of a diagonal matrix $\bar{z}$ which is the component in the field direction (z-direction) of the time-independent vector matric $\bar{\mathfrak{r}}$ representing the electrical centre of the orbit. It can, however, be shown that this matrix $\bar{\mathfrak{r}}$ is connected with the matrix $\mathfrak{A}$ in just the same way as in classical theory, viz. by the relation

$$\bar{\mathfrak{r}} = \frac{3}{2} \frac{Ze^2}{2|E|} \mathfrak{A}, \tag{53}$$

(classically, $Ze^2/2|E|$ represents the semi-major axis a of the Kepler ellipse). Namely, in the first place the same commutation rules apply to $\bar{\mathfrak{r}}$ and $\mathfrak{P}$ as to $\mathfrak{A}$ and $\mathfrak{P}$, from (39) and (II). Further, if we next compare (52) with (III), we obtain the relations

$$\left[\mathfrak{A}, \bar{\mathfrak{r}} - \frac{3}{2} \frac{Ze^2}{2|E|} \mathfrak{A}\right] = 0$$

involving the difference

$$\bar{\mathfrak{r}} - \frac{3}{2} \frac{Ze^2}{2|E|} \mathfrak{A}.$$

1 Quantenmechanik II, Chapter 2, § 2 and Chapter 3.

As is shown by the more detailed discussion based on the solutions of equations (I) to (IV) derived in the next section, these constitute a sufficient number of homogeneous linear equations to permit us to conclude that

$$\bar{\mathfrak{r}} - \frac{3}{2}\,\frac{Ze^2}{2|E|}\,\mathfrak{A}$$

vanishes. If we supplement the electric field in the z-direction by a parallel magnetic field along the same direction, we can use (53) to characterize this case by stipulating that A_z *and* P_z are *diagonal matrices*.

Finally, at the end of the next section we also treat the case of crossed electric and magnetic fields. As explained in detail in § 2, this is of especial interest because of the occurrence of additional exclusion conditions for singular motions, which appeared in the earlier theory.

5. Solution of equations (I) to (IV). Derivation of the Balmer terms

(a) P_z *and* $\mathfrak{P}^2$ *are diagonal matrices*. For this first case, where the degeneracy is removed by superimposing an additional central field and a weak magnetic field in the z-direction, we make the following *Ansatz* in order to satisfy equations (I) and (II). For a given value of $\mathfrak{P}^2$, let the possible values of P_z be

$$P_{z\,k,m}^{\;k,m} = mh/2\pi, \tag{54}$$

where m runs from $-k$ to $+k$:

$$-k \leqq m \leqq k. \tag{54'}$$

Further, let the partial vibrations of $\mathfrak{P}$, which belong to a change in m by ± 1, be left- and right-circular in the (x, y)-plane:

$$P_{y\,k,m\pm 1}^{\;k,m} = \pm\, iP_{x\,k,m\pm 1}^{\;k,m}. \tag{55}$$

It then follows from (I) that

$$|P_{x\,k,m\mp 1}^{\;k,m}|^2 = |P_{y\,k,m\mp 1}^{\;k,m}|^2 = \frac{1}{4}\,\frac{h^2}{4\pi^2}\,[k(k+1) - m(m\mp 1)]$$

$$= \frac{1}{4}\,\frac{h^2}{4\pi^2}\,(k\pm m)(k+1\mp m). \tag{56}$$

$$(\mathfrak{P}^2)_{k,m}^{k,m} = \frac{h^2}{4\pi^2}\,k(k+1). \tag{57}$$

Next we set for the matrix $\mathfrak{A}$,

$$A_{yk',m\pm 1}^{k,m} = \pm\, \mathrm{i} A_{xk',m\pm 1}^{k,m} \qquad (k' = k+1 \quad \text{or} \quad k-1), \tag{58}$$

$$|A_{xk,m\pm 1}^{k+1,m}|^2 = |A_{yk,m\pm 1}^{k+1,m}|^2 = \tfrac{1}{4} C_k^{k+1}(k \mp m)(k \mp m + 1), \tag{59}$$

in accordance with the Hönl–Kronig formula for the intensity of the Zeeman components. When m is replaced by $m-1$ or $m+1$, it further follows that

$$|A_{xk+1,m\pm 1}^{k,m}|^2 = |A_{yk+1,m\pm 1}^{k,m}|^2$$
$$= \tfrac{1}{4} C_{k+1}^{k}(k \pm m + 1)(k \pm m + 2). \tag{59a}$$

Finally, we have for A_z,

$$|A_{zk,m}^{k+1,m}|^2 = C_k^{k+1}[(k+1)^2 - m^2]. \tag{60}$$

It still remains to be seen whether m (and thus also k) is an integer or half-integer; also, the C_k^{k+1} remain for the time being undetermined functions of k which can never take on negative values and which satisfy the symmetry relation

$$C_k^{k+1} = C_{k+1}^{k}. \tag{61}$$

A further remark concerning the signs of $\mathfrak{A}$ relative to those of $\mathfrak{P}$: if P_x and A_z are assumed to be positive and real, then A_x has to be taken as positive or negative (and real) depending on whether one is dealing with transitions that correspond to changes of k and m in the opposite sense (as for $A_{xk+1,m-1}^{k,m}$ and $A_{xk-1,m+1}^{k,m}$) or in the same sense (as for $A_{xk+1,m+1}^{k,m}$ and $A_{xk-1,m-1}^{k,m}$). When the calculations are carried through, it is evident that this approach satisfies equations (I) and (II) of the previous section. Moreover, it follows conversely from considerations by Born, Heisenberg and Jordan[1] that if $\mathfrak{P}^2$ and P_z are assumed to be diagonal matrices, the expression chosen here for $\mathfrak{A}$ and $\mathfrak{P}$ is a necessary consequence of (I) and (II).

In order now to determine the normalization of m and k and the function C_{k+1}^{k}, we make use of equation (III) from the preceding section. It suffices, however, to use just the z-component,

$$\mathsf{A}_x\mathsf{A}_y - \mathsf{A}_y\mathsf{A}_x = \frac{h}{2\pi\mathrm{i}}\,\frac{2}{m_0 Z^2 e^4}\,\mathsf{E}P_z. \tag{62}$$

[1] Quantenmechanik II, Chapter 4, § 1. Cf. also the discussion of the Zeeman effect in Chapter 4, § 2.

Namely, if we form the expression

$$\mathsf{P}_y(\mathsf{A}_x\mathsf{A}_y - \mathsf{A}_y\mathsf{A}_x) - (\mathsf{A}_x\mathsf{A}_y - \mathsf{A}_y\mathsf{A}_x)\mathsf{P}_y = \frac{h}{2\pi i}\,\frac{2}{m_0Z^2e^4}\,\mathsf{E}(\mathsf{P}_y\mathsf{P}_z - \mathsf{P}_z\mathsf{P}_y)$$

and use (I) and (II), we obtain an equation which agrees with the x-component of (III). Similarly, the y-component of (III) also follows from the z-component of this vector equation and equations (I) and (II).

If we form the element of equation (62) which occupies the (k, m) position in the diagonal series, we first obtain, for the left-hand side, from (58) and (59),

$$\begin{aligned}(\mathsf{A}_x\mathsf{A}_y - \mathsf{A}_y\mathsf{A}_x)^{k,m}_{k,m} &= 2i\{|A_{x\,k+1,m-1}^{\;k,m}|^2 - |A_{x\,k+1,m+1}^{\;k,m}|^2 \\ &\quad + |A_{x\,k-1,m-1}^{\;k,m}|^2 - |A_{x\,k-1,m+1}^{\;k,m}|^2\} \\ &= im\{-(2k+3)C_k^{k+1} + (2k-1)C_{k-1}^k\}.\end{aligned}$$

Noting further that E has a negative sign, and introducing the Rydberg constant

$$R = 2\pi^2e^4m_0/h^3 \tag{63}$$

together with the value of P_z given by (54), we see that equation (62) yields the condition

$$m\{-(2k+3)C_k^{k+1} + (2k-1)C_{k-1}^k\} = \frac{|E|}{RhZ^2}\,m. \tag{64}$$

Let us first of all consider the smallest possible value of k for a given $|E|$. Obviously the contribution from the transition $k \to k-1$ on the left-hand side vanishes for this value of k, and the coefficient of m on the left-hand side can therefore certainly not be positive, whereas the coefficient of m on the right-hand side is positive. Hence equation (64) can be satisfied for the minimum value of k only if $m=0$. But according to (54), this means that the minimum value of k must itself vanish, since otherwise m could assume other, non-zero, values. *Hence k and m are necessarily integer*, and k assumes the values

$$k = 0, 1, 2, \ldots n^*, \tag{65}$$

the integer n^* being the largest value of k that can be attained for a given $|E|$. Now (64) implies

$$(2k-1)C_{k-1}^k - (2k+3)C_k^{k+1} = \frac{|E|}{RhZ^2} \quad \text{for} \quad k = 1, \ldots n^*. \tag{64'}$$

Furthermore, we have to set

$$C_{n^*}^{n^*+1} = 0, \tag{64''}$$

since obviously the contribution from the transition $k+1 \rightarrow k$ (second term) disappears for $k=n^*$. Beginning with $k=n^*$ and reducing k stepwise, we can successively calculate the values of

$$C_{n^*-1}^{n^*}, \; C_{n^*-2}^{n^*-1}, \; \ldots, \; C_0^1$$

from (64′). The result can be expressed by the formula

$$\begin{aligned} C_k^{k+1} &= \frac{|E|}{RhZ^2} \frac{n^*(n^*+2) - k(k+2)}{(2k+1)(2k+3)} \\ &= \frac{|E|}{RhZ^2} \frac{(n^*-k)(n^*+k+2)}{(2k+1)(2k+3)}. \end{aligned} \tag{66}$$

Replacing k by $k-1$, we also obtain

$$\begin{aligned} C_{k-1}^{k} &= \frac{|E|}{RhZ^2} \frac{n^*(n^*+2) - (k-1)(k+1)}{(2k-1)(2k+1)} \\ &= \frac{|E|}{RhZ^2} \frac{(n^*-k+1)(n^*+k+1)}{(2k-1)(2k+1)}. \end{aligned} \tag{66'}$$

With the help of these formulae, we can confirm directly that relations (64′) and (64″) are satisfied.

Finally, in order to derive the energy value itself, we make use of the last equation (IV). First of all, we determine the value of $\mathfrak{A}^2$ at the (k, m) position of the diagonal series. Because of (59) and (60), we obtain

$$\begin{aligned} (\mathfrak{A}^2)_{k,m}^{k,m} &= 2|A_{x\,k+1,m+1}^{k,m}|^2 + 2|A_{x\,k+1,m-1}^{k,m}|^2 + |A_{z\,k+1,m}^{k,m}|^2 \\ &\quad + 2|A_{x\,k-1,m+1}^{k,m}|^2 + 2|A_{x\,k-1,m-1}^{k,m}|^2 + |A_{z\,k-1,m}^{k,m}|^2 \\ &= (k+1)(2k+3)C_k^{k+1} + k(2k-1)C_{k-1}^k, \end{aligned}$$

and on substitution from (66), (66′),

$$(\mathfrak{A}^2)_{k,m}^{k,m} = \frac{|E|}{RhZ^2}\,[n^{*2} + 2n^* - k(k+1)]. \tag{67}$$

This expression for $\mathfrak{A}^2$ and the expression (57) for $\mathfrak{P}^2$ now have to be substituted in (IV), giving

$$1 = \frac{|E|}{RhZ^2}(n^{*2} + 2n^* + 1) = \frac{|E|}{RhZ^2}(n^* + 1)^2,$$

and hence

$$|E| = \frac{RhZ^2}{(n^* + 1)^2} = \frac{RhZ^2}{n^2}, \tag{68}$$

(setting $n = n^* + 1$), as in § 2. This demonstrates that the Balmer terms result correctly from the new quantum mechanics and that the weight n^2 is associated with the nth quantum state in the new theory.

(b) *A_z and P_z are diagonal matrices* (*Stark effect*). If a homogeneous electric field of strength F acts in the z-direction, the time-average of the perturbation energy is given by (53),

$$E_1 = \tfrac{3}{2}eF\bar{\mathsf{z}} = \tfrac{3}{2}eF\frac{Ze^2}{2|E|}\mathsf{A}_z. \tag{69}$$

For this case, therefore, we seek a solution of equations (I) to (IV) for which A_z is a diagonal matrix. The added condition that P_z, too, should be a diagonal matrix has the physical meaning that we consider the degeneracy of the secular perturbation of the Stark effect to be removed by an additional weak magnetic field parallel to the electric field.

We shall confine ourselves here simply to quoting the result, without going through the individual calculations in detail and without supplying the proof that the given solution of equations (I) to (IV) is the only one meeting the requirement that A_z and P_z be diagonal matrices. The states which belong to a definite value of the unperturbed energy as given by (68), have to be classified by two quantum numbers s and m, of which the former determines the value of A_z (and of the additional energy E_1), according to

$$A_{z s,m}^{s,m} = s/n, \qquad E_1 = \tfrac{3}{2}eFa_1 ns \qquad (0 \leqq s \leqq n^*), \tag{70}$$

(where $a_1 = h^2/4\pi^2 Ze^2 m_0$) and the latter determines that of P_z, according to

$$P_{z s,m}^{s,m} = mh/2\pi. \tag{71}$$

The range of values taken on by s and m has already been given in § 2 by the relation (13*). The matrices P_x, P_y, A_x, A_y have non-zero elements only in those positions which correspond to a change of ± 1 in s and m. Their values are given by

$$P_{y_{s',m\pm1}}^{s,m} = \pm\, \mathrm{i} P_{x_{s',m\pm1}}^{s,m}, \qquad A_{y_{s',m\pm1}}^{s,m} = \pm\, \mathrm{i} A_{x_{s',m\pm1}}^{s,m} \tag{72}$$

$$(s' = s+1 \quad \text{or} \quad s-1).$$

$$\begin{aligned} A_{x_{s\pm1,m\pm1}}^{s,m} &= +\frac{2\pi}{h}\frac{1}{n} P_{x_{s\pm1,m\pm1}}^{s,m}, \\ A_{x_{s\mp1,m\pm1}}^{s,m} &= -\frac{2\pi}{h}\frac{1}{n} P_{x_{s\mp1,m\pm1}}^{s,m}, \end{aligned} \tag{73}$$

(in these last relations either the upper or the lower sign is to be taken throughout), and

$$\begin{aligned} |P_{x_{s-1,m-1}}^{s,m}|^2 &= |P_{y_{s-1,m-1}}^{s,m}|^2 \\ &= \frac{1}{16}\frac{h^2}{4\pi^2}[n^* + 2 - (m+s)][n^* + (m+s)], \\ |P_{x_{s+1,m-1}}^{s,m}|^2 &= |P_{y_{s+1,m-1}}^{s,m}|^2 \\ &= \frac{1}{16}\frac{h^2}{4\pi^2}[n^* + 2 - (m-s)][n^* + (m-s)]. \end{aligned} \tag{74}$$

It is easy to verify that equations (I) to (IV) are indeed satisfied when expressions (70) to (74) are used.

(c) *Crossed fields.* If the vectors $\mathfrak{E}$ and $\mathfrak{H}$ represent the strengths of the external electric and magnetic field respectively, the time-average of the perturbation energy when both fields are simultaneously present is given by

$$E_1 = \tfrac{3}{2} ea(\mathfrak{E}\mathfrak{A}) + \frac{e}{2m_0 c}(\mathfrak{H}\mathfrak{P}). \tag{75}$$

The quantity a, which in the earlier theory represented the semi-axis of the Kepler ellipse, is now to be regarded simply as an abbreviation for

$$a = \frac{Ze^2}{2|E|}. \tag{76}$$

We introduce the vectors $\mathfrak{o}_F$ and $\mathfrak{o}_H$, which are respectively parallel to $\mathfrak{E}$ and $\mathfrak{H}$ and whose magnitudes are equal to the secular frequencies that would be obtained if just one of the homogeneous external fields were to act by itself, so that (cf. (9) and (10))

$$\mathfrak{o}_H = \frac{e\mathfrak{H}}{4\pi m_0 c}, \qquad \mathfrak{o}_F = \frac{3}{4\pi}\sqrt{\frac{a}{Zm_0}}\,\mathfrak{E} = \frac{3}{4\pi}\,\frac{e\mathfrak{E}}{\sqrt{(2m_0|E|)}}. \tag{77}$$

We can then write (75) in the form

$$E_1 = \sqrt{\frac{Z^2Rh}{|E|}}\,(\mathfrak{A}\mathfrak{o}_F)h + 2\pi(\mathfrak{P}\mathfrak{o}_H). \tag{75a}$$

It is now expedient[1] to introduce vector matrices $\mathfrak{J}_1$ and $\mathfrak{J}_2$ which are defined by

$$\begin{aligned} 2\mathfrak{J}_1 &= \frac{2\pi}{h}\,\mathfrak{P} + \sqrt{\frac{Z^2Rh}{|E|}}\,\mathfrak{A}, \\ 2\mathfrak{J}_2 &= \frac{2\pi}{h}\,\mathfrak{P} - \sqrt{\frac{Z^2Rh}{|E|}}\,\mathfrak{A}, \end{aligned} \tag{78}$$

so that

$$\frac{2\pi}{h}\,\mathfrak{P} = \mathfrak{J}_1 + \mathfrak{J}_2, \qquad \sqrt{\frac{Z^2Rh}{|E|}}\,\mathfrak{A} = \mathfrak{J}_1 - \mathfrak{J}_2, \tag{78a}$$

and to introduce the two vectors

$$\mathfrak{o}_1 = \mathfrak{o}_H + \mathfrak{o}_F, \qquad \mathfrak{o}_2 = \mathfrak{o}_H - \mathfrak{o}_F, \tag{79}$$

whose absolute magnitudes have in § 2 been denoted by ω_1 and ω_2 (cf. equation (18)). The perturbation energy (75a) can then simply be written as

$$E_1 = (\mathfrak{J}_1\mathfrak{o}_1)h + (\mathfrak{J}_2\mathfrak{o}_2)h. \tag{80}$$

Similarly, we introduce into equations (I) to (IV) the new vector matrices $\mathfrak{J}_1$ and $\mathfrak{J}_2$ (as given by (78a)) in place of $\mathfrak{A}$ and $\mathfrak{P}$. A simple calculation then yields the following relations:

$$I_{1x}I_{2x} = I_{2x}I_{1x}, \ldots, \quad I_{1x}I_{2y} = I_{2y}I_{1x}, \qquad I_{2x}I_{1y} = I_{1y}I_{2x}, \ldots \tag{81}$$

$$[\mathfrak{J}_1\mathfrak{J}_1] = i\mathfrak{J}_1, \qquad [\mathfrak{J}_2\mathfrak{J}_2] = i\mathfrak{J}_2, \tag{82}$$

$$\mathfrak{J}_1^2 = \mathfrak{J}_2^2 = \tfrac{1}{4}(RhZ^2/|E| - 1) = \tfrac{1}{4}(n^2 - 1) = \tfrac{1}{2}n^*(\tfrac{1}{2}n^* + 1). \tag{83}$$

Relations (81) state that each component of $\mathfrak{J}_1$ commutes with each component of $\mathfrak{J}_2$. Relations (82) are quite analogous in structure to

[1] The following should be compared with the papers of Klein and Lenz quoted in footnotes [1]) and [2]), following eqs. (17), § 2.

equations (I); in the last part of (83), the energy values (68) were used.

It follows from (80) that the case of crossed fields is characterized by the fact that $(\mathfrak{J}_1\mathfrak{o}_1)$ and $(\mathfrak{J}_2\mathfrak{o}_2)$ or, what amounts to the same, the *components of* $\mathfrak{J}_1$ *and* $\mathfrak{J}_2$ *parallel to* $\mathfrak{o}_1$ *and* $\mathfrak{o}_2$ (written as $\mathfrak{J}_{1\parallel}$ and $\mathfrak{J}_{2\parallel}$), *are diagonal matrices.* The solution of equations (82) is in this case completely analogous to the solution of equations (I) if we assume P^2 and P_z to be diagonal matrices. Here, $\frac{1}{2}n^*$ appears instead of k, and we have to replace m (following the notation of § 2, eqs. (15) and (16)) by the numbers $\frac{1}{2}n^*-n_1$ and $\frac{1}{2}n^*-n_2$, which can run from $-\frac{1}{2}n^*$ to $+\frac{1}{2}n^*$. We then obtain

$$\begin{gathered}(\mathfrak{J}_{1\parallel})^{n_1}_{n_1} = \tfrac{1}{2}n^* - n_1, \qquad (\mathfrak{J}_{2\parallel})^{n_2}_{n_2} = \tfrac{1}{2}n^* - n_2, \\ 0 \leqq n_1 \leqq n^*, \qquad 0 \leqq n_2 \leqq n^*, \\ E_1 = (\tfrac{1}{2}n^* - n_1)\omega_1 h + (\tfrac{1}{2}n^* - n_2)\omega_2 h \\ (\omega_1 = |\mathfrak{o}_1|, \quad \omega_2 = |\mathfrak{o}_2|).\end{gathered} \tag{84}$$

The projections $\mathfrak{J}_{1\perp}$ and $\mathfrak{J}_{2\perp}$ in the planes perpendicular to the directions of $\mathfrak{o}_1$ and $\mathfrak{o}_2$ respectively, describe circular oscillations and are therefore represented by matrices which are analogous to those described by (56) for P_x and P_y (one has to replace k by $\frac{1}{2}n^*$ and m by $\frac{1}{2}n^*-n_1$ or $\frac{1}{2}n^*-n_2$; $\mathfrak{J}^2_\perp$ corresponds to the sum of P_x^2 and P_y^2):

$$\begin{aligned}|\mathfrak{J}_{1\perp}{}^{n_1}_{n_1+1}|^2 &= \tfrac{1}{2}(n_1+1)(n^*-n_1), \\ |\mathfrak{J}_{2\perp}{}^{n_2}_{n_2+1}|^2 &= \tfrac{1}{2}(n_2+1)(n^*-n_2).\end{aligned} \tag{85}$$

Equations (81) and (82) are thereby satisfied, and since, analogously to (57),

$$(\mathfrak{J}_1^2)^{n_1}_{n_1} = (\mathfrak{J}_2^2)^{n_2}_{n_2} = \tfrac{1}{2}n^*(\tfrac{1}{2}n^*+1),$$

equation (83), too, is satisfied by the energy values (68).

Thus all the results given in § 2 have been derived from the new mechanics.

6. On the relationship between the hydrogen spectrum and the alkali spectra

It has already been mentioned in § 2 that the modifications to the basic foundations of the new quantum mechanics which would still be necessary in order to interpret the anomalous Zeeman effects might possibly make themselves felt even in the case of atoms with a single electron; in particular, the result that the ground state of such an atom should be non-magnetic might well not be regarded as conclusive. A suggestion of a special kind, aimed at taking account of the anomalous Zeeman effect, has recently been put forward by Goudsmit and Uhlenbeck.[1] According to this suggestion, the electron is no longer regarded as a point charge, but instead has a preferred axis, angular momentum and (doubly anomalous) magnetism associated with it. Whether this assumption, when combined with the new quantum mechanics, suffices to explain all the empirical results, will probably be decided only when the calculation of relativistic fine structure, too, has been carried out on the basis of the new mechanics. This calculation had to be left, for the present, because we have as yet been unable to effect the requisite evaluation of the time-average $\overline{1/r^2}$.

Independently of the conception of any particular model one is, however, prompted to ask whether the hydrogen spectrum (including the fine-structure and the influence of external fields) could be regarded as a limiting case of the alkali spectra or X-ray spectra for a vanishing central force exerted by the rest of the atom on the valency electron or, respectively, for vanishingly small screening numbers (so that the levels forming a screening doublet coincide).[2] The fine structure of the Balmer lines would then differ from that predicted by the earlier theory, not by the position of the energy levels and line components, but by their intensities: instead of the selection rule $\Delta k = \pm 1$, we should now have the selection rule $\Delta j = 0, \pm 1$, which permits the occurrence of components that had been forbidden in the earlier theory. Goudsmit and Uhlenbeck[1] were able to show that the observed results would make such an alteration of the selection rule appear quite probable. At the same time, however, they draw attention to the fact that the following difficulty stands in the way of a complete analogy between the hydrogen spectrum and alkali spectra: the Zeeman effect for spectra of single-electron atoms in magnetic fields that are weak (relative to the fine structure) does not at all seem to resemble that for alkali spectra according to the available observations.

Although therefore the question as to how far one may pursue the above-mentioned relationship between hydrogen- and alkali-spectra cannot yet be regarded as resolved, one may nevertheless be justified

[1] S. A. Goudsmit and G. E. Uhlenbeck, Naturwiss. **13** (1925) 953.

[2] S. Goudsmit and G. E. Uhlenbeck, Physica **5** (1925) 266. Similar arguments were communicated to me in a letter by Mr. A. Landé a considerable time ago.

in letting oneself be guided by this analogy, at least for all those cases in which the relativistic (or doublet) fine structure can be left out of consideration. This would lead us to assume that, in magnetic fields in which the Zeeman splitting is large compared with the separation of the fine-structure components, the magnetic energy levels in the spectra of single-electron atoms agree with the Paschen–Back terms of the alkali elements, as far as the number and position of these levels is concerned. One would then have to assign twice as many states to the hydrogen atom in an external field of force as were derived in the preceding sections on the present basis of the new quantum mechanics (i.e., $2n^2$ states instead of n^2). In an external magnetic field, to each value of the quantum number m (lying between $-n^*$ and $+n^*$) would have to belong the two magnetic energy values $(m\pm 1)o_H h$ (where o_H=Larmor frequency); in the same way, for the case of crossed fields, each state characterized by n^*, n_1, n_2 would have to split into two states whose energy values would differ from those given by (84) by an amount $\pm o_H h$. According to the correspondence principle, only those transitions would then occur which leave the sign of the additional term $\pm o_H h$ unchanged.

One possible way of differentiating between the term manifold derived in the previous section (for which the ground state of the hydrogen atom is non-magnetic) and the manifold considered here in analogy with the Paschen–Back terms of the alkali elements (for which energy values $\pm o_H h$ are assigned to the ground state of a hydrogen atom in a magnetic field) may be offered by investigations of the Stern–Gerlach type on the deflection of atomic hydrogen beams in an inhomogeneous magnetic field.

Received January 22, 1926

17

QUANTUM MECHANICS AND A PRELIMINARY INVESTIGATION OF THE HYDROGEN ATOM

P. A. M. DIRAC

1. The algebraic laws governing dynamical variables

Although the classical electrodynamic theory meets with a considerable amount of success in the description of many atomic phenomena, it fails completely on certain fundamental points. It has long been thought that the way out of this difficulty lies in the fact that there is one basic assumption of the classical theory which is false, and that if this assumption were removed and replaced by something more general, the whole of atomic theory would follow quite naturally. Until quite recently, however, one has had no idea of what this assumption could be.

A recent paper by Heisenberg* provides the clue to the solution of this question, and forms the basis of a new quantum theory. According to Heisenberg, if x and y are two functions of the co-ordinates and momenta of a dynamical system, then in general xy is not equal to yx. Instead of the commutative law of multiplication, the canonical variables $q_r\, p_r$ $(r=1 \ldots u)$ of a system of u degrees of freedom satisfy the quantum conditions, which were given by the author† in the form

Editor's note. This paper was published as Proc. Roy. Soc. A **110** (1926) 561–569. Sections 5–7 (pp. 570–579) of the original paper, in which the hydrogen problem is partly solved, are not reproduced here since, 5 days earlier, the complete solution was given by Pauli (cf. paper **16** in this volume). Dirac at the time was 1851 Exhibition Senior Research Student of St. John's College, Cambridge. The paper was communicated by R. H. Fowler, F.R.S.

* W. Heisenberg, Zs. f. Phys. **33** (1925) 879.

† P. A. M. Dirac, Roy. Soc. Proc. A **109** (1925) 642. These quantum conditions have been obtained independently by Born, Heisenberg and Jordan, Zs. f. Phys. **35** (1926) 557.

$$\begin{aligned} q_r q_s - q_s q_r &= 0 \\ p_r p_s - p_s p_r &= 0 \\ q_r p_s - p_s q_r &= 0 \qquad (r \neq s) \\ q_r p_r - p_r q_r &= ih \end{aligned} \tag{1}$$

where i is a root of -1 and h is a real universal constant, equal to $(2\pi)^{-1}$ times the usual Planck's constant. These equations are just sufficient to enable one to calculate $xy - yx$ when x and y are given functions of the p's and q's, and are therefore capable of replacing the classical commutative law of multiplication. They appear to be the simplest assumptions one could make which would give a workable theory.

The fact that the variables used for describing a dynamical system do not satisfy the commutative law means, of course, that they are not numbers in the sence of the word previously used in mathematics. To distinguish the two kinds of numbers, we shall call the quantum variables q-numbers and the numbers of classical mathematics which satisfy the commutative law c-numbers, while the word number alone will be used to denote either a q-number or a c-number. When $xy = yx$ we shall say that x commutes with y.

At present one can form no picture of what a q-number is like. One cannot say that one q-number is greater or less than another. All one knows about q-numbers is that if z_1 and z_2 are two q-numbers, or one q-number and one c-number, there exist the numbers $z_1 + z_2$, $z_1 z_2$, $z_2 z_1$, which will in general be q-numbers but may be c-numbers. One knows nothing of the processes by which the numbers are formed except that they satisfy all the ordinary laws of algebra, excluding the commutative law of multiplication, i.e.,

$$z_1 + z_2 = z_2 + z_1,$$

$$(z_1 + z_2) + z_3 = z_1 + (z_2 + z_3),$$

$$(z_1 z_2) z_3 = z_1 (z_2 z_3),$$

$$z_1(z_2 + z_3) = z_1 z_2 + z_1 z_3, \qquad (z_1 + z_2) z_3 = z_1 z_3 + z_2 z_3,$$

and if

$$z_1 z_2 = 0,$$

either

$$z_1 = 0 \quad \text{or} \quad z_2 = 0;$$

but

$$z_1 z_2 \neq z_2 z_1,$$

in general, except when z_1 or z_2 is a c-number. One may define further numbers, x say, by means of equations involving x and the z's, such as $x^2=z$, which defines $z^{\frac{1}{2}}$, or $xz=1$, which defines z^{-1}. There may be more than one value of x satisfying such an equation, but this is not so for the equation $xz=1$, since if $x_1z=1$ and $x_2z=1$ then $(x_1-x_2)z=0$, which gives $x_1=x_2$ provided $z\neq 0$.

A function $f(z)$ of a q-number z cannot be defined in a manner analogous to the general definition of a function of a real c-number variable, but can be defined only by an algebraic relation connecting $f(z)$ with (z). When this relation does not involve any q-number that does not commute with z and $f(z)$, one can define $\partial f/\partial z$ without ambiguity by the same algebraic relation as when z is a c-number, e.g., if $f(z)=z^n$, then $\partial f/\partial z=nz^{n-1}$ where n is a c-number.

In order to be able to get results comparable with experiment from our theory, we must have some way of representing q-numbers by means of c-numbers, so that we can compare these c-numbers with experimental values. The representation must satisfy the condition that one can calculate the c-numbers that represent $x+y$, xy and yx when one is given the c-numbers that represent x and y. If a q-number x is a function of the co-ordinates and momenta of a multiply periodic system, and if it is itself multiply periodic, then it will be shown that the aggregate of all its values for all values of the action variables of the system can be represented by a set of harmonic components of the type $x(nm) \exp i\omega(nm)t$, where $x(nm)$ and $\omega(nm)$ are c-numbers, each associated with two sets of values of the action variables denoted by the labels n and m, and t is the time, also a c-number. This representation was taken as defining a q-number in the previous papers on the new theory.* It seems preferable though to take the above algebraic laws and the general conditions (1) as defining the properties of q-numbers, and to deduce from them that a q-number can be represented by c-numbers in this manner when it has the necessary periodic properties. A q-number thus still has a meaning and can be used in the analysis when it is not multiply periodic, although there is at present no way of representing it by c-numbers.

* See particularly, Born and Jordan, Zs. f. Phys. **34** (1925) 858. Also Born, Heisenberg and Jordan, loc. cit.

2. The Poisson bracket expressions

If x and y are two numbers, we define their Poisson bracket expression $[x, y]$ by

$$xy - yx = ih[x, y]. \tag{2}$$

It has the following properties, which follow at once from the definition and make it analogous to the Poisson bracket of classical mechanics.

(i) It contains no reference to any particular set of canonical variables.

(ii) It satisfies the laws

$$[x_1+x_2, y] = [x_1, y] + [x_2, y],$$
$$[x_1x_2, y] = x_1[x_2, y] + [x_1, y]x_2,$$
$$[x, y] = -[y, x].$$

(iii) It satisfies the identity

$$[[x, y], z] + [[y, z], x] + [[z, x], y] = 0.$$

(iv) The elementary P.B.'s (Poisson brackets) are given, from (1), by

$$[p_r, p_s] = 0, \qquad [q_r, q_s] = 0,$$
$$[q_r, p_s] = 0 \quad (r \neq s), \quad \text{or} \quad 1 \quad (r = s),$$

and also

$$[p_r, c] = [q_r, c] = 0,$$

when c is a c-number.

If x and y are given functions of the p's and q's, then, by successive applications of the laws (ii) the P.B. $[x, y]$ can be expressed in terms of the elementary P.B.'s occurring in (iv), and thus evaluated. It is often more convenient to evaluate a P.B. in this way than by the direct use of (2). For example, to evaluate $[q^2, p^2]$ we have

$$[q^2, p^2] = q[q, p^2] + [q, p^2]q,$$

and

$$[q, p^2] = p[q, p] + [q, p]p = 2p,$$

so that

$$[q^2, p^2] = 2qp + 2pq.$$

One may greatly reduce the labour of evaluating P.B.'s of functions of the p's and q's in certain special cases by observing that the classical theory expression for the P.B. $[x, y]$, namely

$$\sum_r \left(\frac{\partial x}{\partial q_r} \frac{\partial y}{\partial p_r} - \frac{\partial y}{\partial q_r} \frac{\partial x}{\partial p_r} \right)$$

may usually be taken over directly into the quantum theory when this does not give rise to any ambiguity concerning order of factors of products, e.g., we can say at once that

$$[f(x), x] = 0,$$

when $f(x)$ does not involve any number that does not commute with x, and also

$$[f(q_r), p_r] = \partial f / \partial q_r, \tag{3}$$

when $f(q_r)$ does not involve any number that does not commute with q_r.

The conditions that a set of variables Q_r, P_r shall be canonical are defined to be that from the relations connecting the Q_r, P_r with the q_r, p_r (which are given to be canonical) one can deduce the equations

$$[Q_r, Q_s] = 0, \qquad [P_r, P_s] = 0,$$
$$[Q_r, P_s] = 0 \quad (r \neq s) \quad \text{or} \quad 1 \quad (r = s).$$

One could evaluate the P.B. of two functions of the Q_r, P_r, either by working entirely in the variables Q_r, P_r, or by first substituting for these variables in terms of the q_r, p_r. The relations connecting the Q_r, P_r with the q_r, p_r may be put in the form

$$Q_r = b q_r b^{-1}, \qquad P_r = b p_r b^{-1},$$

where b is a q-number which determines the transformation, but these formulae do not appear to be of great practical value.

A dynamical system is determined on the classical theory by a Hamiltonian H, which is a certain function of the p's and q's, and the classical equations of motion may be written

$$\dot{x} = [x, H]. \tag{4}$$

We assume that the equations of motion on the quantum theory are also of the form (4), where the Hamiltonian H is now a q-number, and is for the present an unknown function of the p's and q's. The representation of a q-number by c-numbers when it is multiply periodic must be such that if x is represented by the harmonic components

$x(nm) \exp i\omega(nm)t$, $\dot{x}$ defined by (4) has the components

$$i\omega(nm)\, x(nm) \exp i\omega(nm)t.$$

3. Some elementary algebraic theorems

In all previous descriptions of natural phenomena the two roots of -1 have always played symmetrical parts. The occurrence of a root of -1 in the fundamental equations (1) means that this is not so in the present theory. For mathematical convenience we shall continually be using in the analysis a root of -1, j say, which is independent of the i in (1), that is to say, from any equation one can obtain another equation by writing $-$j for j without at the same time changing the sign of i. From these two equations one can obtain two more equations by reversing the order of the factors of all products occurring in them and at the same time writing $-h$ for h, since if this operation is applied to equations (1) it will give correct results, so that it must still give correct results when applied to any equation derivable from (1). To avoid having two symbols i and j, both denoting roots of -1, we shall take j$=$i, and must then modify the above rules to read: From any equation one may obtain another equation by writing $-$i for i wherever it occurs and at the same time writing $-h$ for h, or by reversing the order of all factors and writing $-h$ for h, or by applying the two previous operations together, which reduces to reversing the order of all factors and writing $-$i for i. This third operation applied to any number gives what may be defined as the conjugate imaginary number. A number is defined to be real if it is equal to its conjugate imaginary.

The remainder of this section will be devoted to some simple analytical rules which will be of use in the subsequent work.

When forming the reciprocal of a quantity composed of two or more factors, one must reverse their order, i.e.,

$$\frac{1}{(xy)} = \frac{1}{y} \cdot \frac{1}{x}. \tag{5}$$

This equation may be verified by multiplying each side by xy either in front or behind.

To differentiate the reciprocal of a quantity x one must proceed as follows:

$$\frac{d}{dt}\left(\frac{1}{x}\cdot x\right) = \frac{d}{dt}(1) = [1, H] = 0.$$

$$0 = \frac{d}{dt}\left(\frac{1}{x}\cdot x\right) = \frac{d}{dt}\left(\frac{1}{x}\right)\cdot x + \frac{1}{x}\,\dot{x}.$$

Hence, dividing by x behind, one gets

$$\frac{d}{dt}\left(\frac{1}{x}\right) = -\frac{1}{x}\,\dot{x}\,\frac{1}{x}.$$

The binomial expansion for $(1+x)^n$ when n is a c-number is the same as in ordinary algebra. Also one defines e^x by the same power series as in ordinary algebra. The ordinary exponential law, however, is not valid, i.e., e^{x+y} is not in general equal to $e^x e^y$, except when x commutes with y.

If (αq) denotes $\sum_r (\alpha_r q_r)$, where the α_r $(r=1\ldots u)$ are c-numbers, then from (3)

$$[e^{i(\alpha q)}, p_r] = i\alpha_r\, e^{i(\alpha q)}.$$

Hence, since

$$e^{i(\alpha q)}p_r - p_r\, e^{i(\alpha q)} = ih\,[e^{i(\alpha q)}, p_r],$$

we have

$$e^{i(\alpha q)}p_r = (p_r - \alpha_r h)\, e^{i(\alpha q)}.$$

More generally, if $f(q_r, p_r)$ is any function of the q's and p's,

$$\begin{aligned} e^{i(\alpha q)}f(q_r, p_r) &= f(q_r, p_r - \alpha_r h)\, e^{i(\alpha q)}, \\ f(q_r, p_r)\, e^{i(\alpha q)} &= e^{i(\alpha q)}f(q_r, p_r + \alpha_r h). \end{aligned} \tag{6}$$

To prove this result, we observe that if it is true for any two functions f, f_1 and f_2, say, it must also be true for (f_1+f_2) and $f_1 f_2$. Now we have proved it true when $f=p_r$, and it is obviously true when $f=q_r$ since the q's commute with each other. Hence it is true when f is any power series in the p's and q's so that we may take it to be generally true.

Equations (6) show the law of interchange of any function of the p's and q's with a quantity of the form $e^{i(\alpha q)}$. They are of great value in the theory of multiply periodic systems. There are, of course, corresponding equations for any set of canonical variables, Q_r, P_r.

4. Multiply periodic systems

A dynamical system is multiply periodic on the quantum theory when there exists a set of uniformising variables J_r, w_r having the following properties:

(i) They are canonical variables, i.e.,

$$[J_r, J_s] = 0, \qquad [w_r, w_s] = 0,$$
$$[w_r, J_s] = 0 \quad (r \neq s), \quad \text{or} \quad 1 \quad (r = s).$$

(ii) The Hamiltonian H is a function of the J's only.*

(iii) The original p's and q's that describe the system are multiply periodic functions of the w's of period 2π, the condition for this being defined to be that a p or q can be expanded in either of the forms

$$\Sigma_\alpha C_\alpha \exp i(\alpha_1 w_1 + \alpha_2 w_2 + \ldots + \alpha_u w_u) = \Sigma_\alpha C_\alpha \exp i(\alpha w)$$

or

$$\Sigma_\alpha \exp i(\alpha_1 w_1 + \alpha_2 w_2 + \ldots + \alpha_u w_u) C'_\alpha = \Sigma_\alpha \exp i(\alpha w) C'_\alpha.$$

C'_α, where the C_α's and C'_α's are functions of the J's only and the α's are integers. We have taken the w's 2π times as great and the J's $1/2\pi$ times as great as the usual uniformising variables in order to save writing.

We have at once

$$\dot{J}_r = [J_r, H] = 0$$

from (ii), and

$$\dot{w}_r = [w_r, H] = \partial H/\partial J_r,$$

using (3). The quantities $\dot{w}_r$ are, therefore, constants and may be called the frequencies. There are, however, other quantities that have claims to be called frequencies. We have

$$\frac{d}{dt} e^{i(\alpha w)} = [e^{i(\alpha w)}, H] = \frac{e^{i(\alpha w)} H - H e^{i(\alpha w)}}{ih}.$$

From (6) applied to the J's and w's,

$$e^{i(\alpha w)} H(J_r) = H(J_r - \alpha_r h)\, e^{i(\alpha w)},$$

* H is not necessarily the same function of the J's as on the classical theory with the present definition of the J's.

and

$$H(J_r)\, e^{i(\alpha w)} = e^{i(\alpha w)} H(J_r + \alpha_r h).$$

Hence

$$\frac{d}{dt}\, e^{i(\alpha w)} = i(\alpha\omega)\, e^{i(\alpha w)} = ie^{i(\alpha w)}(\alpha\omega)',$$

where

$$\begin{aligned}(\alpha\omega)h &= H(J_r) - H(J_r - \alpha_r h),\\ (\alpha\omega)'h &= H(J_r + \alpha_r h) - H(J_r),\end{aligned} \tag{7}$$

The quantities $\dot{w}_r$ correspond to the orbital frequencies on Bohr's theory, while the $(\alpha\omega)$ and $(\alpha\omega)'$ correspond, when the α's are integers, to the transition frequencies. It must be remembered though that the w_r, $(\alpha\omega)$ and $(\alpha\omega)'$ are q-numbers, and, therefore, they cannot be equated to Bohr's frequencies, which are c-numbers. They are merely the same functions of the present J's, which are q-numbers, as Bohr's frequencies are of his J's, which are c-numbers.

Suppose x can be expanded in the form

$$x = \Sigma_\alpha\, x_\alpha\, e^{i(\alpha w)} = \Sigma_\alpha\, e^{i(\alpha w)} x'_\alpha, \tag{8}$$

where the α's are integers and the x_α, x'_α are functions of the J's only. From (6)

$$x'_\alpha(J_r) = x_\alpha(J_r + \alpha_r h).$$

Also

$$\dot{x} = \Sigma_\alpha\, x_\alpha\, i(\alpha\omega)\, e^{i(\alpha w)} = \Sigma_\alpha\, e^{i(\alpha w)} i(\alpha\omega)' x'_\alpha \tag{9}$$

If x and the J's are real and if $\bar{x}_\alpha$ denotes the conjugate imaginary of x_α, then by equating the conjugate imaginaries of both sides of (8) we get

$$x = \Sigma_\alpha\, e^{-i(\alpha w)} \bar{x}_\alpha(J_r) = \Sigma_\alpha\, \bar{x}_\alpha(J_r + \alpha_r h)\, e^{-i(\alpha w)}.$$

Comparing this with equation (8) we find that

$$\bar{x}_\alpha(J_r + \alpha_r h) = x_{-\alpha}(J_r).$$

This relation is brought out more clearly if we change the notation. For $x_\alpha(J_r)$ write $x(J, J-\alpha h)$. Then

$$\bar{x}(J + \alpha h, J) = x(J, J + \alpha h),$$

which shows that there is some kind of symmetry in the way in which the amplitude $x(J, J-\alpha h)$ is related to the two sets of variables to which it explicitly refers. Our expansion for x is now

$$x = \Sigma_\alpha\, x(J, J - \alpha h)\, e^{i(\alpha w)} = \Sigma_\alpha\, e^{i(\alpha w)} x(J + \alpha h, J).$$

The expressions (7) for the transition frequencies suggest that we should put

$$(\alpha\omega)(J) = \omega(J, J - \alpha h),$$

and

$$(\alpha\omega)'(J) = \omega(J + \alpha h, J).$$

We should then have from (9)

$$\begin{aligned} \dot{x} &= \Sigma\, x(J, J - \alpha h) i\omega(J, J - \alpha h)\, e^{i(\alpha w)} \\ &= \Sigma\, e^{i(\alpha w)} i\omega(J + \alpha h, J) x(J + \alpha h, J). \end{aligned} \qquad (10)$$

Suppose y can also be expanded in the form

$$y = \Sigma_\beta\, y(J, J - \beta h)\, e^{i(\beta w)}.$$

Then

$$\begin{aligned} xy &= \Sigma_{\alpha\beta}\, x(J, J - \alpha h)\, e^{i(\alpha w)} y(J, J - \beta h)\, e^{i(\beta w)}, \\ &= \Sigma_{\alpha\beta}\, x(J, J - \alpha h) \cdot y(J - \alpha h, J - \alpha h - \beta h)\, e^{i[(\alpha+\beta) w]}, \end{aligned}$$

by again using (6), and the fact that the w's commute; or, the amplitudes of xy are given by

$$xy(J, J - \gamma h) = \Sigma_\alpha\, x(J, J - \alpha h) \cdot y(J - \alpha h, J - \gamma h). \qquad (11)$$

These formulae provide a way of representing q-numbers by means of c-numbers. Suppose that in the expressions $x(J, J - \alpha h)$ and $\omega(J, J - \alpha h)$, considered merely as functions of the J's, we substitute for each J_r the c-number $n_r h$, and denote the resulting c-numbers by $x(n, n-\alpha)$ and $\omega(n, n-\alpha)$. We may consider the aggregate of all the c-numbers $x(n, n-\alpha) \exp i\omega(n, n-\alpha)t$, in which it is sufficient (but not necessary) for the n to take a series of values differing successively by unity as representing the values of the q-number x for all values of the q-numbers J_r. Equation (10) shows that

$$\dot{x}(n, n - \alpha) = i\omega(n, n - \alpha) x(n, n - \alpha),$$

while equation (11) gives

$$xy(n, n - \gamma) = \sum_\alpha x(n, n - \alpha)y(n - \alpha, n - \gamma),$$

which is just Heisenberg's law of multiplication. Also we have obviously

$$(x + y)(n, n - \alpha) = x(n, n - \alpha) + y(n, n - \alpha).$$

Our representation thus satisfies the conditions mentioned in §§ 1 and 2, which proves the sufficiency of this discrete set of n's.

One gets different representations of the q-numbers x by c-numbers $x(nm) \exp i\omega(nm)t$ by taking different values for the c-numbers, η_r, say, by which the n_r's differ from integers. Only one of these representations, though, is of physical importance, this being the one (assumed to exist) for which every $x(nm)$ vanishes when an m_r is less than a certain value, n_{or}, say, which fixes the normal state of the system on Bohr's theory, and each $n_r \geqq n_{or}$. This requires that every coefficient $x(J, J-\alpha h)$ in the expansion of x shall vanish when for each J_r is substituted the c-number $(n_{or}+m_r h)$, where the m_r are integers not less than zero, at least one of which is less than the corresponding α_r.

NAME INDEX

Italic numbers refer to the first page of a paper written by the author concerned; subsequent pages of the same paper are not listed.

SOME DOVER SCIENCE BOOKS

CHANCE, LUCK AND STATISTICS: THE SCIENCE OF CHANCE,
Horace C. Levinson

Theory of probability and science of statistics in simple, non-technical language. Part I deals with theory of probability, covering odd superstitions in regard to "luck," the meaning of betting odds, the law of mathematical expectation, gambling, and applications in poker, roulette, lotteries, dice, bridge, and other games of chance. Part II discusses the misuse of statistics, the concept of statistical probabilities, normal and skew frequency distributions, and statistics applied to various fields—birth rates, stock speculation, insurance rates, advertising, etc. "Presented in an easy humorous style which I consider the best kind of expository writing," Prof. A. C. Cohen, Industry Quality Control. Enlarged revised edition. Formerly titled *The Science of Chance*. Preface and two new appendices by the author. Index. xiv + 365pp. 5⅜ x 8. Paperbound $2.00

BASIC ELECTRONICS,
prepared by the U.S. Navy Training Publications Center

A thorough and comprehensive manual on the fundamentals of electronics. Written clearly, it is equally useful for self-study or course work for those with a knowledge of the principles of basic electricity. Partial contents: Operating Principles of the Electron Tube; Introduction to Transistors; Power Supplies for Electronic Equipment; Tuned Circuits; Electron-Tube Amplifiers; Audio Power Amplifiers; Oscillators; Transmitters; Transmission Lines; Antennas and Propagation; Introduction to Computers; and related topics. Appendix. Index. Hundreds of illustrations and diagrams. vi + 471pp. 6½ x 9¼.
Paperbound $2.75

BASIC THEORY AND APPLICATION OF TRANSISTORS,
prepared by the U.S. Department of the Army

An introductory manual prepared for an army training program. One of the finest available surveys of theory and application of transistor design and operation. Minimal knowledge of physics and theory of electron tubes required. Suitable for textbook use, course supplement, or home study. Chapters: Introduction; fundamental theory of transistors; transistor amplifier fundamentals; parameters, equivalent circuits, and characteristic curves; bias stabilization; transistor analysis and comparison using characteristic curves and charts; audio amplifiers; tuned amplifiers; wide-band amplifiers; oscillators; pulse and switching circuits; modulation, mixing, and demodulation; and additional semiconductor devices. Unabridged, corrected edition. 240 schematic drawings, photographs, wiring diagrams, etc. 2 Appendices. Glossary. Index. 263pp. 6½ x 9¼. Paperbound $1.25

GUIDE TO THE LITERATURE OF MATHEMATICS AND PHYSICS,
N. G. Parke III

Over 5000 entries included under approximately 120 major subject headings of selected most important books, monographs, periodicals, articles in English, plus important works in German, French, Italian, Spanish, Russian (many recently available works). Covers every branch of physics, math, related engineering. Includes author, title, edition, publisher, place, date, number of volumes, number of pages. A 40-page introduction on the basic problems of research and study provides useful information on the organization and use of libraries, the psychology of learning, etc. This reference work will save you hours of time. 2nd revised edition. Indices of authors, subjects, 464pp. 5⅜ x 8.
Paperbound $2.75

THE RISE OF THE NEW PHYSICS (formerly THE DECLINE OF MECHANISM), *A. d'Abro*

This authoritative and comprehensive 2-volume exposition is unique in scientific publishing. Written for intelligent readers not familiar with higher mathematics, it is the only thorough explanation in non-technical language of modern mathematical-physical theory. Combining both history and exposition, it ranges from classical Newtonian concepts up through the electronic theories of Dirac and Heisenberg, the statistical mechanics of Fermi, and Einstein's relativity theories. "A must for anyone doing serious study in the physical sciences," *J. of Franklin Inst.* 97 illustrations. 991pp. 2 volumes.

Vol. 1 Paperbound $2.25, Vol. 2 Paperbound $2.25, The set $4.50

THE STRANGE STORY OF THE QUANTUM, AN ACCOUNT FOR THE GENERAL READER OF THE GROWTH OF IDEAS UNDERLYING OUR PRESENT ATOMIC KNOWLEDGE, *B. Hoffmann*

Presents lucidly and expertly, with barest amount of mathematics, the problems and theories which led to modern quantum physics. Dr. Hoffmann begins with the closing years of the 19th century, when certain trifling discrepancies were noticed, and with illuminating analogies and examples takes you through the brilliant concepts of Planck, Einstein, Pauli, de Broglie, Bohr, Schroedinger, Heisenberg, Dirac, Sommerfeld, Feynman, etc. This edition includes a new, long postscript carrying the story through 1958. "Of the books attempting an account of the history and contents of our modern atomic physics which have come to my attention, this is the best," H. Margenau, Yale University, in *American Journal of Physics.* 32 tables and line illustrations. Index. 275pp. 5⅜ x 8.

Paperbound $1.75

GREAT IDEAS AND THEORIES OF MODERN COSMOLOGY, *Jagjit Singh*

The theories of Jeans, Eddington, Milne, Kant, Bondi, Gold, Newton, Einstein, Gamow, Hoyle, Dirac, Kuiper, Hubble, Weizsäcker and many others on such cosmological questions as the origin of the universe, space and time, planet formation, "continuous creation," the birth, life, and death of the stars, the origin of the galaxies, etc. By the author of the popular *Great Ideas of Modern Mathematics.* A gifted popularizer of science, he makes the most difficult abstractions crystal-clear even to the most non-mathematical reader. Index. xii + 276pp. 5⅜ x 8½.

Paperbound $2.00

GREAT IDEAS OF MODERN MATHEMATICS: THEIR NATURE AND USE, *Jagjit Singh*

Reader with only high school math will understand main mathematical ideas of modern physics, astronomy, genetics, psychology, evolution, etc., better than many who use them as tools, but comprehend little of their basic structure. Author uses his wide knowledge of non-mathematical fields in brilliant exposition of differential equations, matrices, group theory, logic, statistics, problems of mathematical foundations, imaginary numbers, vectors, etc. Original publications, 2 appendices. 2 indexes. 65 illustr. 322pp. 5⅜ x 8. Paperbound $2.00

THE MATHEMATICS OF GREAT AMATEURS, *Julian L. Coolidge*

Great discoveries made by poets, theologians, philosophers, artists and other non-mathematicians: Omar Khayyam, Leonardo da Vinci, Albrecht Dürer, John Napier, Pascal, Diderot, Bolzano, etc. Surprising accounts of what can result from a non-professional preoccupation with the oldest of sciences. 56 figures. viii + 211pp. 5⅜ x 8½.

Paperbound $1.50

COLLEGE ALGEBRA, *H. B. Fine*
Standard college text that gives a systematic and deductive structure to algebra; comprehensive, connected, with emphasis on theory. Discusses the commutative, associative, and distributive laws of number in unusual detail, and goes on with undetermined coefficients, quadratic equations, progressions, logarithms, permutations, probability, power series, and much more. Still most valuable elementary-intermediate text on the science and structure of algebra. Index. 1560 problems, all with answers. x + 631pp. 5⅜ x 8. Paperbound $2.75

HIGHER MATHEMATICS FOR STUDENTS OF CHEMISTRY AND PHYSICS, *J. W. Mellor*
Not abstract, but practical, building its problems out of familiar laboratory material, this covers differential calculus, coordinate, analytical geometry, functions, integral calculus, infinite series, numerical equations, differential equations, Fourier's theorem, probability, theory of errors, calculus of variations, determinants. "If the reader is not familiar with this book, it will repay him to examine it," *Chem. & Engineering News*. 800 problems. 189 figures. Bibliography. xxi + 641pp. 5⅜ x 8. Paperbound $2.50

TRIGONOMETRY REFRESHER FOR TECHNICAL MEN, *A. A. Klaf*
A modern question and answer text on plane and spherical trigonometry. Part I covers plane trigonometry: angles, quadrants, trigonometrical functions, graphical representation, interpolation, equations, logarithms, solution of triangles, slide rules, etc. Part II discusses applications to navigation, surveying, elasticity, architecture, and engineering. Small angles, periodic functions, vectors, polar coordinates, De Moivre's theorem, fully covered. Part III is devoted to spherical trigonometry and the solution of spherical triangles, with applications to terrestrial and astronomical problems. Special time-savers for numerical calculation. 913 questions answered for you! 1738 problems; answers to odd numbers. 494 figures. 14 pages of functions, formulae. Index. x + 629pp. 5⅜ x 8. Paperbound $2.00

CALCULUS REFRESHER FOR TECHNICAL MEN, *A. A. Klaf*
Not an ordinary textbook but a unique refresher for engineers, technicians, and students. An examination of the most important aspects of differential and integral calculus by means of 756 key questions. Part I covers simple differential calculus: constants, variables, functions, increments, derivatives, logarithms, curvature, etc. Part II treats fundamental concepts of integration: inspection, substitution, transformation, reduction, areas and volumes, mean value, successive and partial integration, double and triple integration. Stresses practical aspects! A 50 page section gives applications to civil and nautical engineering, electricity, stress and strain, elasticity, industrial engineering, and similar fields. 756 questions answered. 556 problems; solutions to odd numbers. 36 pages of constants, formulae. Index. v + 431pp. 5⅜ x 8. Paperbound $2.00

INTRODUCTION TO THE THEORY OF GROUPS OF FINITE ORDER, *R. Carmichael*
Examines fundamental theorems and their application. Beginning with sets, systems, permutations, etc., it progresses in easy stages through important types of groups: Abelian, prime power, permutation, etc. Except 1 chapter where matrices are desirable, no higher math needed. 783 exercises, problems. Index. xvi + 447pp. 5⅜ x 8. Paperbound $3.00

NUMERICAL SOLUTIONS OF DIFFERENTIAL EQUATIONS,
H. Levy & E. A. Baggott

Comprehensive collection of methods for solving ordinary differential equations of first and higher order. All must pass 2 requirements: easy to grasp and practical, more rapid than school methods. Partial contents: graphical integration of differential equations, graphical methods for detailed solution. Numerical solution. Simultaneous equations and equations of 2nd and higher orders. "Should be in the hands of all in research in applied mathematics, teaching," *Nature.* 21 figures. viii + 238pp. 5⅜ x 8. Paperbound $1.85

ELEMENTARY STATISTICS, WITH APPLICATIONS IN MEDICINE AND THE BIOLOGICAL SCIENCES, *F. E. Croxton*

A sound introduction to statistics for anyone in the physical sciences, assuming no prior acquaintance and requiring only a modest knowledge of math. All basic formulas carefully explained and illustrated; all necessary reference tables included. From basic terms and concepts, the study proceeds to frequency distribution, linear, non-linear, and multiple correlation, skewness, kurtosis, etc. A large section deals with reliability and significance of statistical methods. Containing concrete examples from medicine and biology, this book will prove unusually helpful to workers in those fields who increasingly must evaluate, check, and interpret statistics. Formerly titled "Elementary Statistics with Applications in Medicine." 101 charts. 57 tables. 14 appendices. Index. vi + 376pp. 5⅜ x 8. Paperbound $2.00

INTRODUCTION TO SYMBOLIC LOGIC,
S. Langer

No special knowledge of math required — probably the clearest book ever written on symbolic logic, suitable for the layman, general scientist, and philosopher. You start with simple symbols and advance to a knowledge of the Boole-Schroeder and Russell-Whitehead systems. Forms, logical structure, classes, the calculus of propositions, logic of the syllogism, etc. are all covered. "One of the clearest and simplest introductions," *Mathematics Gazette.* Second enlarged, revised edition. 368pp. 5⅜ x 8. Paperbound $2.00

A SHORT ACCOUNT OF THE HISTORY OF MATHEMATICS,
W. W. R. Ball

Most readable non-technical history of mathematics treats lives, discoveries of every important figure from Egyptian, Phoenician, mathematicians to late 19th century. Discusses schools of Ionia, Pythagoras, Athens, Cyzicus, Alexandria, Byzantium, systems of numeration; primitive arithmetic; Middle Ages, Renaissance, including Arabs, Bacon, Regiomontanus, Tartaglia, Cardan, Stevinus, Galileo, Kepler; modern mathematics of Descartes, Pascal, Wallis, Huygens, Newton, Leibnitz, d'Alembert, Euler, Lambert, Laplace, Legendre, Gauss, Hermite, Weierstrass, scores more. Index. 25 figures. 546pp. 5⅜ x 8. Paperbound $2.25

INTRODUCTION TO NONLINEAR DIFFERENTIAL AND INTEGRAL EQUATIONS,
Harold T. Davis

Aspects of the problem of nonlinear equations, transformations that lead to equations solvable by classical means, results in special cases, and useful generalizations. Thorough, but easily followed by mathematically sophisticated reader who knows little about non-linear equations. 137 problems for student to solve. xv + 566pp. 5⅜ x 8½. Paperbound $2.00

AN INTRODUCTION TO THE GEOMETRY OF N DIMENSIONS,
D. H. Y. Sommerville
An introduction presupposing no prior knowledge of the field, the only book in English devoted exclusively to higher dimensional geometry. Discusses fundamental ideas of incidence, parallelism, perpendicularity, angles between linear space; enumerative geometry; analytical geometry from projective and metric points of view; polytopes; elementary ideas in analysis situs; content of hyper-spacial figures. Bibliography. Index. 60 diagrams. 196pp. 5⅜ x 8.
Paperbound $1.50

ELEMENTARY CONCEPTS OF TOPOLOGY, *P. Alexandroff*
First English translation of the famous brief introduction to topology for the beginner or for the mathematician not undertaking extensive study. This unusually useful intuitive approach deals primarily with the concepts of complex, cycle, and homology, and is wholly consistent with current investigations. Ranges from basic concepts of set-theoretic topology to the concept of Betti groups. "Glowing example of harmony between intuition and thought," David Hilbert. Translated by A. E. Farley. Introduction by D. Hilbert. Index. 25 figures. 73pp. 5⅜ x 8. Paperbound $1.00

ELEMENTS OF NON-EUCLIDEAN GEOMETRY,
D. M. Y. Sommerville
Unique in proceeding step-by-step, in the manner of traditional geometry. Enables the student with only a good knowledge of high school algebra and geometry to grasp elementary hyperbolic, elliptic, analytic non-Euclidean geometries; space curvature and its philosophical implications; theory of radical axes; homothetic centres and systems of circles; parataxy and parallelism; absolute measure; Gauss' proof of the defect area theorem; geodesic representation; much more, all with exceptional clarity. 126 problems at chapter endings provide progressive practice and familiarity. 133 figures. Index. xvi + 274pp. 5⅜ x 8. Paperbound $2.00

INTRODUCTION TO THE THEORY OF NUMBERS, *L. E. Dickson*
Thorough, comprehensive approach with adequate coverage of classical literature, an introductory volume beginners can follow. Chapters on divisibility, congruences, quadratic residues & reciprocity. Diophantine equations, etc. Full treatment of binary quadratic forms without usual restriction to integral coefficients. Covers infinitude of primes, least residues. Fermat's theorem. Euler's phi function, Legendre's symbol, Gauss's lemma, automorphs, reduced forms, recent theorems of Thue & Siegel, many more. Much material not readily available elsewhere. 239 problems. Index. I figure. viii + 183pp. 5⅜ x 8.
Paperbound $1.75

MATHEMATICAL TABLES AND FORMULAS,
compiled by Robert D. Carmichael and Edwin R. Smith
Valuable collection for students, etc. Contains all tables necessary in college algebra and trigonometry, such as five-place common logarithms, logarithmic sines and tangents of small angles, logarithmic trigonometric functions, natural trigonometric functions, four-place antilogarithms, tables for changing from sexagesimal to circular and from circular to sexagesimal measure of angles, etc. Also many tables and formulas not ordinarily accessible, including powers, roots, and reciprocals, exponential and hyperbolic functions, ten-place logarithms of prime numbers, and formulas and theorems from analytical and elementary geometry and from calculus. Explanatory introduction. viii + 269pp. 5⅜ x 8½. Paperbound $1.25

A SOURCE BOOK IN MATHEMATICS,
D. E. Smith

Great discoveries in math, from Renaissance to end of 19th century, in English translation. Read announcements by Dedekind, Gauss, Delamain, Pascal, Fermat, Newton, Abel, Lobachevsky, Bolyai, Riemann, De Moivre, Legendre, Laplace, others of discoveries about imaginary numbers, number congruence, slide rule, equations, symbolism, cubic algebraic equations, non-Euclidean forms of geometry, calculus, function theory, quaternions, etc. Succinct selections from 125 different treatises, articles, most unavailable elsewhere in English. Each article preceded by biographical introduction. Vol. I: Fields of Number, Algebra. Index. 32 illus. 338pp. 5⅜ x 8. Vol. II: Fields of Geometry, Probability, Calculus, Functions, Quaternions. 83 illus. 432pp. 5⅜ x 8.

Vol. 1 Paperbound $2.00, Vol. 2 Paperbound $2.00,
The set $4.00

FOUNDATIONS OF PHYSICS,
R. B. Lindsay & H. Margenau

Excellent bridge between semi-popular works & technical treatises. A discussion of methods of physical description, construction of theory; valuable for physicist with elementary calculus who is interested in ideas that give meaning to data, tools of modern physics. Contents include symbolism; mathematical equations; space & time foundations of mechanics; probability; physics & continua; electron theory; special & general relativity; quantum mechanics; causality. "Thorough and yet not overdetailed. Unreservedly recommended," *Nature* (London). Unabridged, corrected edition. List of recommended readings. 35 illustrations. xi + 537pp. 5⅜ x 8. Paperbound $3.00

FUNDAMENTAL FORMULAS OF PHYSICS,
ed. by D. H. Menzel

High useful, full, inexpensive reference and study text, ranging from simple to highly sophisticated operations. Mathematics integrated into text—each chapter stands as short textbook of field represented. Vol. 1: Statistics, Physical Constants, Special Theory of Relativity, Hydrodynamics, Aerodynamics, Boundary Value Problems in Math, Physics, Viscosity, Electromagnetic Theory, etc. Vol. 2: Sound, Acoustics, Geometrical Optics, Electron Optics, High-Energy Phenomena, Magnetism, Biophysics, much more. Index. Total of 800pp. 5⅜ x 8.

Vol. 1 Paperbound $2.25, Vol. 2 Paperbound $2.25,
The set $4.50

THEORETICAL PHYSICS,
A. S. Kompaneyets

One of the very few thorough studies of the subject in this price range. Provides advanced students with a comprehensive theoretical background. Especially strong on recent experimentation and developments in quantum theory. Contents: Mechanics (Generalized Coordinates, Lagrange's Equation, Collision of Particles, etc.), Electrodynamics (Vector Analysis, Maxwell's equations, Transmission of Signals, Theory of Relativity, etc.), Quantum Mechanics (the Inadequacy of Classical Mechanics, the Wave Equation, Motion in a Central Field, Quantum Theory of Radiation, Quantum Theories of Dispersion and Scattering, etc.), and Statistical Physics (Equilibrium Distribution of Molecules in an Ideal Gas, Boltzmann Statistics, Bose and Fermi Distribution. Thermodynamic Quantities, etc.). Revised to 1961. Translated by George Yankovsky, authorized by Kompaneyets. 137 exercises. 56 figures. 529pp. 5⅜ x 8½.

Paperbound $2.50

MATHEMATICAL PHYSICS, *D. H. Menzel*
Thorough one-volume treatment of the mathematical techniques vital for classical mechanics, electromagnetic theory, quantum theory, and relativity. Written by the Harvard Professor of Astrophysics for junior, senior, and graduate courses, it gives clear explanations of all those aspects of function theory, vectors, matrices, dyadics, tensors, partial differential equations, etc., necessary for the understanding of the various physical theories. Electron theory, relativity, and other topics seldom presented appear here in considerable detail. Scores of definition, conversion factors, dimensional constants, etc. "More detailed than normal for an advanced text . . . excellent set of sections on Dyadics, Matrices, and Tensors," *Journal of the Franklin Institute.* Index. 193 problems, with answers. x + 412pp. 5⅜ x 8. Paperbound $2.50

THE THEORY OF SOUND, *Lord Rayleigh*
Most vibrating systems likely to be encountered in practice can be tackled successfully by the methods set forth by the great Nobel laureate, Lord Rayleigh. Complete coverage of experimental, mathematical aspects of sound theory. Partial contents: Harmonic motions, vibrating systems in general, lateral vibrations of bars, curved plates or shells, applications of Laplace's functions to acoustical problems, fluid friction, plane vortex-sheet, vibrations of solid bodies, etc. This is the first inexpensive edition of this great reference and study work. Bibliography, Historical introduction by R. B. Lindsay. Total of 1040pp. 97 figures. 5⅜ x 8. Vol. 1 Paperbound $2.50, Vol. 2 Paperbound $2.50, The set $5.00

HYDRODYNAMICS, *Horace Lamb*
Internationally famous complete coverage of standard reference work on dynamics of liquids & gases. Fundamental theorems, equations, methods, solutions, background, for classical hydrodynamics. Chapters include Equations of Motion, Integration of Equations in Special Gases, Irrotational Motion, Motion of Liquid in 2 Dimensions, Motion of Solids through Liquid-Dynamical Theory, Vortex Motion, Tidal Waves, Surface Waves, Waves of Expansion, Viscosity, Rotating Masses of Liquids. Excellently planned, arranged; clear, lucid presentation. 6th enlarged, revised edition. Index. Over 900 footnotes, mostly bibliographical. 119 figures. xv + 738pp. 6⅛ x 9¼. Paperbound $4.00

DYNAMICAL THEORY OF GASES, *James Jeans*
Divided into mathematical and physical chapters for the convenience of those not expert in mathematics, this volume discusses the mathematical theory of gas in a steady state, thermodynamics, Boltzmann and Maxwell, kinetic theory, quantum theory, exponentials, etc. 4th enlarged edition, with new material on quantum theory, quantum dynamics, etc. Indexes. 28 figures. 444pp. 6⅛ x 9¼. Paperbound $2.75

THERMODYNAMICS, *Enrico Fermi*
Unabridged reproduction of 1937 edition. Elementary in treatment; remarkable for clarity, organization. Requires no knowledge of advanced math beyond calculus, only familiarity with fundamentals of thermometry, calorimetry. Partial Contents: Thermodynamic systems; First & Second laws of thermodynamics; Entropy; Thermodynamic potentials: phase rule, reversible electric cell; Gaseous reactions: van't Hoff reaction box, principle of LeChatelier; Thermodynamics of dilute solutions: osmotic & vapor pressures, boiling & freezing points; Entropy constant. Index. 25 problems. 24 illustrations. x + 160pp. 5⅜ x 8. Paperbound $1.75

CELESTIAL OBJECTS FOR COMMON TELESCOPES,
Rev. T. W. Webb

Classic handbook for the use and pleasure of the amateur astronomer. Of inestimable aid in locating and identifying thousands of celestial objects. Vol I, The Solar System: discussions of the principle and operation of the telescope, procedures of observations and telescope-photography, spectroscopy, etc., precise location information of sun, moon, planets, meteors. Vol. II, The Stars: alphabetical listing of constellations, information on double stars, clusters, stars with unusual spectra, variables, and nebulae, etc. Nearly 4,000 objects noted. Edited and extensively revised by Margaret W. Mayall, director of the American Assn. of Variable Star Observers. New Index by Mrs. Mayall giving the location of all objects mentioned in the text for Epoch 2000. New Precession Table added. New appendices on the planetary satellites, constellation names and abbreviations, and solar system data. Total of 46 illustrations. Total of xxxix + 606pp. 5⅜ x 8. Vol. 1 Paperbound $2.25, Vol. 2 Paperbound $2.25
The set $4.50

PLANETARY THEORY,
E. W. Brown and C. A. Shook

Provides a clear presentation of basic methods for calculating planetary orbits for today's astronomer. Begins with a careful exposition of specialized mathematical topics essential for handling perturbation theory and then goes on to indicate how most of the previous methods reduce ultimately to two general calculation methods: obtaining expressions either for the coordinates of planetary positions or for the elements which determine the perturbed paths. An example of each is given and worked in detail. Corrected edition. Preface. Appendix. Index. xii + 302pp. 5⅜ x 8½. Paperbound $2.25

STAR NAMES AND THEIR MEANINGS,
Richard Hinckley Allen

An unusual book documenting the various attributions of names to the individual stars over the centuries. Here is a treasure-house of information on a topic not normally delved into even by professional astronomers; provides a fascinating background to the stars in folk-lore, literary references, ancient writings, star catalogs and maps over the centuries. Constellation-by-constellation analysis covers hundreds of stars and other asterisms, including the Pleiades, Hyades, Andromedan Nebula, etc. Introduction. Indices. List of authors and authorities. xx + 563pp. 5⅜ x 8½. Paperbound $2.50

A SHORT HISTORY OF ASTRONOMY, *A. Berry*

Popular standard work for over 50 years, this thorough and accurate volume covers the science from primitive times to the end of the 19th century. After the Greeks and the Middle Ages, individual chapters analyze Copernicus, Brahe, Galileo, Kepler, and Newton, and the mixed reception of their discoveries. Post-Newtonian achievements are then discussed in unusual detail: Halley, Bradley, Lagrange, Laplace, Herschel, Bessel, etc. 2 Indexes. 104 illustrations, 9 portraits. xxxi + 440pp. 5⅜ x 8. Paperbound $2.75

SOME THEORY OF SAMPLING, *W. E. Deming*

The purpose of this book is to make sampling techniques understandable to and useable by social scientists, industrial managers, and natural scientists who are finding statistics increasingly part of their work. Over 200 exercises, plus dozens of actual applications. 61 tables. 90 figs. xix + 602pp. 5⅜ x 8½.
Paperbound $3.50

PRINCIPLES OF STRATIGRAPHY,
A. W. Grabau

Classic of 20th century geology, unmatched in scope and comprehensiveness. Nearly 600 pages cover the structure and origins of every kind of sedimentary, hydrogenic, oceanic, pyroclastic, atmoclastic, hydroclastic, marine hydroclastic, and bioclastic rock; metamorphism; erosion; etc. Includes also the constitution of the atmosphere; morphology of oceans, rivers, glaciers; volcanic activities; faults and earthquakes; and fundamental principles of paleontology (nearly 200 pages). New introduction by Prof. M. Kay, Columbia U. 1277 bibliographical entries. 264 diagrams. Tables, maps, etc. Two volume set. Total of xxxii + 1185pp. 5⅜ x 8. Vol. 1 Paperbound $2.50, Vol. 2 Paperbound $2.50, The set $5.00

SNOW CRYSTALS, *W. A. Bentley and W. J. Humphreys*

Over 200 pages of Bentley's famous microphotographs of snow flakes—the product of painstaking, methodical work at his Jericho, Vermont studio. The pictures, which also include plates of frost, glaze and dew on vegetation, spider webs, windowpanes; sleet; graupel or soft hail, were chosen both for their scientific interest and their aesthetic qualities. The wonder of nature's diversity is exhibited in the intricate, beautiful patterns of the snow flakes. Introductory text by W. J. Humphreys. Selected bibliography. 2,453 illustrations. 224pp. 8 x 10¼. Paperbound $3.25

THE BIRTH AND DEVELOPMENT OF THE GEOLOGICAL SCIENCES,
F. D. Adams

Most thorough history of the earth sciences ever written. Geological thought from earliest times to the end of the 19th century, covering over 300 early thinkers & systems: fossils & their explanation, vulcanists vs. neptunists, figured stones & paleontology, generation of stones, dozens of similar topics. 91 illustrations, including medieval, renaissance woodcuts, etc. Index. 632 footnotes, mostly bibliographical. 511pp. 5⅜ x 8. Paperbound $2.75

ORGANIC CHEMISTRY, *F. C. Whitmore*

The entire subject of organic chemistry for the practicing chemist and the advanced student. Storehouse of facts, theories, processes found elsewhere only in specialized journals. Covers aliphatic compounds (500 pages on the properties and synthetic preparation of hydrocarbons, halides, proteins, ketones, etc.), alicyclic compounds, aromatic compounds, heterocyclic compounds, organophosphorus and organometallic compounds. Methods of synthetic preparation analyzed critically throughout. Includes much of biochemical interest. "The scope of this volume is astonishing," *Industrial and Engineering Chemistry*. 12,000-reference index. 2387-item bibliography. Total of x + 1005pp. 5⅜ x 8. Two volume set, paperbound $4.50

THE PHASE RULE AND ITS APPLICATION,
Alexander Findlay

Covering chemical phenomena of 1, 2, 3, 4, and multiple component systems, this "standard work on the subject" (*Nature*, London), has been completely revised and brought up to date by A. N. Campbell and N. O. Smith. Brand new material has been added on such matters as binary, tertiary liquid equilibria, solid solutions in ternary systems, quinary systems of salts and water. Completely revised to triangular coordinates in ternary systems, clarified graphic representation, solid models, etc. 9th revised edition. Author, subject indexes. 236 figures. 505 footnotes, mostly bibliographic. xii + 494pp. 5⅜ x 8. Paperbound $2.75

A Course in Mathematical Analysis,
Edouard Goursat
Trans. by E. R. Hedrick, O. Dunkel, H. G. Bergmann. Classic study of fundamental material thoroughly treated. Extremely lucid exposition of wide range of subject matter for student with one year of calculus. Vol. 1: Derivatives and differentials, definite integrals, expansions in series, applications to geometry. 52 figures, 556pp. Paperbound \$2.50. Vol. 2, Part 1: Functions of a complex variable, conformal representations, doubly periodic functions, natural boundaries, etc. 38 figures, 269pp. Paperbound \$1.85. Vol. 2, Part 2: Differential equations, Cauchy-Lipschitz method, nonlinear differential equations, simultaneous equations, etc. 308pp. Paperbound \$1.85. Vol. 3, Part 1: Variation of solutions, partial differential equations of the second order. 15 figures, 339pp. Paperbound \$3.00. Vol. 3, Part 2: Integral equations, calculus of variations. 13 figures, 389pp. Paperbound \$3.00

Planets, Stars and Galaxies,
A. E. Fanning
Descriptive astronomy for beginners: the solar system; neighboring galaxies; seasons; quasars; fly-by results from Mars, Venus, Moon; radio astronomy; etc. all simply explained. Revised up to 1966 by author and Prof. D. H. Menzel, former Director, Harvard College Observatory. 29 photos, 16 figures. 189pp. 5⅜ x 8½. Paperbound \$1.50

Great Ideas in Information Theory, Language and Cybernetics,
Jagjit Singh
Winner of Unesco's Kalinga Prize covers language, metalanguages, analog and digital computers, neural systems, work of McCulloch, Pitts, von Neumann, Turing, other important topics. No advanced mathematics needed, yet a full discussion without compromise or distortion. 118 figures. ix + 338pp. 5⅜ x 8½. Paperbound \$2.00

Geometric Exercises in Paper Folding,
T. Sundara Row
Regular polygons, circles and other curves can be folded or pricked on paper, then used to demonstrate geometric propositions, work out proofs, set up well-known problems. 89 illustrations, photographs of actually folded sheets. xii + 148pp. 5⅜ x 8½. Paperbound \$1.00

Visual Illusions, Their Causes, Characteristics and Applications,
M. Luckiesh
The visual process, the structure of the eye, geometric, perspective illusions, influence of angles, illusions of depth and distance, color illusions, lighting effects, illusions in nature, special uses in painting, decoration, architecture, magic, camouflage. New introduction by W. H. Ittleson covers modern developments in this area. 100 illustrations. xxi + 252pp. 5⅜ x 8. Paperbound \$1.50

Atoms and Molecules Simply Explained,
B. C. Saunders and R. E. D. Clark
Introduction to chemical phenomena and their applications: cohesion, particles, crystals, tailoring big molecules, chemist as architect, with applications in radioactivity, color photography, synthetics, biochemistry, polymers, and many other important areas. Non technical. 95 figures. x + 299pp. 5⅜ x 8½. Paperbound \$1.50

THE PRINCIPLES OF ELECTROCHEMISTRY,
D. A. MacInnes

Basic equations for almost every subfield of electrochemistry from first principles, referring at all times to the soundest and most recent theories and results; unusually useful as text or as reference. Covers coulometers and Faraday's Law, electrolytic conductance, the Debye-Hueckel method for the theoretical calculation of activity coefficients, concentration cells, standard electrode potentials, thermodynamic ionization constants, pH, potentiometric titrations, irreversible phenomena. Planck's equation, and much more. 2 indices. Appendix. 585-item bibliography. 137 figures. 94 tables. ii + 478pp. 5⅝ x 8⅜.
Paperbound $2.75

MATHEMATICS OF MODERN ENGINEERING,
E. G. Keller and R. E. Doherty

Written for the Advanced Course in Engineering of the General Electric Corporation, deals with the engineering use of determinants, tensors, the Heaviside operational calculus, dyadics, the calculus of variations, etc. Presents underlying principles fully, but emphasis is on the perennial engineering attack of set-up and solve. Indexes. Over 185 figures and tables. Hundreds of exercises, problems, and worked-out examples. References. Two volume set. Total of xxxiii + 623pp. 5⅜ x 8. Two volume set, paperbound $3.70

AERODYNAMIC THEORY: A GENERAL REVIEW OF PROGRESS,
William F. Durand, editor-in-chief

A monumental joint effort by the world's leading authorities prepared under a grant of the Guggenheim Fund for the Promotion of Aeronautics. Never equalled for breadth, depth, reliability. Contains discussions of special mathematical topics not usually taught in the engineering or technical courses. Also: an extended two-part treatise on Fluid Mechanics, discussions of aerodynamics of perfect fluids, analyses of experiments with wind tunnels, applied airfoil theory, the nonlifting system of the airplane, the air propeller, hydrodynamics of boats and floats, the aerodynamics of cooling, etc. Contributing experts include Munk, Giacomelli, Prandtl, Toussaint, Von Karman, Klemperer, among others. Unabridged republication. 6 volumes. Total of 1,012 figures, 12 plates, 2,186pp. Bibliographies. Notes. Indices. 5⅜ x 8½.
Six volume set, paperbound $13.50

FUNDAMENTALS OF HYDRO- AND AEROMECHANICS,
L. Prandtl and O. G. Tietjens

The well-known standard work based upon Prandtl's lectures at Goettingen. Wherever possible hydrodynamics theory is referred to practical considerations in hydraulics, with the view of unifying theory and experience. Presentation is extremely clear and though primarily physical, mathematical proofs are rigorous and use vector analysis to a considerable extent. An Engineering Society Monograph, 1934. 186 figures. Index. xvi + 270pp. 5⅜ x 8.
Paperbound $2.00

APPLIED HYDRO- AND AEROMECHANICS,
L. Prandtl and O. G. Tietjens

Presents for the most part methods which will be valuable to engineers. Covers flow in pipes, boundary layers, airfoil theory, entry conditions, turbulent flow in pipes, and the boundary layer, determining drag from measurements of pressure and velocity, etc. Unabridged, unaltered. An Engineering Society Monograph. 1934. Index. 226 figures, 28 photographic plates illustrating flow patterns. xvi + 311pp. 5⅜ x 8. Paperbound $2.00